Lectures on BSDEs, Stochastic Control, and Stochastic Differential Games with Financial Applications

Financial Mathematics

Carmona, René, *Lectures on BSDEs, Stochastic Control, and Stochastic Differential Games with Financial Applications*

RENÉ CARMONA

Princeton University
Princeton, New Jersey

Lectures on BSDEs, Stochastic Control, and Stochastic Differential Games with Financial Applications

Society for Industrial and Applied Mathematics
Philadelphia

Copyright © 2016 by the Society for Industrial and Applied Mathematics

10 9 8 7 6 5 4 3 2 1

All rights reserved. Printed in the United States of America. No part of this book may be reproduced, stored, or transmitted in any manner without the written permission of the publisher. For information, write to the Society for Industrial and Applied Mathematics, 3600 Market Street, 6th Floor, Philadelphia, PA 19104-2688 USA.

Trademarked names may be used in this book without the inclusion of a trademark symbol. These names are used in an editorial context only; no infringement of trademark is intended.

Royalties from the sale of this book will be donated to the SIAG/FME to be used for a prize.

Publisher	David Marshall
Acquisitions Editor	Elizabeth Greenspan
Developmental Editor	Gina Rinelli
Managing Editor	Kelly Thomas
Production Editor	Ann Manning Allen
Copy Editor	Nicola Howcroft
Production Manager	Donna Witzleben
Production Coordinator	Cally Shrader
Compositor	Techsetters, Inc.
Graphic Designer	Lois Sellers

Library of Congress Cataloging-in-Publication Data

Names: Carmona, R. (René)
Title: Lectures on BSDEs, stochastic control, and stochastic differential games with financial applications / René Carmona, Princeton University, Princeton, New Jersey.
Other titles: Lectures on backward stochastic differential equations, stochastic control, and stochastic differential games with financial applications
Description: Philadelphia : Society for Industrial and Applied Mathematics, [2016] | Series: Financial mathematics ; 01 | Includes bibliographical references and index.
Identifiers: LCCN 2015038344 | ISBN 9781611974232
Subjects: LCSH: Stochastic differential equations. | Stochastic control theory. | Business mathematics.
Classification: LCC QA274.23 .C27 2016 | DDC 519.2/7--dc23
LC record available at http://lccn.loc.gov/2015038344

 is a registered trademark.

Contents

Preface   vii

List of Notation   ix

I   Stochastic Calculus Preliminaries   1

1   Stochastic Differential Equations   3
    1.1   Notation and First Definitions 3
    1.2   Existence and Uniqueness of Strong Solutions: The Lipschitz Case . . 4
    1.3   SDEs of McKean–Vlasov Type 11
    1.4   Conditional Propagation of Chaos 18
    1.5   Notes & Complements 26

2   Backward Stochastic Differential Equations   27
    2.1   Introduction and First Definitions 27
    2.2   Mean-Field BSDEs 35
    2.3   Reflected Backward Stochastic Differential Equations (RBSDEs) . . . 37
    2.4   Forward-Backward Stochastic Differential Equations (FBSDEs) . . . 40
    2.5   Existence and Uniqueness of Solutions 46
    2.6   The Affine Case 58
    2.7   Notes & Complements 63

II   Stochastic Control   65

3   Continuous Time Stochastic Optimization and Control   67
    3.1   Optimization of Stochastic Dynamical Systems 67
    3.2   First Financial Applications 75
    3.3   Dynamic Programming and the HJB Equation 79
    3.4   Infinite Horizon Case 85
    3.5   Constraints and Singular Control Problems 87
    3.6   Viscosity Solutions of HJB Equations and QVIs 101
    3.7   Impulse Control Problems 107
    3.8   Ergodic Control 112

4   Probabilistic Approaches to Stochastic Control   119
    4.1   BSDEs and Stochastic Control 119
    4.2   Pontryagin Stochastic Maximum Principle 125
    4.3   Linear-Quadratic (LQ) Models 136

		4.4	Optimal Control of McKean–Vlasov Dynamics 141
		4.5	Notes & Complements 160

III Stochastic Differential Games   163

5 Stochastic Differential Games   165
    5.1 Introduction and First Definitions 165
    5.2 Specific Examples 176
    5.3 Weak Formulation and the Case of Uncontrolled Volatility 182
    5.4 Game Versions of the Stochastic Maximum Principle 186
    5.5 A Simple Model for Systemic Risk 198
    5.6 A Predatory Trading Game Model 204
    5.7 Notes & Complements 216

6 Mean-Field Games   219
    6.1 Introduction and First Definitions 219
    6.2 A Full Solution Without the Common Noise 226
    6.3 Propagation of Chaos and Approximate Nash Equilibriums 236
    6.4 Applications and Open Problems 243
    6.5 Notes & Complements 250

Bibliography   253

Author Index   261

Subject Index   263

Preface

This book grew out of the lecture notes I prepared for a graduate class I taught at Princeton University in 2011–12, and again in 2012–13. My goal was to introduce the students to stochastic analysis tools, which play an increasing role in the probabilistic approach to optimization problems, including stochastic control and stochastic differential games. I had invested quite a bit of effort in trying to understand the groundbreaking works of Lasry and Lions on mean field games, and of Caines, Huang, and Malhamé on Nash certainty equivalence. These initial results were intriguing, and definitely screaming for a probabilistic interpretation. While the tools of optimal control of stochastic differential systems are taught in many graduate programs in applied mathematics and operations research, I was intrigued by the fact that game theory, and especially the theory of stochastic differential games, are rarely taught in these programs. In fact, I was shocked by the lack of published literature in book form, a sad state of affairs which prompted me to write lecture notes for the class.

This would have been the end of the story if I hadn't shared some of my notes with a friend and co-author of mine, Jean-Pierre Fouque, who decided to convince me to turn these lecture notes into a book. His perseverance, together with my desire to help those applied mathematicians trying to learn the theory of stochastic differential games despite the lack of sources in textbook form, helped me to find the time to clean up my original class notes. Still, this short preface should contain a clear disclaimer emphasizing the fact that the present manuscript is more a set of lecture notes than a polished and exhaustive textbook on the subject matter. Many experts on the subject could produce better treatises. But until then, I hope these notes will be helpful to young researchers and newcomers to stochastic analysis, eager to understand the subtleties of the modern probabilistic approach to stochastic control and stochastic differential games.

Acknowledgments. I would like to thank the students whose patience allowed me to meander through the maze of concepts and theories I was discovering while teaching. Special thanks are due to Dan Lacker, Kevin Webster and Geoffrey Zhu, whose regular feedback helped me keep the lectures on track. Last but not least, I want to thank François Delarue, not only for a couple of enlightening guest lectures, but for an enjoyable collaboration which taught me the subtle intricacies of forward-backward stochastic differential equations—nothing can rival learning from the master!

Important Aside. The royalties generated by the sales (if any) of this book will be used, in their entirety, to fund a student prize awarded every other year during the SIAM Conference on Financial Mathematics.

René Carmona
Princeton, NJ
August 6, 2015

List of Notation

$BSDE(\Psi, \xi)$, 27
$C([0,T]; \mathbb{R}^n)$, 69
$C_b^2(\mathbb{R}^d)$, 7
$C^{1,2}([0,T] \times \mathbb{R}^d)$, 8
D, 95
M_h^f, 8
$W^{(0)}$, 11, 12
$W^{(2)}$, 11, 12
$\mathbb{F}^x$, 69
$\mathbb{H}^2(A)$, 123
$\mathbb{P}_X$, 13
$\mathbb{Q}$, 33
Ξ, 244
$\mathcal{A}$, 68, 167
$\mathcal{A}^i$, 167
$\mathcal{B}(E)$, 68
$\mathcal{C}$, 69
$\mathcal{E}$, 33, 74
$\mathcal{L}(X)$, 13
$\mathcal{M}(\mathbb{R}^d)$, 84
$\mathcal{P}$, 27, 68, 69
$\mathcal{P}_p(E)$, 142
δ_x, 16, 220
$\langle \varphi, \mu \rangle$, 10
$\mathbb{B}^k$, 3
$\mathbb{H}^{0,k}$, 3
$\mathbb{H}^{2,k}$, 3

$\mathbb{S}^2$, 3
$\mathcal{P}^{W^0}$, 19
μ^N, 16
μ_x^N, 220
$\mathscr{L}_s$, 8, 10
$\mathscr{L}_t$, 10
$\mathscr{L}_t^*$, 10
$\sigma^\dagger$, 71
$\text{var}\{\eta\}$, 73
$\tilde{H}$, 71
$a = (a^1, \cdots, a^N)$, 167
$a = (a_t)_{t \in [0,T]}$, 68
g^{conv}, 80
v^*, 102
v_*, 102
(**E**), 101

BAU, 76
BSDE, 27

CL, 68
CLFFNE, 170
CLNE, 170
CLPS, 167
CRRA, 89, 98, 132, 246

DNE, 170

DPP, 81

FBSDE, 41, 42
FPS, 167

HJB, 67, 82, 119

LQ, 63, 136, 179

MFG, 172, 216
MPS, 167
MRT, 27

NCE, 251

ODE, 84
OL, 68, 167
OLNE, 170

PDE, 7, 42

QVI, 44, 87, 94, 109

RBSDE, 37

SDE, 3, 16, 42
SPDE, 222

Part I

Stochastic Calculus Preliminaries

Chapter 1
Stochastic Differential Equations

This introductory chapter is devoted to the analysis of stochastic differential equations of Itô type. In preparation for the study of stochastic control problems and stochastic differential games, we prove existence and uniqueness for equations whose coefficients can be random. Next, we discuss the Markov property and the connection with partial differential equations when the coefficients do not depend upon the past. The last two sections are not classical in nature and they can be skipped in a first reading. They are devoted to McKean–Vlasov equations, whose coefficients depend upon the distribution of the solution and to their connection with the so-called propagation of chaos. The last section considers the case when the marginal distributions are conditioned by the knowledge of a separate Itô diffusion.

1.1 ▪ Notation and First Definitions

We assume that $(\Omega, \mathcal{F}, \mathbb{F}, \mathbb{P})$ is a stochastic basis where the filtration $\mathbb{F} = (\mathcal{F}_t)_{0 \leq t \leq T}$ supports an m-dimensional $\mathbb{F}$-Brownian motion $\mathbf{W} = (W_t)_{0 \leq t \leq T}$ in $\mathbb{R}^m$. For each integer $k \geq 1$, we denote by $\mathbb{H}^{0,k}$ the collection of all $\mathbb{R}^k$-valued progressively measurable processes on $[0, T] \times \mathbb{R}$, and we introduce the subspaces

$$\mathbb{H}^{2,k} := \left\{ Z \in \mathbb{H}^{0,k};\ \mathbb{E} \int_0^T |Z_s|^2 ds < \infty \right\} \text{ and } \mathbb{S}^2 := \left\{ Y \in \mathbb{H}^{0,k};\ \mathbb{E} \sup_{0 \leq t \leq T} |Y_s|^2 < \infty \right\}.$$

We also use the notation $\mathbb{B}^k$ for the subspace of bounded processes, namely,

$$\mathbb{B}^k := \left\{ Z \in \mathbb{H}^{0,k};\ \sup_{0 \leq t \leq T} |Z_t| < \infty,\ \mathbb{P}-a.s. \right\}.$$

When dealing with several possible filtrations, we add a subscript to specify the filtration with respect to which the progressive measurability needs to be understood. In the case of scalar processes, when $k = 1$, we skip the exponent k from our notation. We are interested in stochastic differential equations (SDEs) of the form

$$dX_t = b(t, X_t)dt + \sigma(t, X_t)dW_t, \tag{1.1}$$

where the coefficients b and σ

$$(b, \sigma) : [0, T] \times \Omega \times \mathbb{R}^d \to \mathbb{R}^d \times \mathbb{R}^{d \times m}$$

satisfy the following assumptions.

(A1) For each $x \in \mathbb{R}^d$, the processes $(b(t,x))_{0 \le t \le T}$ and $(\sigma(t,x))_{0 \le t \le T}$ are in $\mathbb{H}^{2,d}$ and $\mathbb{H}^{2,dm}$, respectively.

(A2) $\exists c > 0, \forall t \in [0,T], \forall \omega \in \Omega, \forall x, x' \in \mathbb{R}^d$,

$$|(b,\sigma)(t,\omega,x) - (b,\sigma)(t,\omega,x')| \le c|x-x'|.$$

We shall follow the standard practice of not making the dependence upon $\omega \in \Omega$ explicit in the formulas whenever possible.

Definition 1.1. *We say that an $\mathbb{F}$-progressively measurable process $\mathbf{X} = (X_t)_{0 \le t \le T}$ is a strong solution of the SDE (1.1) if*

- $\int_0^T (|b(t,X_t)| + |\sigma(t,X_t)|^2)dt < \infty$ $\mathbb{P}$-almost surely,
- $X_t = X_0 + \int_0^t b(s,X_s)ds + \int_0^T \sigma(s,X_s)dW_s$, $\quad 0 \le t \le T$.

1.2 ▪ Existence and Uniqueness of Strong Solutions: The Lipschitz Case

Theorem 1.2. *Let us assume that $X_0 \in L^2$ is independent of $\mathbf{W}$ and that the coefficients b and σ satisfy the assumptions (A1) and (A2) above. Then, there exists a unique solution of (1.1) in $\mathbb{H}^{2,d}$, and for some $c > 0$ depending only upon T and the Lipschitz constant of b and σ, this solution satisfies*

$$\mathbb{E} \sup_{0 \le t \le T} |X_t|^2 \le c(1 + \mathbb{E}|X_0|^2)e^{cT}. \tag{1.2}$$

Here and in the following, we use the letter c for a generic constant whose value can change from line to line.

Proof. For each $\mathbf{X} \in \mathbb{H}^{2,d}$, the space of square integrable progressively measurable processes, we define the process $U(\mathbf{X})$ by

$$U(\mathbf{X})_t = X_0 + \int_0^t b(s,X_s)ds + \int_0^t \sigma(s,X_s)dW_s. \tag{1.3}$$

First we prove that $U(\mathbf{X}) \in \mathbb{H}^{2,d}$, and then, since $\mathbf{X}$ is a solution of the SDE (1.1) if and only if $U(\mathbf{X}) = \mathbf{X}$, we prove that U is a strict contraction in the Hilbert space $\mathbb{H}^{2,d}$. By definition of the norm of $\mathbb{H}^{2,d}$, we have

$$\|U(\mathbf{X})\|^2 \le (i) + (ii) + (iii)$$

with

$$(i) = 3T\mathbb{E}|X_0|^2 < \infty,$$

and (ii) and (iii) defined below. Using the fact that

$$|b(t,x)|^2 \le c(1 + |b(t,0)|^2 + |x|^2),$$

1.2. Existence and Uniqueness of Strong Solutions: The Lipschitz Case

implied by our Lipschitz assumption, we have

$$(ii) = 3\mathbb{E}\int_0^T \left|\int_0^t b(s, X_s)ds\right|^2 dt$$
$$\leq 3\mathbb{E}\int_0^T t\left(\int_0^t |b(s, X_s)|^2 ds\right) dt$$
$$\leq 3c\mathbb{E}\int_0^T t\left(\int_0^t (1 + |b(s,0)|^2 + |X_s|^2)ds\right) dt$$
$$\leq 3cT^2\left(1 + \|b(\cdot, 0)\|^2 + \mathbb{E}\sup_{0\leq t\leq T} |X_t|^2\right)$$
$$< \infty.$$

Finally, in order to estimate

$$(iii) = 3\mathbb{E}\int_0^T \left|\int_0^t \sigma(s, X_s)dW_s\right|^2 dt$$

we use Doob's maximal inequality together with the fact that

$$|\sigma(t, x)|^2 \leq c(1 + |\sigma(t, 0)|^2 + |x|^2),$$

implied by our Lipschitz assumption on σ. Doing so, we have

$$(iii) \leq 3T\mathbb{E}\sup_{0\leq t\leq T}\left|\int_0^t \sigma(s, X_s)dW_s\right|^2 dt$$
$$\leq 12T\mathbb{E}\int_0^T |\sigma(s, X_s)|^2 ds$$
$$\leq 12Tc\mathbb{E}\int_0^T (1 + |\sigma(s,0)|^2 + |X_s|^2)ds$$
$$\leq 12cT^2\left(1 + |\sigma(\cdot, 0)|^2 + \mathbb{E}\sup_{0\leq t\leq T} |X_t|^2\right)$$
$$< \infty.$$

Again, remember that the value of the generic constant c can change from line to line. Now that we know that U maps $\mathbb{H}^{2,d}$ into itself, we prove that it is a strict contraction. In order to do so, we find it convenient to change the norm of the Hilbert space $\mathbb{H}^{2,d}$ to an equivalent norm. For each $\alpha > 0$ we define a norm on the space $\mathbb{H}^{2,d}$ by

$$\|\boldsymbol{\xi}\|_\alpha^2 = \mathbb{E}\int_0^T e^{-\alpha t}|\xi_t|^2 dt.$$

The norm $\|\cdot\|_\alpha$ and the original norm $\|\cdot\|$ (which correspond to $\alpha = 0$) are equivalent and define the same topology. If $\mathbf{X}$ and $\mathbf{Y}$ are generic elements of $\mathbb{H}^{2,d}$ with $X_0 = Y_0$, we have

$$\mathbb{E}|U(X)_t - U(Y)_t|^2$$
$$\leq 2\mathbb{E}\left|\int_0^t [b(s,X_s) - b(s,Y_s)]ds\right|^2 + 2\mathbb{E}\left|\int_0^t [\sigma(s,X_s) - \sigma(s,Y_s)]dW_s\right|^2$$
$$\leq 2t\mathbb{E}\int_0^t |b(s,X_s) - b(s,Y_s)|^2 ds + 2\mathbb{E}\int_0^t |\sigma(s,X_s) - \sigma(s,Y_s)|^2 ds$$
$$\leq c(t+1)\int_0^t \mathbb{E}|X_s - Y_s|^2 ds$$

if we use the Lipschitz property of the coefficients. Consequently,

$$\|U(\mathbf{X}) - U(\mathbf{Y})\|_\alpha^2 = \int_0^T e^{-\alpha t}\mathbb{E}|U(\mathbf{X})_t - U(\mathbf{Y})_t|^2 dt$$
$$\leq cT\int_0^T e^{-\alpha t}\int_0^t \mathbb{E}|X_s - Y_s|^2 ds\, dt$$
$$\leq cT\int_0^T \mathbb{E}|X_s - Y_s|^2 ds \int_t^T e^{-\alpha t} dt$$
$$\leq \frac{cT}{\alpha}\|\mathbf{X} - \mathbf{Y}\|_\alpha^2,$$

and U is indeed a strict contraction if $\alpha > cT$ is large enough! Finally, we prove the estimate (1.2) for the solution. For $t \in [0,T]$ fixed we have

$$\mathbb{E}\sup_{0\leq s\leq t}|X_s|^2 = \mathbb{E}\sup_{0\leq s\leq t}\left|X_0 + \int_0^s b(r,X_r)dr + \int_0^s \sigma(r,X_r)dW_r\right|^2$$
$$\leq 3\left(\mathbb{E}|X_0|^2 + t\mathbb{E}\int_0^t |b(s,X_s)|^2 ds + 4\mathbb{E}\int_0^t |\sigma(s,X_s)|^2 ds\right)$$
$$\leq c\left(1 + \mathbb{E}|X_0|^2 + \int_0^t \mathbb{E}\sup_{0\leq r\leq s}|X_r|^2 dr\right),$$

where the constant c depends only upon T, $\|b(\,\cdot\,,0)\|^2$, and $\|\sigma(\,\cdot\,,0)\|^2$, and finally we conclude using Gronwall's inequality. $\square$

Remark 1.3. *Gronwall's inequality. Since we will use Gronwall's inequality repeatedly in the following, we state it for later reference:*

$$\varphi(t) \leq \alpha + \int_0^t \beta(s)\varphi(s)ds \implies \varphi(t) \leq \alpha e^{\int_0^t \beta(s)ds}. \tag{1.4}$$

1.2.1 ▪ Flow and Markov Solutions of SDEs

In this subsection we consider SDEs for which the coefficients are given by deterministic functions of the desired solution X_t. So we now assume that

$$(b,\sigma) : [0,T] \times \mathbb{R}^d \to \mathbb{R}^d \times \mathbb{R}^{d\times d}$$

are measurable deterministic functions which are Lipschitz in the space variable uniformly in time, i.e.,

(**L**) $\quad \exists c > 0, \forall t \in [0,T], \forall x,y \in \mathbb{R}^d, \quad |(b,\sigma)(t,x) - (b,\sigma)(t,y)| \leq c|x-y|.$

1.2. Existence and Uniqueness of Strong Solutions: The Lipschitz Case

For each $(t,x) \in [0,T] \times \mathbb{R}^d$ we denote by $(X_s^{t,x})_{t \leq s \leq T}$ the unique solution on the interval $[t,T]$, which is equal to x at time t. In other words,

$$X_s^{t,x} = x + \int_t^s b(r, X_r^{t,x})dr + \int_t^s \sigma(r, X_r^{t,x})dW_r, \qquad s \in [t,T].$$

A closer look at the proof of Theorem 1.2 shows that there exists a constant $c > 0$ for which, for all (t,x) and (t',x') in $[0,T] \times \mathbb{R}^d$ such that $t \leq t'$, we have

$$\mathbb{E} \sup_{t \leq s \leq t'} |X_s^{t,x} - X_s^{t',x'}|^2 \leq c e^{ct'} |x - x'|^2. \tag{1.5}$$

This makes it possible to define the processes $X^{t,x}$ simultaneously for all $(t,x) \in [0,T] \times \mathbb{R}^d$. More precisely, it is possible to show that this solution can be constructed pathwise in the sense that there exists a function

$$F: [0,T] \times \mathbb{R}^d \times [0,T] \times C([t,T]; \mathbb{R}^d) \ni (t,x,s,w) \hookrightarrow F(t,x,s,w) \in \mathbb{R}^d$$

such that the solution at time $s \in [t,T]$ is a deterministic function of the increments of the Wiener process between time t and s:

$$X_s^{t,x} = F(t, x, s, (W_r - W_t)_{t \leq r \leq s}) \qquad \text{a.s.}$$

Together with pathwise uniqueness, this implies that

$$X_s^{t,x} = X_s^{r, X_r^{t,x}} \qquad \text{whenever } t \leq r \leq s \leq T.$$

In words, at time $s \geq t$, the solution which started from x at time t is equal to the solution which started at time $r \in [t,s]$ from the point it got to after starting from x at time t, namely $X_r^{t,x}$. This form of the Markov property is not limited to deterministic times t, s, and r. It can easily be extended to stopping times.

1.2.2 ▪ First Connection with Partial Differential Equations (PDEs)

If f is a twice continuously differentiable real-valued function on $\mathbb{R}^d$ with bounded derivatives ($f \in C_b^2(\mathbb{R}^d)$ in notation), Itô's formula gives

$$f(X_{t+h}) = f(X_t) + \int_t^{t+h} \left[\frac{1}{2} \text{trace}(\sigma \sigma^\dagger D^2 f)(s, X_s) + b(s, X_s) Df(X_s) \right] ds$$
$$+ \int_t^{t+h} Df(X_s) \sigma(s, X_s) dW_s,$$

where we used the notation $\sigma^\dagger$ to denote the transpose of the matrix σ. Taking expectations on both sides, the expectation of the stochastic integral vanishes because Df is bounded and σ is linear growth, so we get

$$\frac{1}{h}[\mathbb{E}f(X_{t+h}) - f(x)] = \frac{1}{h} \int_t^{t+h} \mathbb{E}[\mathscr{L}_s f](s, X_s) ds,$$

where we use the notation $\mathscr{L}_s$ for the time-dependent partial differential operator defined by

$$[\mathscr{L}_s f](x) = \frac{1}{2} \text{trace}(\sigma \sigma^\dagger D^2 f)(s, x) + b(s, x) Df(x)$$
$$= \frac{1}{2} \sum_{i,j=1}^d a_{i,j}(s, x) \partial_{i,j}^2 f(x) + \sum_{i=1}^d b_i(s, x) \partial_i f(x), \tag{1.6}$$

where the entries of the diffusion matrix $a = \sigma\sigma^\dagger$ are given by

$$a_{i,j}(t,x) = \sum_{k=1}^{d} \sigma_{i,k}(t,x)\sigma_{j,k}(t,x).$$

A closer look at the above argument shows that for every $f \in C_b^2(\mathbb{R}^d)$, the process M^f defined by

$$M_h^f = f(X_{t+h}) - f(X_t) - \int_t^{t+h} [\mathscr{L}_s f](s, X_s) ds$$

is a martingale starting from 0 since $M_0^f = 0$. This identifies $\mathscr{L}_s$ as the generator of the diffusion given by the drift coefficient b and the volatility σ.

Theorem 1.4. *Let us assume as before that b and σ are deterministic functions which are Lipschitz in $x \in \mathbb{R}^d$ uniformly in $t \in [0, T]$, and that g is a bounded measurable function on $\mathbb{R}^d$ such that the function u defined by $u(t,x) = \mathbb{E}g(X_T^{t,x})$ is continuously differentiable in t and twice continuously differentiable in x ($u \in C^{1,2}([0,T] \times \mathbb{R}^d)$ in notation). Then u solves the PDE*

$$\partial_t u + \mathscr{L}_t u = 0 \tag{1.7}$$

on $[0, T] \times \mathbb{R}^d$ with terminal condition $u(T, \cdot) = g$.

Proof. Fix $(t,x) \in [0,T] \times \mathbb{R}^d$ and define the stopping time τ_1 by

$$\tau_1 = T \wedge \inf\{s > t;\ |X_s^{t,x} - x| \geq 1\}.$$

For each $s \in [t, T]$ we have

$$u(t,x) = \mathbb{E}g(X_T^{t,x}) = \mathbb{E}g(X_T^{s\wedge\tau_1, X_{s\wedge\tau_1}^{t,x}})$$
$$= \mathbb{E}\left[\mathbb{E}\{g(X_T^{s\wedge\tau_1, X_{s\wedge\tau_1}^{t,x}})|\mathcal{F}_{s\wedge\tau_1}\}\right]$$
$$= \mathbb{E}u(s \wedge \tau_1, X_{s\wedge\tau_1}^{t,x}).$$

Using Itô's formula as before, we get

$$\mathbb{E}u(s \wedge \tau_1, X_{s\wedge\tau_1}^{t,x}) - u(t,x) = \mathbb{E}\int_t^{s\wedge\tau_1} [\partial_s u + \mathscr{L}_s u](r, X_r^{t,x}) dr,$$

or, in other words,

$$0 = \mathbb{E}\int_t^{s\wedge\tau_1} [\partial_s u + \mathscr{L}_s u](r, X_r^{t,x}) dr.$$

We conclude by letting $s \searrow t$ and by Lebesgue's dominated convergence theorem. □

1.2.3 ▪ Second Connection with PDEs: The Feynman–Kac Formula

In this subsection, we look for a probabilistic representation of the solutions to the PDE

$$\begin{cases} \partial_t u(t,x) + \mathscr{L}_t u(t,x) + c(t,x)u(t,x) + d(t,x) = 0, & (t,x) \in [0,T) \times \mathbb{R}^d, \\ u(T,x) = g(x). \end{cases} \tag{1.8}$$

1.2. Existence and Uniqueness of Strong Solutions: The Lipschitz Case

Theorem 1.5. *Let us assume as before that b and σ are deterministic functions which are Lipschitz in $x \in \mathbb{R}^d$ uniformly in $t \in [0, T]$ and such that*

$$\int_0^T (|b(t,0)|^2 + |\sigma(t,0)|^2) dt < \infty. \tag{1.9}$$

Let us assume that c is a measurable function uniformly bounded above (i.e., there exists $\bar{c} > 0$ such that $c(t,x) \leq \bar{c}$ for all $(t,x) \in [0,T] \times \mathbb{R}^d$), and that the function d is measurable and of quadratic growth in x uniformly in $t \in [0,T]$. Now, if $u \in C^{1,2}([0,T] \times \mathbb{R}^d)$ has quadratic growth in x uniformly in $t \in [0,T]$ and solves the PDE (1.8), then u has the following representation:

$$u(t,x) = \mathbb{E}\left[\int_t^T d(s, X_s^{t,x}) e^{-\int_t^s c(r, X_r^{t,x}) dr} ds + g(X_T^{t,x}) e^{-\int_t^T c(r, X_r^{t,x}) dr}\right]. \tag{1.10}$$

Representation (1.10) is often called the Feynman–Kac formula.

Proof. Fix $(t,x) \in [0,T] \times \mathbb{R}^d$ and in order to lighten the notation, set $X_s = X_s^{t,x}$. For each integer $n \geq 1$, we define the stopping time τ_n by

$$\tau_n = T \wedge \inf\{s > t; |X_s - x| \geq n\}.$$

Clearly, $\tau_n \nearrow T$ almost surely. The following computation of the Itô differential of a product of two processes (where the time variable is s since t is fixed) is standard since one of the factors is of bounded variations. Using the fact that u solves (1.10), we have

$$d\left[e^{\int_t^s c(r, X_r) dr} u(s, X_s)\right]$$
$$= c(s, X_s) e^{\int_t^s c(r, X_r) dr} u(s, X_s) ds$$
$$+ e^{\int_t^s c(r, X_r) dr} [\partial_s u(s, X_s) + \mathscr{L}_s u(s, X_s)] ds$$
$$+ e^{\int_t^s c(r, X_r) dr} \partial_x u(s, X_s) \sigma(s, X_s) dW_s$$
$$= e^{\int_t^s c(r, X_r) dr} [-d(s, X_s) ds + \partial_x u(s, X_s) \sigma(s, X_s) dW_s].$$

If we integrate both sides in s between t and τ_n, and if we then take expectations, we get

$$\mathbb{E}\left[e^{\int_t^{\tau_n} c(r, X_r) dr} u(\tau_n, X_{\tau_n})\right] - u(t,x)$$
$$= \mathbb{E}\left[\int_t^{\tau_n} e^{\int_t^s c(r, X_r) dr} [-d(s, X_s) ds + \partial_x u(s, X_s) \sigma(s, X_s) dW_s]\right].$$

All these integrals and expectations are finite. Indeed, using the fact that c is bounded above and u has quadratic growth in x uniformly in t, we see that

$$\mathbb{E}\left[e^{\int_t^{\tau_n} c(r, X_r) dr} |u(\tau_n, X_{\tau_n})|\right] \leq e^{(T-t)\bar{c}} K(|x| + n)^2$$

for some finite constant $K > 0$. The case is similar for the expectations in the right-hand side because of the cut-off provided by the stopping time τ_n. In particular, the expectation of the stochastic integral vanishes and we get

$$u(t,x) = \mathbb{E}\left[\int_t^{\tau_n} d(s, X_s^{t,x}) e^{-\int_t^s c(r, X_r^{t,x}) dr} ds + u(\tau_n, X_{\tau_n}) e^{-\int_t^{\tau_n} c(r, X_r^{t,x}) dr}\right],$$

which is almost what we want except for the fact that we have the stopping time τ_n instead of T and $u(\tau_n, X_{\tau_n})$ instead of $g(X_T)$. Since $\tau_n \nearrow T$ almost surely, this is remedied by taking the limit $n \to \infty$ and using Lebesgue's dominated convergence theorem, since the terms under the expectation are bounded in absolute value by a multiple of $1 + \sup_{t \leq s \leq T} |X_s|^2$ because of the quadratic growth assumptions, and this supremum has finite expectations, as proven in Theorem 1.2. $\square$

1.2.4 ▪ Kolmogorov's Equation

We consider the stochastic differential equation

$$dX_t = b(t, X_t)dt + \sigma(t, X_t)dW_t, \tag{1.11}$$

where

$$(b, \sigma) : [0, T] \times \mathbb{R}^d \to \mathbb{R}^d \times \mathbb{R}^{d \times d}$$

are deterministic functions which are Lipschitz in $x \in \mathbb{R}^d$ uniformly in $t \in [0, T]$ with linear growth. If $\varphi : \mathbb{R}^d \hookrightarrow \mathbb{R}$ is a C^2 function (i.e., twice continuously differentiable) with bounded derivatives, Itô's formula gives

$$\varphi(X_t) = \varphi(X_0) + \int_0^t [\mathscr{L}_s \varphi](X_s) ds + \int_0^t \partial_x \varphi(X_s) \sigma(s, X_s) dW_s,$$

where the partial differential operator $\mathscr{L}_t$ is defined by

$$[\mathscr{L}_t \varphi](x) = b(t, x) \cdot \partial_x \varphi(x) + \frac{1}{2} \text{trace}[\sigma(t, x) \sigma(t, x)^\dagger \partial_{xx}^2 \varphi(t, x)].$$

Taking expectations on both sides and denoting by μ_t the distribution of X_t, i.e., $\mu_t(dx) = \mathbb{P}\{X_t \in dx\}$, we get

$$\langle \varphi, \mu_t \rangle = \langle \varphi, \mu_0 \rangle + \int_0^t \langle \mathscr{L}_s \varphi, \mu_s \rangle \, ds$$

if we use the notation

$$\langle \varphi, \mu \rangle = \int_{\mathbb{R}^d} \varphi(x) \mu(dx)$$

so that

$$\langle \varphi, \mu_t \rangle = \left\langle \varphi, \mu_0 + \int_0^t \mathscr{L}_s^* \mu_s ds \right\rangle$$

if we denote by $\mathscr{L}_s^*$ the adjoint of the operator $\mathscr{L}_s$. So, since this equality is true for all test functions φ, we have proven that μ_t is the solution of the following equation:

$$\frac{d}{dt} \mu_t = \mathscr{L}_t^* \mu_t \tag{1.12}$$

with initial condition μ_0. This equation is most often called the Kolmogorov equation, and sometimes also the Fokker–Planck equation. Assuming that μ_t has a density, say $\mu_t(dx) = m(t, x)dx$, and rewriting

$$\langle \varphi, \mathscr{L}_t^* \mu_t \rangle = \langle \mathscr{L}_t \varphi, \mu_t \rangle = \int_{\mathbb{R}^d} m(t, x) [\mathscr{L}_t \varphi](x) dx$$

using integration by parts, one shows that, like $\mathscr{L}_t$, $\mathscr{L}_t^*$ is also a partial differential operator. It is easy to prove that

$$[\mathscr{L}_t^* m(t,\,\cdot\,)](x) = \frac{1}{2}\sum_{hk=1}^{d} a_{hk}(t,x)\frac{\partial^2 m}{\partial x_h \partial x_k}(t,x) - \text{div}[m(t,\,\cdot\,)b(t,\,\cdot\,)](x)$$
$$+ \sum_{hk=1}^{d}\frac{\partial a_{hk}}{\partial x_k}(t,x)\frac{\partial m}{\partial x_h}(t,x) + \frac{1}{2}\left(\sum_{hk=1}^{d}\frac{\partial^2 a_{hk}}{\partial x_h \partial x_k}(t,x)\right)m(t,x).$$

In particular, when $\sigma(t,x) \equiv \sigma$ is constant we have

$$\mathscr{L}_t\varphi = \frac{\sigma^2}{2}\Delta\varphi + b(t,x)\cdot\nabla\varphi$$

and

$$\mathscr{L}_t^* m = \frac{\sigma^2}{2}\Delta\varphi - \text{div}[m(t,\,\cdot\,)b(t,\,\cdot\,)]$$
$$= \frac{\sigma^2}{2}\Delta\varphi - b(t,x)\cdot\nabla m - -\text{div}[b(t,\,\cdot\,)]m.$$

1.3 ▪ SDEs of McKean–Vlasov Type

In this section, we study stochastic differential equations with coefficients which can depend upon the marginal distributions of the solutions. As we show later in this section, this type of nonlinearity occurs in the asymptotic regime of large stochastic systems with mean-field interactions.

1.3.1 ▪ Notation

For any given measurable space $(E,\mathcal{E})$, we denote by $\mathcal{P}(E)$ the set of probability measures on $(E,\mathcal{E})$. In the following, we shall most often assume that E is a complete separable metric space, e.g., $E = \mathbb{R}^d$ or a separable Banach space with norm $\|\cdot\|$. In such a case, we denote by $\mathcal{P}_2(E)$ the set of probability measures with finite second-order moment (i.e., the set of those $\mu \in \mathcal{P}(E)$ for which $\int_E d(x,x_0)^2 \mu(dx) < \infty$) for some (or equivalently any) $x_0 \in E$. Obviously, we use the notation d for the distance of the metric space E. For μ and ν in $\mathcal{P}_2(E)$, the Wasserstein distance is defined by the formula

$$W^{(2)}(\mu,\nu) = \inf\left\{\left[\int d(x,y)^2 \pi(dx,dy)\right]^{1/2}; \pi \in \mathcal{P}(E\times E) \text{ with marginals } \mu \text{ and } \nu\right\}.$$
(1.13)

This distance induces the topology of weak convergence of measures together with the convergence of all moments of order up to 2. Using a bounded distance such as $d(x,y) = \|x-y\|\wedge 1$, one can extend the definition of the Wasserstein distance to the whole $\mathcal{P}(E)$. Indeed, the modified Wasserstein distance $W^{(0)}$ on $\mathcal{P}(E)$ is defined as

$$W^{(0)}(\mu,\nu) = \inf\left\{\int 1\wedge d(x,y)\,\pi(dx,dy); \pi \in \mathcal{P}(E\times E) \text{ with marginals } \mu \text{ and } \nu\right\}.$$
(1.14)

It induces the topology of weak convergence on $\mathcal{P}(E)$.

In order to formulate the results of this part of the chapter we need to assign meaning to the continuity and Lipschitz properties of functions of probability measures on the space $E = C([0,T];\mathbb{R}^d)$ of $\mathbb{R}^d$-valued bounded continuous functions on $[0,T]$ equipped with the norm of the uniform convergence. A notion of distance on the space of such measures is all we need for now. Later on, we will need a notion of differentiability of these functions, and matters will become more involved. For the sake of definiteness, we rewrite the definitions of the Wasserstein distances $W^{(0)}$ and $W^{(2)}$ in the particular case of the complete metric space $E = C([0,T];\mathbb{R}^d)$.

$$W^{(0)}(m_1, m_2) = \inf \left\{ \int \sup_{0 \le t \le T} 1 \wedge |X_t(w_1) - X_t(w_2)| \, m(dw_1, dw_2); \right.$$
$$\left. m \in \mathcal{P}(C([0,T];\mathbb{R}^d) \times C([0,T];\mathbb{R}^d)) \text{ with marginals } m_1 \text{ and } m_2 \right\},$$

where $(X_t)_{0 \le t \le T}$ is the canonical process $w \hookrightarrow X_t(w) = w(t)$ on $\mathcal{P}(C([0,T];\mathbb{R}^d))$. $W^{(0)}$ is a distance on $\mathcal{P}(C([0,T];\mathbb{R}^d))$ which defines a notion of convergence equivalent to the weak convergence of probability measures. So $\mathcal{P}(C([0,T];\mathbb{R}^d))$ is a complete metric space for $W^{(0)}$. Similarly, we define $\mathcal{P}_2(C([0,T];\mathbb{R}^d)$ as the subset of $\mathcal{P}(C([0,T];\mathbb{R}^d))$ of the measures m with a finite moment of order 2, namely satisfying

$$\int \sup_{0 \le t \le T} |X_t(w)|^2 m(dw) < \infty.$$

The corresponding Wasserstein distance $W^{(2)}$ can be defined as

$$W^{(2)}(m_1, m_2) = \inf \left\{ \left[\int \sup_{0 \le t \le T} |X_t(w_1) - X_t(w_2)|^2 \, m(dw_1, dw_2) \right]^{1/2} ; \right.$$
$$\left. m \in \mathcal{P}_2(C([0,T];\mathbb{R}^d) \times C([0,T];\mathbb{R}^d)) \text{ with marginals } m_1 \text{ and } m_2 \right\}$$

and for later convenience we introduce the notation

$$W_t^{(2)}(m_1, m_2) = \inf \left\{ \left(\int \sup_{0 \le s \le t} |X_s(w_1) - X_s(W^{(2)})|^2 \, m(dw_1, dW^{(2)}) \right)^{1/2} ; \right.$$
$$\left. m \in \mathcal{P}_2(C([0,T];\mathbb{R}^d) \times C([0,T];\mathbb{R}^d)) \text{ with marginals } m_1 \text{ and } m_2 \right\}$$

so that $W^{(2)}(m_1, m_2) = W_T^{(2)}(m_1, m_2)$.

We now introduce the assumptions on the coefficients of the *nonlinear* stochastic differential equations which we study in this section. As in the classical case studied earlier, we assume that $(\Omega, \mathcal{F}, \mathbb{F}, \mathbb{P})$ is a stochastic basis where the filtration $\mathbb{F} = (\mathcal{F}_t)_{0 \le t \le T}$ supports an $\mathbb{F}$-Brownian motion $\mathbf{W} = (W_t)_{0 \le t \le T}$ in $\mathbb{R}^d$. We are interested in stochastic differential equations of the form

$$dX_t = b(t, X_t, \mathcal{L}(X_t))dt + \sigma(t, X_t, \mathcal{L}(X_t))dW_t, \tag{1.15}$$

where the coefficients b and σ

$$(b, \sigma) : [0, T] \times \Omega \times \mathbb{R}^d \times \mathcal{P}(\mathbb{R}^d) \to \mathbb{R}^d \times \mathbb{R}^{d \times m}$$

satisfy the following assumptions.

1.3. SDEs of McKean–Vlasov Type

(A1) For each $x \in \mathbb{R}^d$ and $\mu \in \mathcal{P}(\mathbb{R}^d)$, the processes $(b(t,x,\mu))_{0 \leq t \leq T}$ and $(\sigma(t,x,\mu))_{0 \leq t \leq T}$ are in $\mathbb{H}^{2,d}$ and $\mathbb{H}^{2,dm}$, respectively.

(A2) $\exists c > 0, \forall t \in [0,T], \forall \omega \in \Omega, \forall x, x' \in \mathbb{R}^d, \forall \mu, \mu' \in \mathcal{P}_2(\mathbb{R}^d)$,

$$|(b,\sigma)(t,\omega,x,\mu) - (b,\sigma)(t,\omega,x',\mu')| \leq c(|x-x'| + W^{(2)}(\mu,\mu')).$$

Here and in the following, we use the notation $\mathbb{P}_X$ or $\mathcal{L}(X)$ for the distribution or law of the random element X.

1.3.2 ▪ Examples of Mean-Field Interactions

In practical applications, interactions through the marginal distribution of the process as in the case of the SDE of McKean–Vlasov type (1.15) come in various forms of complexity.

In the simplest case, which we shall call the *mean-field interaction of scalar* type, the dependence upon the distribution degenerates into a dependence upon some moments of this distribution. To be more specific, in the case of interactions of scalar type we have

$$b(t,\omega,x,\mu) = \tilde{b}(t,\omega,x,\langle \varphi, \mu \rangle) \tag{1.16}$$

for some scalar function φ defined on $\mathbb{R}^d$ and some function $\tilde{b}$ defined on $[0,T] \times \Omega \times \mathbb{R}^d \times \mathbb{R}$. As before, we use the angular bracket notation

$$\langle \varphi, \mu \rangle = \int \varphi(x') d\mu(x')$$

for the integral of a function with respect to a measure.

Remark 1.6. *It is clear that scalar interactions can include functions of several, say n, moments of the measure. Still using the same notation and framework, it amounts to considering vector functions φ taking values in a Euclidean space $\mathbb{R}^n$ and having the function $\tilde{b}$ be defined on $[0,T] \times \Omega \times \mathbb{R}^d \times \mathbb{R}^n$.*

We shall also encounter applications in which the dependence upon the distribution is given by means of an auxiliary function $\tilde{b}$ defined on $[0,T] \times \Omega \times \mathbb{R}^d \times \mathbb{R}^d$, the interaction taking the form

$$b(t,\omega,x,\mu) = \int_{\mathbb{R}^d} \tilde{b}(t,\omega,x,x') \mu(dx'). \tag{1.17}$$

This type of mean-field interaction will be called *interaction of order* 1. It is linear in μ. Similarly, one could define mean-field interactions of higher orders. For example, a mean-field interaction of order 2 (which is quadratic in μ) should be of the form

$$b(t,\omega,x,\mu) = \int\int_{\mathbb{R}^d \times \mathbb{R}^d} \tilde{b}(t,\omega,x,x',x'') \mu(dx') \mu(dx''). \tag{1.18}$$

1.3.3 ▪ Existence and Uniqueness of Solutions: The Lipschitz Case

Theorem 1.7. *Let us assume that $X_0 \in L^2$ is independent of $\mathbf{W}$, and that the coefficients b and σ satisfy the assumptions* **(A1)** *and* **(A2)** *stated above. Then, there exists a unique solution to (1.15) in $\mathbb{H}^{2,d}$, and for some $c > 0$ depending only upon T and the Lipschitz constant of b and σ, this solution satisfies*

$$\mathbb{E} \sup_{0 \leq t \leq T} |X_t|^2 \leq c(1 + \mathbb{E}|X_0|^2)e^{cT}. \tag{1.19}$$

Proof. Let $m \in \mathcal{P}_2(C([0,T];\mathbb{R}^d))$ be temporarily fixed, and let us denote by m_t its time marginals, i.e., the push-forward image of the measure m by X_t viewed as a map from $C([0,T];\mathbb{R}^d)$ into $\mathbb{R}^d$. By Lebesgue's dominated convergence theorem, the inequality

$$W^{(2)}(m_s, m_t)^2 \leq \int |X_s(\omega) - X_t(\omega)|^2 \, m(d\omega)$$

implies that the map $[0,T] \ni t \hookrightarrow m_t \in \mathcal{P}_2(\mathbb{R}^d)$ is continuous for the Wasserstein distance $W^{(2)}$. Hence, substituting momentarily m_t for $\mathcal{L}(X_t)$ for all $t \in [0,T]$ in (1.15), since X_0 is given, Theorem 1.2 gives the existence and uniqueness of a strong solution of the classical stochastic differential equation

$$dX_t = b(t, X_t, m_t)dt + \sigma(t, X_t, m_t)dW_t \qquad (1.20)$$

with random coefficients, and we denote its solution by $\mathbf{X}^m = (X_t^m)_{0 \leq t \leq T}$. We first notice that, because of the upper bound proven in Theorem 1.2, the law of $\mathbf{X}^m$ is of order 2. We then define the mapping $\Phi : \mathcal{P}_2(C([0,T];\mathbb{R}^d)) \ni m \hookrightarrow \Phi(m) = \mathcal{L}(\mathbf{X}^m) = \mathbb{P}_{\mathbf{X}^m} \in \mathcal{P}_2(C([0,T];\mathbb{R}^d))$. Since a process $\mathbf{X} = (X_t)_{0 \leq t \leq T}$ satisfying $\mathbb{E} \sup_{0 \leq t \leq T} |X_t|^2 < \infty$ is a solution of (1.15) if and only if its law is a fixed point of Φ, we prove the existence and uniqueness result of the theorem by proving that the mapping Φ has a unique fixed point. Let us choose m and m' in $\mathcal{P}_2(C([0,T];\mathbb{R}^d))$. Then, since $\mathbf{X}^m$ and $\mathbf{X}^{m'}$ have the same initial conditions, for each $t \in [0,T]$, using Doob's maximal inequality and the Lipschitz assumption, we have

$$\mathbb{E} \sup_{0 \leq s \leq t} |X_s^m - X_s^{m'}|^2 \leq 2\mathbb{E} \sup_{0 \leq s \leq t} \left| \int_0^s [b(r, X_r^m, m_r) - b(r, X_r^{m'}, m'_r)] dr \right|^2$$

$$+ 2\mathbb{E} \sup_{0 \leq s \leq t} \left| \int_0^s [\sigma(r, X_r^m, m_r) - \sigma(r, X_r^{m'}, m'_r)] dW_r \right|^2$$

$$\leq c(T+1) \left(\int_0^t \mathbb{E} \sup_{0 \leq r \leq s} |X_r^m - X_r^{m'}|^2 ds + \int_0^t W^{(2)}(m_s, m'_s) ds \right.$$

$$+ \mathbb{E} \int_0^t |\sigma(r, X_r^m, m_r) - \sigma(r, X_r^{m'}, m'_r)|^2 dr \Bigg)$$

$$\leq cT \left(\int_0^t \mathbb{E} \sup_{0 \leq r \leq s} |X_r^m - X_r^{m'}|^2 ds + \int_0^t W^{(2)}(m_s, m'_s) ds \right).$$

As usual, and except for the dependence upon T which we keep track of, we use the same notation even though the value of the constant c can change from line to line. Using Gronwall's inequality one concludes that

$$\mathbb{E} \sup_{0 \leq s \leq t} |X_s^m - X_s^{m'}|^2 \leq c(T) \int_0^t W^{(2)}(m_s, m'_s)^2 \, ds \qquad (1.21)$$

with $c(T) = cTe^{cT}$. Notice that

$$W_t^{(2)}(\Phi(m), \Phi(m'))^2 \leq \mathbb{E} \sup_{0 \leq s \leq t} |X_s^m - X_s^{m'}|^2$$

and that $W^{(2)}(m_s, m'_s) \leq W_s^{(2)}(m, m')$ so that (1.21) implies

$$W_t^{(2)}(\Phi(m), \Phi(m'))^2 \leq c(T) \int_0^t W_s^{(2)}(m, m')^2 \, ds.$$

1.3. SDEs of McKean–Vlasov Type

Iterating this inequality and denoting by Φ^k the kth composition of the mapping Φ with itself, we get that for any integer $k > 1$

$$W_T^{(2)}(\Phi^k(m), \Phi^k(m'))^2 \leq c(T)^k \int_0^T \frac{(T-s)^{k-1}}{(k-1)!} W_s^{(2)}(m, m')^2 \, ds$$
$$\leq \frac{c^k T^k}{k!} W_T^{(2)}(m, m')^2,$$

which shows that for k large enough, Φ^k is a strict contraction, and hence Φ admits a unique fixed point. □

Remark 1.8. *Let us fix $\varphi \in C_b^2(\mathbb{R}^d)$ and let us apply Itô's formula to $\varphi(X_t)$. We get*

$$\varphi(X_t) = \varphi(X_0) + \int_0^t \left[\frac{1}{2} \text{trace}[\sigma(s, X_s, \mathcal{L}(X_s))^\dagger \sigma(s, X_s, \mathcal{L}(X_s)) D^2 \varphi(X_s)] \right.$$
$$\left. + b(s, X_s, \mathcal{L}(X_s)) D\varphi(X_s) \right] ds + \int_0^t D\varphi(X_s) \sigma(s, X_s, \mathcal{L}(X_s)) \, dW_s.$$

Taking expectations of both sides and using the notation $\mu_t = \mathcal{L}(X_t)$, we get

$$\langle \varphi, \mu_t \rangle = \langle \varphi, \mu_0 \rangle + \int_0^t \left[\frac{1}{2} \langle \text{trace}[\sigma(s, \cdot, \mu_s)^\dagger \sigma(s, \cdot, \mu_s) D^2 \varphi(\cdot)], \mu_s \rangle \right.$$
$$\left. + \langle b(s, \cdot, \mu_s) D\varphi(\cdot), \mu_s \rangle \right] ds$$

and after integration by parts

$$\langle \varphi, \mu_t \rangle = \langle \varphi, \mu_0 \rangle + \int_0^t \left\langle \frac{1}{2} \text{trace}[\sigma(s, \cdot, \mu_s)^\dagger \sigma(s, \cdot, \mu_s) D^2 \mu_s] - \text{div}(b(s, \cdot, \mu_s)\mu_s), \varphi \right\rangle ds,$$

or in differential form

$$\partial_t \mu_t = L_t(\mu)\mu_t,$$

where for each $t \in [0, T]$ and $\mu \in \mathcal{P}(\mathbb{R}^d)$, the second-order partial differential operator $L_t(\mu)$ is defined by

$$L_t(\mu) f = \frac{1}{2} \text{trace}[\sigma(t, \cdot, \mu)^\dagger \sigma(t, \cdot, \mu) D^2 f] - \text{div}[b(t, \cdot, \mu) f].$$

This is a form of the (nonlinear) Kolmogorov equation for $\boldsymbol{\mu} = (\mu_t)_{0 \leq t \leq T}$.

1.3.4 ▪ Particle Approximations and Propagation of Chaos

Our next step is to study pathwise particle approximations of the solution of the McKean–Vlasov SDE (1.15). In particular, this will provide a proof of the original propagation of chaos result, which was stated in terms of convergence of laws instead of a pathwise behavior. Let $((X_0^i, \mathbf{W}^i))_{i \geq 1}$ be a sequence of independent copies of $(X_0, \mathbf{W})$. For each

$i \geq 1$, we let $\mathbf{X}^i = (X_t^i)_{0 \leq t \leq T}$ denote the solution of (1.15) constructed in Theorem 1.7 starting from X_0^i and driven by the Wiener process $\mathbf{W}^i$. It satisfies

$$X_t^i = X_0^i + \int_0^t b(s, X_s^i, \mathcal{L}(X_s^i))ds + \int_0^t \sigma(s, X_s^i, \mathcal{L}(X_s^i))dW_s^i. \qquad (1.22)$$

Notice that the probability measures $\mathcal{L}(X_s^i)$ do not depend upon i. Clearly, all the processes $\mathbf{X}^i$ are independent by construction. We show that they can be approximated by finite systems of classical Itô processes (which we often call particles) depending upon each other through specific interactions. For each integer $N \geq 1$, we consider the *particle processes* $\mathbf{X}^{i,N}$ for $i = 1, \cdots, N$ solving the system of standard SDEs

$$X_t^{i,N} = X_0^i + \int_0^t b\left(s, X_s^{i,N}, \frac{1}{N}\sum_{d=1}^N \delta_{X_s^{d,N}}\right) ds + \int_0^t \sigma\left(s, X_s^{i,N}, \frac{1}{N}\sum_{d=1}^N \delta_{X_s^{d,N}}\right) dW_s^i \qquad (1.23)$$

for $i = 1, \cdots, N$, where we use the standard notation δ_x for the unit (Dirac) point mass at x. Notice that the coupling between these N SDEs is obtained by replacing in the McKean–Vlasov dynamics (1.15) the distributions $\mathcal{L}(X_t^i)$, whose presence creates the nonlinearity in the form of a self-interaction, by the empirical distributions of the particles $X_t^{1,N}, \cdots, X_t^{N,N}$. The hope is that a form of the law of large numbers will prove that the impact of this substitution on the solutions will be minimal. A simple application of the definition of the Wasserstein distance to the case of point measures shows that if $\mathbf{x} = (x_1, \cdots, x_N)$ and $\mathbf{y} = (y_1, \cdots, y_N)$ are generic elements of $\mathbb{R}^{dN}$, one has

$$W^{(2)}\left(\frac{1}{N}\sum_{i=1}^N \delta_{x_i}, \frac{1}{N}\sum_{i=1}^N \delta_{y_i}\right) \leq \left(\frac{1}{N}\sum_{i=1}^N |x_i - y_i|^2\right)^{1/2} = \frac{1}{\sqrt{N}}|\mathbf{x} - \mathbf{y}|. \qquad (1.24)$$

This implies, for fixed N, the uniform Lipschitz property for the coefficients of the system (1.23) and, in turn, the existence and uniqueness of a strong solution. The main result of this section establishes pathwise propagation of chaos for the interacting particle system (1.23). The following lemma will be useful in the control of the particle approximation.

Lemma 1.9. *Let $\mu \in \mathcal{P}_2(\mathbb{R}^d)$, let $(\xi_i)_{i \geq 1}$ be a sequence of independent random variables with common law μ, and for each integer $N \geq 1$, let μ^N denote the empirical distribution of $\xi_1, \cdots, \xi_N$, namely $\mu^N = \frac{1}{N}\sum_{i=1}^N \delta_{\xi_i}$. Then, for each $N \geq 1$, we have*

$$\mathbb{E}W^{(2)}(\mu^N, \mu)^2 \leq 4\int_{\mathbb{R}^d} |x|^2 \mu(dx) \quad \text{and} \quad \lim_{N \to \infty} \mathbb{E}W^{(2)}(\mu^N, \mu)^2 = 0.$$

Proof. By the strong law of large numbers, μ^N converges weakly toward μ almost surely as $N \to \infty$. Similarly, for each $1 \leq i, j \leq d$,

$$\lim_{N \to \infty} \int_{\mathbb{R}^d} x_i \mu^N(dx) = \int_{\mathbb{R}^d} x_i \mu(dx) \quad \text{and} \quad \lim_{N \to \infty} \int_{\mathbb{R}^d} x_i x_j \mu^N(dx) = \int_{\mathbb{R}^d} x_i x_j \mu(dx)$$

almost surely. Since the Wasserstein distance $W^{(2)}$ induces the topology of weak con-

1.3. SDEs of McKean–Vlasov Type

vergence together with the convergence of all the moments up to order 2, one concludes that $W^{(2)}(\mu^N, \mu)$ converges toward 0 almost surely as $N \to \infty$. Moreover, the sequence $(W^{(2)}(\mu^N, \mu)^2)_{N \geq 1}$ of random variables is uniformly integrable. Indeed, for any coupling π of μ^N and μ,

$$W^{(2)}(\mu^N, \mu)^2 \leq \int_{\mathbb{R}^d \times \mathbb{R}^d} |x-y|^2 \, \pi(dx, dy)$$

$$\leq 2 \int_{\mathbb{R}^d \times \mathbb{R}^d} (|x|^2 + |y|^2) \, \pi(dx, dy)$$

$$= 2 \left(\int_{\mathbb{R}^d} |x|^2 \, \mu^N(dx) + \int_{\mathbb{R}^d} |y|^2 \, \mu(dy) \right)$$

$$= \frac{2}{N} \sum_{i=1}^{N} |\xi_i|^2 + 2 \int_{\mathbb{R}^d} |x|^2 \mu(dx),$$

which is nonnegative and converges almost surely toward $4 \int_{\mathbb{R}^d} |x|^2 \mu(dx)$, which is a finite constant. Since the limit is a constant, the convergence is also in the sense of L^1 from which we conclude. □

Theorem 1.10. *Under the above assumptions we have*

$$\lim_{N \to \infty} \sup_{1 \leq i \leq N} \mathbb{E} \sup_{0 \leq t \leq T} |X_t^{i,N} - X_t^i|^2 = 0. \tag{1.25}$$

Moreover, if $\mathbb{E}|X_0|^{d+5} < \infty$, *then*

$$\sup_{1 \leq i \leq N} \mathbb{E} \sup_{0 \leq t \leq T} |X_t^{i,N} - X_t^i|^2 \leq C N^{-2/(d+4)}, \tag{1.26}$$

and in the case of interaction of the first order given by Lip-1 functions, we also have

$$\sup_{1 \leq i \leq N} \mathbb{E} \sup_{0 \leq t \leq T} |X_t^{i,N} - X_t^i|^2 \leq \frac{C}{N}, \tag{1.27}$$

where the constant C does not depend upon N.

Usually, propagation of chaos is a statement about distributions rather than a pathwise statement. It says that for fixed $k \geq 1$, if we let $N \nearrow \infty$, then the law of $(X_t^{i,N})_{t \in [0,T]}^{i=1,\cdots,k}$ converges toward the probability distribution of $(X_t^i)_{t \in [0,T]}^{i=1,\cdots,k}$, implying that the k particles $\mathbf{X}^{1,N}, \cdots, \mathbf{X}^{k,N}$ become independent and acquire the same distribution as given by the solution of the McKean–Vlasov SDE (1.15). Theorem 1.10, especially statement (1.25), gives a pathwise form of this statement by constructing the finite particle systems and their chaotic limits on the same probability space and proving pathwise convergence via a mean-field coupling.

Proof.

$$\mathbb{E}\sup_{0\le s\le t}|X_s^{i,N}-X_s^i|^2 \le c\int_0^t \mathbb{E}|b\left(s,X_s^{i,N},\frac{1}{N}\sum_{k=1}^N \delta_{X_s^{k,N}}\right) - b\left(s,X_s^i,\frac{1}{N}\sum_{k=1}^N \delta_{X_s^k}\right)|^2 ds$$

$$+ c\int_0^t \mathbb{E}|b\left(s,X_s^i,\frac{1}{N}\sum_{k=1}^N \delta_{X_s^k}\right) - b(s,X_s^i,\mathcal{L}(X_s^i))|^2 ds$$

$$+ c\int_0^t \mathbb{E}|\sigma\left(s,X_s^{i,N},\frac{1}{N}\sum_{k=1}^N \delta_{X_s^{k,N}}\right) - \sigma\left(s,X_s^i,\frac{1}{N}\sum_{k=1}^N \delta_{X_s^k}\right)|^2 ds$$

$$+ c\int_0^t \mathbb{E}|\sigma\left(s,X_s^i,\frac{1}{N}\sum_{k=1}^N \delta_{X_s^k}\right) - \sigma(s,X_s^i,\mathcal{L}(X_s^i))|^2 ds.$$

Using the Lipschitz property of the coefficients b and σ, the elementary estimate (1.24) controlling the distance between two empirical measures, and the exchangeability of the couples $(X^i, X^{i,N})_{1\le i\le N}$, one sees that the first and third terms of the above right-hand side are bounded from above by $c\int_0^t \mathbb{E}\sup_{0\le r\le s}|X_r^{i,N}-X_r^i|^2 ds$. Using Gronwall's inequality and again the Lipschitz property of the coefficients b and σ, we get

$$\mathbb{E}\sup_{0\le s\le t}|X_s^{i,N}-X_s^i|^2 \le c\int_0^t \mathbb{E}|b\left(s,X_s^i,\frac{1}{N}\sum_{k=1}^N \delta_{X_s^{k,N}}\right) - b(s,X_s^i,\mathcal{L}(X_s^i))|^2 ds$$

$$+ c\int_0^t \mathbb{E}|\sigma\left(s,X_s^i,\frac{1}{N}\sum_{k=1}^N \delta_{X_s^{k,N}}\right) - \sigma(s,X_s^i,\mathcal{L}(X_s^i))|^2 ds$$

$$\le c\int_0^t \mathbb{E}W^{(2)}\left(\frac{1}{N}\sum_{k=1}^N \delta_{X_s^{k,N}},\mathcal{L}(X_s^i)\right)^2 ds.$$

The first claim of the theorem now follows from Lemma 1.9 and Lebesgue's dominated convergence theorem. We will not prove the second assertion about the rate of convergence to 0. The third one follows from the remark that

$$\mathbb{E}|b\left(s,X_s^i,\frac{1}{N}\sum_{k=1}^N \delta_{X_s^{k,N}}\right) - b(s,X_s^i,\mathcal{L}(X_s^i))|^2 = \sum_{k=1}^d \frac{1}{N^2}\sum_{j,\ell=1}^d,$$

$$\mathbb{E}\left(\left[\tilde{b}_k(s,X_s^i,X_s^j) - \int_{\mathbb{R}^d} \tilde{b}_k(s,X_s^i,y)\mathbb{P}_{X_s^i}(dy)\right]\left[\tilde{b}_k(s,X_s^i,X_s^\ell) - \int_{\mathbb{R}^d} \tilde{b}_k(s,X_s^i,y)\mathbb{P}_{X_s^i}(dy)\right]\right),$$

which is enough to conclude, since by independence of the random variables $X_s^1,\cdots,X_s^N$, which have the common law $\mathbb{P}_{X_s^i}$, the expectation in the above summation vanishes whenever $j \ne \ell$, and a similar argument can be used to control the terms involving σ. □

1.4 ▪ Conditional Propagation of Chaos

In this section we generalize the propagation of chaos result derived earlier to a conditional framework. This generalization is motivated by the analysis of large stochastic systems for which the mean-field interaction will not be enough to produce deterministic limits when the size of the population goes to infinity. While the ideas and tools are very similar to those

1.4. Conditional Propagation of Chaos

presented in the previous section, the *conditional* nature of quantities of interest makes the mathematical analysis much more involved, and some readers may want to skip some of the technicalities of this section if they are not interested in the particular applications we use as motivation.

To be specific, we consider a system of $(N+1)$ interacting particles with stochastic dynamics:

$$\begin{cases} dX_t^{0,N} = b_0(t, X_t^{0,N}, \mu_t^N)dt + \sigma_0(t, X_t^{0,N}, \mu_t^N)dW_t^0, \\ dX_t^{i,N} = b(t, X_t^{i,N}, \mu_t^N, X_t^{0,N})dt + \sigma(t, X_t^{i,N}, \mu_t^N, X_t^{0,N})dW_t^i, \quad i = 1, 2, \ldots, N, \\ X_0^{0,N} = x_0^0, \quad X_0^{i,N} = x_0, \quad i = 1, 2, \ldots, N, \end{cases} \quad (1.28)$$

on a probability space $(\Omega, \mathcal{F}, \mathbb{P})$, where the empirical measure μ_t^N is defined by

$$\mu_t^N = \frac{1}{N} \sum_{i=1}^N \delta_{X_t^{i,N}}.$$

Here $(\boldsymbol{W}^i)_{i \geq 0}$ is a sequence of independent Wiener processes, $\boldsymbol{W}^0$ being m_0-dimensional and $\boldsymbol{W}^i$ being m-dimensional for $i \geq 1$. The major-particle process $\boldsymbol{X}^{0,N}$ is d_0-dimensional, and the minor-particle processes $\boldsymbol{X}^{i,N}$ are d-dimensional for $i \geq 1$. The coefficient functions

$$(b_0, \sigma_0) : [0, T] \times \Omega \times \mathbb{R}^{d_0} \times \mathcal{P}_2(\mathbb{R}^d) \to \mathbb{R}^{d_0} \times \mathbb{R}^{d_0 \times m_0},$$
$$(b, \sigma) : [0, T] \times \Omega \times \mathbb{R}^d \times \mathcal{P}_2(\mathbb{R}^d) \times \mathbb{R}^{d_0} \to \mathbb{R}^d \times \mathbb{R}^{d \times m}$$

are allowed to be random. We make the following assumptions.

(A1.1) The functions b_0 and σ_0 (resp., b and σ) are $\mathcal{P}^{W^0} \otimes \mathcal{B}(\mathbb{R}^{d_0}) \otimes \mathcal{B}(\mathcal{P}_2(\mathbb{R}^d))$-measurable (resp., $\mathcal{P}^{W^0} \otimes \mathcal{B}(\mathbb{R}^d) \otimes \mathcal{B}(\mathcal{P}_2(\mathbb{R}^d)) \otimes \mathcal{B}(\mathbb{R}^{d_0})$-measurable), where $\mathcal{P}^{W^0}$ is the progressive σ-field associated with the filtration $\mathbb{F}^0 = (\mathcal{F}_t^0)$ on $[0, T] \times \Omega$ generated by the Wiener process $\boldsymbol{W}^0$, and $\mathcal{B}(\mathcal{P}_2(\mathbb{R}^d))$ is the Borel σ-field generated by the metric $W^{(2)}$.

(A1.2) There exists a constant $K > 0$ such that for all $t \in [0,T]$, $\omega \in \Omega$, $x, x' \in \mathbb{R}^d$, $x_0, x_0' \in \mathbb{R}^{d_0}$, and $\mu, \mu' \in \mathcal{P}_2(\mathbb{R}^d)$,

$$|(b_0, \sigma_0)(t, \omega, x_0, \mu) - b_0(t, \omega, x_0', \mu')| \leq K[|x_0 - x_0'| + W^{(2)}(\mu, \mu')],$$
$$|(b, \sigma)(t, \omega, x, \mu, x_0) - b(t, \omega, x', \mu', x_0')| \leq K[|x - x'| + |x_0 - x_0'| + W^{(2)}(\mu, \mu')].$$

(A1.3) We have

$$\mathbb{E}\left[\int_0^T |(b_0, \sigma_0)(t, 0, \delta_0)|^2 + |(b, \sigma)(t, 0, \delta_0, 0)|^2 dt\right] < \infty.$$

Our goal is to study the limiting behavior of the solution of the system (1.28) when N tends to infinity. The limit will be given by the so-called *limiting nonlinear processes*, but before defining it, we need to introduce notations and definitions for the regular versions of conditional probabilities which we use throughout.

1.4.1 ▪ Regular Conditional Distributions and Optional Projections

We assume that $(\Omega, \mathcal{F})$ is a standard measurable space, $\mathcal{F}$ being the Borel σ-field for us to be able to use regular conditional distributions for any sub-σ-field of $\mathcal{F}$. In fact, if $\mathbb{G} = (\mathcal{G}_t)$ is a right continuous filtration, we make use of the existence of a map $\Pi^{\mathbb{G}} : [0, \infty) \times \Omega \hookrightarrow \mathcal{P}(\Omega)$ which is $(\mathcal{O}, \mathcal{B}(\mathcal{P}(\Omega)))$-measurable and such that for each $t \geq 0$,

$\{\Pi_t^\mathbb{G}(\omega, A); \omega \in \Omega, A \in \mathcal{F}\}$ is a regular version of the conditional probability of $\mathbb{P}$ given the σ-field $\mathcal{G}_t$. Here $\mathcal{O}$ denotes the optional σ-field of the filtration $\mathbb{G} = (\mathcal{G}_t)$. This result is a direct consequence of Proposition 1 in [118] applied to the process $\boldsymbol{X} = (X_t)$ given by the identity map of Ω and the constant filtration $\mathcal{F}_t \equiv \mathcal{F}$. For each $t \geq 0$, we define the probability measures $\mathbb{P} \otimes \Pi_t^\mathbb{G}$ and $\Pi_t^\mathbb{G} \otimes \mathbb{P}$ on $\Omega^2 = \Omega \times \Omega$ via the formulas

$$\mathbb{P} \otimes \Pi_t^\mathbb{G}(A \times B) = \int_A \Pi_t^\mathbb{G}(\omega, B)\mathbb{P}(d\omega) \quad \text{and} \quad \Pi_t^\mathbb{G} \otimes \mathbb{P}(A \times B) = \int_B \Pi_t^\mathbb{G}(\omega, A)\mathbb{P}(d\omega). \tag{1.29}$$

It is easy to check that integrals of functions of the form $\Omega^2 \ni (\omega, \tilde{\omega}) \hookrightarrow \varphi(\omega)\psi(\tilde{\omega})$ with respect to these two measures are equal. This shows that these two measures are the same. We will use this result in the following way: if X is measurable and bounded on Ω^2, we can interchange ω and $\tilde{\omega}$ in the integrand of

$$\int_{\Omega^2} X(\omega, \tilde{\omega})\Pi_t^\mathbb{G}(\omega, d\tilde{\omega})\mathbb{P}(d\omega)$$

without changing the value of the integral.

In this section, we often use the notation $\mathbb{E}^{\mathcal{G}_t}$ for the expectation with respect to the transition kernel $\Pi_t^\mathbb{G}$, i.e., for all random variables $X : \Omega^2 \ni (\omega, \tilde{\omega}) \hookrightarrow X(\omega, \tilde{\omega}) \in \mathbb{R}$, we define

$$\mathbb{E}^{\mathcal{G}_t}[X(\omega, \tilde{\omega})] = \int_\Omega X(\omega, \tilde{\omega})\Pi_t^\mathbb{G}(\omega, d\tilde{\omega}),$$

which, as a function of ω, is a random variable on Ω. Also, we still use $\mathbb{E}$ to denote the expectation with respect to the first argument, i.e.,

$$\mathbb{E}[X] = \int_\Omega X(\omega, \tilde{\omega})\mathbb{P}(d\omega),$$

which, as a function of $\tilde{\omega}$, is a random variable on Ω. Finally, whenever we have a random variable X defined on Ω, we define the random variable $\tilde{X}$ on Ω^2 via the formula $\tilde{X}(\omega, \tilde{\omega}) = X(\tilde{\omega})$.

1.4.2 • Conditional McKean–Vlasov SDEs

We first derive a few technical properties of the conditional distributions of a process with respect to a filtration. We assume that the filtration $\mathbb{G} = (\mathcal{G}_t)$ is a subfiltration of a right continuous filtration $\mathbb{F} = (\mathcal{F}_t)$, in particular $\mathcal{G}_t \subseteq \mathcal{F}_t$ for all $t \geq 0$, and that $\boldsymbol{X} = (X_t)$ is an $\mathbb{F}$-adapted continuous process taking values in a Polish space $(E, \mathcal{E})$. Defining $\mu_t^X(\omega)$ as the distribution of the random variable X_t under the probability measure $\Pi_t^\mathbb{G}(\omega, \cdot)$, we obtain the following result which we state as a lemma for future reference.

Lemma 1.11. *There exists a stochastic measure flow $\mu^X : [0, \infty) \times \Omega \to \mathcal{P}(E)$ such that*

1. *μ^X is $\mathcal{P}/\mathcal{B}(\mathcal{P}(E))$-measurable, where $\mathcal{P}$ is the progressive σ-field associated to $\mathbb{G}$ on $[0, \infty) \times \Omega$, and $\mathcal{B}(\mathcal{P}(E))$, the Borel σ-field of the weak topology on $\mathcal{P}(E)$.*
2. *$\forall t \geq 0$, μ_t^X is a regular conditional distribution of X_t given $\mathcal{G}_t$.*

Our goal is to prove the well-posedness of the SDE

$$dX_t = b(t, X_t, \mathcal{L}(X_t|\mathcal{G}_t))dt + \sigma(t, X_t, \mathcal{L}(X_t|\mathcal{G}_t))dW_t. \tag{1.30}$$

We say that this SDE is of the conditional McKean–Vlasov type because the conditional distribution of X_t with respect to $\mathcal{G}_t$ enters the dynamics. Note that when $\mathcal{G}_t$ is the trivial

1.4. Conditional Propagation of Chaos

σ-field, (1.30) reduces to a classical McKean–Vlasov SDE. In the following, when writing $\mathcal{L}(X_t|\mathcal{G}_t)$, we always mean μ_t^X for the stochastic flow μ^X, whose existence is given in Lemma 1.11.

The analysis of the SDE (1.30) is done under the following assumptions. W is an m-dimensional Wiener process on a probability space $(\Omega, \mathcal{F}, \mathbb{P})$, $\mathbb{F}^0 = (\mathcal{F}_t^0)$ its (raw) filtration, $\mathbb{F} = \mathbb{F}^W$ its usual $\mathbb{P}$-augmentation (we shall often write $\mathcal{F}_t$ for $\mathcal{F}_t^W$), and $\mathbb{G} = (\mathcal{G}_t)$ a subfiltration of $\mathcal{F}_t$ also satisfying the usual conditions. We impose the following standard assumptions on b and σ.

(B1.1) The function

$$(b, \sigma) : [0, T] \times \Omega \times \mathbb{R}^n \times \mathcal{P}(\mathbb{R}^n) \ni (t, \omega, x, \mu) \hookrightarrow (b(t, \omega, x, \mu), \sigma(t, \omega, x, \mu)) \in \mathbb{R}^n \times \mathbb{R}^{n \times m}$$

is $\mathcal{P}^{\mathbb{G}} \otimes \mathcal{B}(\mathbb{R}^n) \otimes \mathcal{B}(\mathcal{P}(\mathbb{R}^n))$-measurable, where $\mathcal{P}^{\mathbb{G}}$ is the progressive σ-field associated with the filtration $\mathbb{G}$ on $[0, T] \times \Omega$.

(B1.2) There exists $K > 0$ such that for all $t \in [0, T]$, $\omega \in \Omega$, $x, x' \in \mathbb{R}^n$, and $\mu, \mu' \in \mathcal{P}_2(\mathbb{R}^n)$, we have

$$|b(t, \omega, x, \mu) - b(t, \omega, x', \mu')| + |\sigma(t, \omega, x, \mu) - \sigma(t, \omega, x', \mu')| \leq K(|x - x'| + W^{(2)}(\mu, \mu')).$$

(B1.3) It holds that

$$\mathbb{E}\left[\int_0^T |b(t, 0, \delta_0)|^2 + |\sigma(t, 0, \delta_0)|^2 dt\right] < \infty.$$

Definition 1.12. *By a (strong) solution of (1.30) we mean an $\mathcal{F}_t$-adapted continuous process X taking values in $\mathbb{R}^n$ such that for all $t \in [0, T]$,*

$$X_t = x_0 + \int_0^t b(s, X_s, \mathcal{L}(X_s|\mathcal{G}_s))ds + \int_0^t \sigma(s, X_s, \mathcal{L}(X_s|\mathcal{G}_s))dW_s, \quad a.s.$$

In order to establish the well-posedness of (1.30) we need some form of control on the 2-Wasserstein distance $W^{(2)}$ between two conditional distributions. We shall use the following dual representation.

Proposition 1.13. *If $\mu, \nu \in \mathcal{P}_2(E)$ where E is a Euclidean space, then*

$$W^{(2)}(\mu, \nu)^2 = \sup_{\phi \in \mathcal{C}_b^{\text{Lip}}(E)} \left(\int_E \phi^* d\mu - \int_E \phi d\nu\right),$$

where $\phi^*(x) := \inf_{z \in E} \phi(z) + |x - z|^2$.

We shall use the following consequences of this representation.

Lemma 1.14. *If X and Y are two random variables of order 2 taking values in a Euclidean space, and $\mathcal{G}$ is a sub-σ-field of $\mathcal{F}$, then for all $p \geq 2$ we have*

$$W^{(2)}(\mathcal{L}(X|\mathcal{G}), \mathcal{L}(Y|\mathcal{G}))^p \leq \mathbb{E}[|X - Y|^p|\mathcal{G}], a.s.$$

By taking expectations on both sides we further have

$$\mathbb{E}\left[W^{(2)}(\mathcal{L}(X|\mathcal{G}), \mathcal{L}(Y|\mathcal{G}))^p\right] \leq \mathbb{E}[|X - Y|^p].$$

Proof. By using the above dual representation formula and the characteristic equation for conditional distributions, we get

$$W^{(2)}(\mathcal{L}(X|\mathcal{G}), \mathcal{L}(Y|\mathcal{G}))^2 = \sup_{\phi \in \mathcal{C}_b^{\text{Lip}}(E)} \mathbb{E}[\phi^*(X) - \phi(Y)|\mathcal{G}] \leq \mathbb{E}[|X - Y|^2|\mathcal{G}].$$

The first inequality follows Jensen's inequality for conditional expectations. □

We then have the following well-posedness result.

Proposition 1.15. *If* **(B1.1)**, **(B1.2)**, *and* **(B1.3)** *hold, the conditional McKean–Vlasov SDE (1.30) has a unique strong solution. Moreover, for all $p \geq 2$, if we replace the assumption* **(B1.3)** *by*

$$\mathbb{E} \int_0^T |b(t, 0, \delta_0)|^p + |\sigma(t, 0, \delta_0)|^p dt < \infty,$$

the solution of (1.30) satisfies

$$\mathbb{E}\left[\sup_{0 \leq t \leq T} |X_t|^p\right] < \infty.$$

Proof. The proof is an application of the contraction mapping theorem which mimics the existence proofs already given in this chapter. For each $c > 0$, we consider the Banach space $\mathbb{H}_c^2$ of $\mathbb{F}$-progressively measurable processes satisfying

$$\|\boldsymbol{X}\|_c^2 := \mathbb{E}\left[\int_0^T e^{-ct}|X_t|^2 dt\right] < \infty.$$

For each $\boldsymbol{X} \in \mathbb{H}_c^2$, we have

$$\mathcal{L}(X_t|\mathcal{G}_t) \in \mathcal{P}_2(\mathbb{R}^n), \quad \text{a.s., a.e.,}$$

and we can define

$$U_t = x_0 + \int_0^t b(s, X_s, \mathcal{L}(X_s|\mathcal{G}_s)) ds + \int_0^t \sigma(s, X_s, \mathcal{L}(X_s|\mathcal{G}_s)) dW_s.$$

It is easy to see that $\mathbf{U} \in \mathbb{H}_c^2$. Moreover, if $\boldsymbol{X}, \boldsymbol{X}' \in \mathbb{H}_c^2$ and $\mathbf{U}$ and $\mathbf{U}'$ are the processes defined via the above equality from $\boldsymbol{X}$ and $\boldsymbol{X}'$ respectively, we have

$$\mathbb{E}\left[\left|\int_0^t b(s, X_s', \mathcal{L}(X_s'|\mathcal{G}_s)) - b(s, X_s, \mathcal{L}(X_s|\mathcal{G}_s)) ds\right|^2\right]$$

$$\leq 2TK^2 \mathbb{E}\left[\int_0^t |X_s' - X_s|^2 + W_2^2(\mathcal{L}(X_s'|\mathcal{G}_s), \mathcal{L}(X_s|\mathcal{G}_s)) ds\right]$$

$$\leq 2TK^2 \mathbb{E}\left[\int_0^t |X_s' - X_s|^2 ds\right],$$

and we have the same type of estimate for the stochastic integral term by replacing the

Cauchy–Schwarz inequality with the Itô isometry. This yields

$$\|\mathbf{U}' - \mathbf{U}\|_c^2 = \mathbb{E}\left[\int_0^T e^{-ct}|U_t' - U_t|^2 dt\right]$$

$$\leq 2(T+1)K^2 \mathbb{E}\left[\int_0^T e^{-ct}\left(\int_0^t |X_s' - X_s|^2 ds\right) dt\right]$$

$$\leq \frac{2(T+1)K^2}{c}\|\mathbf{X}' - \mathbf{X}\|_c^2,$$

and this proves that the map $\mathbf{X} \to \mathbf{U}$ is a strict contraction in the Banach space $\mathbb{H}_c^2$ if we choose c sufficiently large. As in the unconditional case, the fact that the solution possesses finite moments can be obtained by using standard estimates and Lemma 1.14. We omit the details here. $\square$

In the above discussion, $\mathbb{G}$ is a rather general subfiltration of the Brownian filtration $\mathbb{F}^W$. From now on, we shall restrict ourselves to subfiltrations $\mathbb{G}$ equal to the Brownian filtration generated by the first r components of $\mathbf{W}$ for some $r < m$. We rewrite (1.30) as

$$dX_t = b(t, X_t, \mathcal{L}(X_t|\mathcal{G}_t^W))dt + \sigma(t, X_t, \mathcal{L}(X_t|\mathcal{G}_t^W))dW_t, \tag{1.31}$$

and we expect that the solution of the SDE (1.31) is given by a deterministic functional of the Brownian paths. In order to prove this fact in a rigorous way, we need the following notion.

Definition 1.16. *By a setup we mean a 4-tuple $(\Omega, \mathcal{F}, \mathbb{P}, \mathbf{W})$, where $(\Omega, \mathcal{F}, \mathbb{P})$ is a probability space with an m-dimensional Wiener process $\mathbf{W}$. We use the notation $\mathbb{F}^W$ for the natural filtration of $\mathbf{W}$ and $\mathbb{G}^W$ for the natural filtration of the first r components of $\mathbf{W}$. By the canonical setup we mean $(\Omega^c, \mathcal{F}^c, \mathbb{W}, B)$, where $\Omega^c = C([0,T]; \mathbb{R}^m)$, $\mathcal{F}^c$ is the Borel σ-field associated with the uniform topology, $\mathbb{W}$ is the Wiener measure, and B_t is the coordinate (marginal) projection.*

Proposition 1.15 basically states that the SDE (1.31) is uniquely solvable on any setup, and in particular that it is uniquely solvable on the canonical setup. The solution in the canonical setup, denoted by $\mathbf{X}^c$, gives us a measurable functional from $C([0,T]; \mathbb{R}^d)$ to $C([0,T]; \mathbb{R}^n)$. Because of the important role played by this functional, in the following we use Φ (instead of X^c) to denote it.

Lemma 1.17. *Let $\psi : C([0,T]; \mathbb{R}^m) \to \mathbb{R}^n$ be $\mathcal{F}_t^B$-measurable; then we have*

$$\mathcal{L}(\psi|\mathcal{G}_t^B)(\mathbf{W}.) = \mathcal{L}(\psi(\mathbf{W}.)|\mathcal{G}_t^W).$$

Proof. By the very definition of conditional distributions, it suffices to prove that for all bounded measurable functions $f : \mathbb{R}^n \to \mathbb{R}^+$, we have

$$\mathbb{E}\left[f(\psi(\mathbf{W}.))|\mathcal{G}_t^W\right] = \mathbb{E}\left[f(\psi)|\mathcal{G}_t^B\right](\mathbf{W}.),$$

and by using the definition of conditional expectations the above equality can be easily proved. $\square$

With the help of Lemma 1.17, we can state and prove the following.

Proposition 1.18. *On any setup* $(\Omega, \mathcal{F}, \mathbb{P}, \boldsymbol{W})$, *the solution of* (1.31) *is given by*

$$\boldsymbol{X}. = \Phi(\boldsymbol{W}.).$$

Proof. We are going to check directly that $\Phi(\boldsymbol{W}.)$ is a solution of (1.31). By the definition of Φ as the solution of (1.31) on the canonical setup, we have

$$\Phi(\mathtt{w}) = x_0 + \int_0^t b(s, \Phi(\mathtt{w})_s, \mathcal{L}(\Phi(\cdot)_s|\mathcal{G}_s^B)(\mathtt{w}))ds$$
$$+ \int_0^t \sigma(s, \Phi(\mathtt{w})_s, \mathcal{L}(\Phi(\cdot)_s|\mathcal{G}_s^B)(\mathtt{w}))dB_s, \mathbb{W}\text{-a.s.},$$

where $\mathtt{w}$ stands for a generic element in the canonical space $C([0,T];\mathbb{R}^m)$. By using Lemma 1.17 we thus have

$$\Phi(\boldsymbol{W}.) = x_0 + \int_0^t b(s, \Phi(\boldsymbol{W}.)_s, \mathcal{L}(\Phi(\boldsymbol{W}.)_s|\mathcal{G}_s^B))ds$$
$$+ \int_0^t \sigma(s, \Phi(\boldsymbol{W}.)_s, \mathcal{L}(\Phi(\boldsymbol{W}.)_s|\mathcal{G}_s^B))dW_s,$$

$\mathbb{P}$-a.s., which proves the desired result. □

1.4.3 ▪ The Nonlinear Processes

The limiting nonlinear process associated with the particle system (1.28) is defined as the solution of

$$\begin{cases} dX_t^0 = b_0(t, X_t^0, \mathcal{L}(X_t^1|\mathcal{F}_t^0))dt + \sigma_0(t, X_t^0, \mathcal{L}(X_t^1|\mathcal{F}_t^0))dW_t^0, \\ dX_t^i = b(t, X_t^i, \mathcal{L}(X_t^i|\mathcal{F}_t^0), X_t^0)dt + \sigma(t, X_t^i, \mathcal{L}(X_t^i|\mathcal{F}_t^0), X_t^0)dW_t^i, \quad i \geq 1, \\ X_0^0 = x_0^0, \qquad X_0^i = x_0, \quad i \geq 1. \end{cases} \quad (1.32)$$

Under the assumptions **(A1.1)–(A1.3)**, the unique solvability of this system is ensured by Proposition 1.15. Due to the strong symmetry among the processes $(\boldsymbol{X}^i)_{i \geq 1}$, we first prove the following proposition.

Proposition 1.19. *For all $i \geq 1$, the solution of* (1.32) *solves the conditional McKean–Vlasov SDE*

$$\begin{cases} dX_t^0 = b_0(t, X_t^0, \mathcal{L}(X_t^i|\mathcal{F}_t^0))dt + \sigma_0(t, X_t^0, \mathcal{L}(X_t^i|\mathcal{F}_t^0))dW_t^0, \\ dX_t^i = b(t, X_t, \mathcal{L}(X_t^i|\mathcal{F}_t^0), X_t^0)dt + \sigma(t, X_t^i, \mathcal{L}(X_t^i|\mathcal{F}_t^0), X_t^0)dW_t^i, \end{cases}$$

and for all fixed $t \in [0, T]$, the random variables $(X_t^i)_{i \geq 1}$ are $\mathcal{F}_t^0$-conditionally i.i.d.

Proof. This is an immediate consequence of Proposition 1.18. □

Now that the nonlinear processes are well-defined, in the next subsection we prove that these processes give the limiting behavior of (1.28) when N tends to infinity.

1.4.4 ▪ Conditional Propagation of Chaos

We extend the result of the unconditional theory to the conditional case involving the influence of a major individual in the population. As in the classical case, we prove the (strong) pathwise version of the propagation of chaos.

Theorem 1.20. *There exists a constant C such that*

$$\max_{0 \leq i \leq N} \mathbb{E}\left[\sup_{0 \leq t \leq T} |X_t^{i,N} - X_t^i|^2\right] \leq CN^{-2/(d+4)},$$

where C only depends on T, the Lipschitz constants of b_0 and b, and $\eta = \mathbb{E}[\int_0^T |X_t^1|^{d+5} dt]$.

Proof. We first note that, by the SDEs satisfied by X^0 and $X^{0,N}$ and the Lipschitz conditions on the coefficients,

$$|X_t^{0,N} - X_t^0|^2 = \left| \int_0^t \left[b_0\left(s, X_s^{0,N}, \frac{1}{N}\sum_{j=1}^N \delta_{X_s^{j,N}}\right) - b_0(s, X_s^0, \mu_s) \right] ds \right.$$

$$\left. + \int_0^t \left[\sigma_0\left(s, X_s^{0,N}, \frac{1}{N}\sum_{j=1}^N \delta_{X_s^{j,N}}\right) - \sigma_0(s, X_s^0, \mu_s) \right] dW_s^0 \right|^2$$

$$\leq K \left[\int_0^t |X_s^{0,N} - X_s^0|^2 ds + \int_0^t W^{(2)}\left(\frac{1}{N}\sum_{j=1}^N \delta_{X_s^{j,N}}, \frac{1}{N}\sum_{j=1}^N \delta_{X_s^j}\right)^2 ds \right.$$

$$\left. + \int_0^t W^{(2)}\left(\frac{1}{N}\sum_{j=1}^N \delta_{X_s^j}, \mu_s\right)^2 ds \right]$$

$$\leq K \left[\int_0^t |X_s^{0,N} - X_s^0|^2 ds + \int_0^t \frac{1}{N}\sum_{j=1}^N |X_s^{j,N} - X_s^j|^2 ds \right.$$

$$\left. + \int_0^t W^{(2)}\left(\frac{1}{N}\sum_{j=1}^N \delta_{X_s^j}, \mu_s\right)^2 ds \right].$$

If we take the supremum and the expectation on both sides, by the exchangeability we get

$$\mathbb{E}\left[\sup_{0 \leq s \leq t} |X_s^{0,N} - X_s^0|^2\right]$$

$$\leq K \left[\int_0^t \mathbb{E}\left[\sup_{0 \leq u \leq s} |X_u^{0,N} - X_u^0|^2\right] ds + \int_0^t \mathbb{E}[(X_s^{1,N} - X_s^1)^2] ds \right.$$

$$\left. + \int_0^t \mathbb{E}\left[W^{(2)}\left(\frac{1}{N}\sum_{j=1}^N \delta_{X_s^j}, \mu_s\right)^2\right] ds \right]$$

$$\leq K \left[\int_0^t \mathbb{E}\left[\sup_{0 \leq u \leq s} |X_u^{0,N} - X_u^0|^2\right] ds + \int_0^t \mathbb{E}\left[\sup_{0 \leq u \leq s} |X_u^{1,N} - X_u^1|^2\right] ds \right.$$

$$\left. + \int_0^t \mathbb{E}\left[W^{(2)}\left(\frac{1}{N}\sum_{j=1}^N \delta_{X_s^j}, \mu_s\right)^2\right] ds \right].$$

Following the same strategy, we readily obtain a similar estimate for $X^{1,N} - X^1$:

$$\mathbb{E}\left[\sup_{0\leq s\leq t} |X_s^{1,N} - X_s^1|^2\right]$$
$$\leq K'\left[\int_0^t \mathbb{E}\left[\sup_{0\leq u\leq s} |X_u^{0,N} - X_u^0|^2\right]ds + \int_0^t \mathbb{E}\left[\sup_{0\leq u\leq s} |X_u^{1,N} - X_u^1|^2\right]ds\right.$$
$$\left. + \int_0^t \mathbb{E}\left[W^{(2)}\left(\frac{1}{N}\sum_{j=1}^N \delta_{X_s^j}, \mu_s\right)^2\right]ds\right],$$

and by summing the above two inequalities and using Gronwall's lemma we get

$$\mathbb{E}\left[\sup_{0\leq t\leq T} |X_t^{0,N} - X_t^0|^2\right] + \mathbb{E}\left[\sup_{0\leq t\leq T} |X_t^{1,N} - X_t^1|^2\right]$$
$$\leq K\int_0^T \mathbb{E}\left[W^{(2)}\left(\frac{1}{N}\sum_{j=1}^N \delta_{X_t^j}, \mu_t\right)^2\right]ds$$
$$\leq K\mathbb{E}\left[\int_0^T |X_t^1|^{d+5}\right] N^{-2/(d+4)},$$

where the second inequality comes from a direct application of Lemma 1.9, with the help of Lemma 1.14. This completes the proof of the desired result. □

1.5 ▪ Notes & Complements

The theory of stochastic differential equations is a well-established field of probability and applied mathematics. It occupies the lion's share of the literature on stochastic analysis. For this reason, we have only presented the basic existence and uniqueness result in the case of Lipschitz coefficients. We limited ourselves to the notion of strong solutions and we emphasized the possible randomness of the coefficients, as this is a crucial element of the formulation of stochastic control and stochastic game models which we study in these lectures. Still, later we shall consider the weak formulations of these models, relying on tools developed for the theory of weak solutions of stochastic differential equations. We used Touzi's monograph [114] as a source to which we refer the reader interested in the subject. We chose to discuss the Kolmogorov forward equation because it will play an important role in our discussion of mean-field games in Chapter 6. The second half of the chapter is devoted to SDEs of mean-field type, also known as McKean–Vlasov SDEs. They will appear in Chapter 4 as a new breed of stochastic dynamics whose optimal control can be handled with the methods presented in these lectures, and the propagation of chaos result will be instrumental in the construction of approximate Nash equilibriums in Chapter 6. This material rarely appears in textbook form. Our presentation is modeled after the lecture notes of Sznitman [113] which have been very influential. The interested reader may want to look at [72] by Jourdain, Méléard, and Woyczynski for an extension of these results to SDEs driven by Lévy processes. The discussion of conditional McKean–Vlasov equations and the corresponding propagation of chaos results are borrowed from the paper [42] by Carmona and Zhu, in which the theory was developed to be applied to mean-field games with major and minor players. See Chapter 6 for details and complements. This is also presented in the book by Carmona and Delarue [35], in which the authors go to great lengths to prepare for the analysis of mean-field games with a common noise, a topic which will not be discussed in Chapter 6. Also, [35] provides a precise rate of convergence for the second claim of Lemma 1.9. This rate is sharper than the rate used in Theorems 1.10 and 1.20.

Chapter 2
Backward Stochastic Differential Equations

2.1 ▪ Introduction and First Definitions

We use the same notation as in the first chapter. Moreover, $\mathcal{P}$ stands for the σ-field on $[0,T] \times \Omega$ generated by the $\mathbb{F}$-progressively measurable bounded processes. Throughout the chapter we assume that the filtration $\mathbb{F}$ is generated by a Wiener process $\mathbf{W}$ with values in $\mathbb{R}^m$.

2.1.1 ▪ First Definitions and Examples

The data of a backward stochastic differential equation (BSDE) comprise:

- a coefficient or driver which is a function $\Psi : [0,T] \times \Omega \times \mathbb{R}^p \times \mathbb{R}^{pm} \hookrightarrow \mathbb{R}^p$ such that $(\Psi(t, \cdot, y, z))_{0 \leq t \leq T}$ is $\mathcal{P}$-measurable for each $(y, z) \in \mathbb{R}^p \times \mathbb{R}^{pm}$;
- a terminal condition $\xi \in L^2(\Omega, \mathcal{F}, \mathbb{P}; \mathbb{R}^p)$.

We often denote the corresponding backward stochastic differential equation by $BSDE(\Psi, \xi)$.

Definition 2.1. *A solution of the BSDE with driver Ψ and terminal condition ξ is a pair $(\mathbf{Y}, \mathbf{Z})$ of $\mathcal{P}$-measurable processes with values in $\mathbb{R}^p$ and $\mathbb{R}^{pm}$ respectively, satisfying*

1. $\mathbb{E}\{\sup_{0 \leq t \leq T} |Y_t|^2 + \int_0^T |Z_t|^2 dt\} < \infty$;
2. $\forall t \leq T, \ Y_t = \xi + \int_t^T \Psi(s, Y_s, Z_s) ds - \int_t^T Z_s dW_s,$ *a.s. on $\mathbb{R}^p$.*

Using the spaces introduced in Chapter 1, condition 1 says that $\mathbf{Y} \in \mathbb{S}^{2,p}$ and $\mathbf{Z} \in \mathbb{H}^{2,pm}$.

Examples

$\Psi \equiv 0$—**Martingale Representation Theorem (MRT):** Since we are in a Brownian filtration, for any $\xi \in L^2(\Omega, \mathcal{F}_T, \mathbb{P})$, if we set $Y_t = \mathbb{E}[\xi|\mathcal{F}_t]$, there exists a unique $Z \in \mathbb{H}^{2,pm}$ such that $Y_t = \mathbb{E}\xi + \int_0^t Z_s dW_s$ and so that

$$Y_t = \xi - \int_t^T Z_s dW_s,$$

proving the existence and uniqueness of a solution to the BSDE driven by $\Psi \equiv 0$.

Affine Driver: For the sake of simplicity, we consider only the scalar case $p = m = 1$. We assume that
$$\Psi(t, y, z) = \alpha_t + \beta_t y + \gamma_t z, \qquad y, z \in \mathbb{R},$$
for some progressively measurable processes $\boldsymbol{\alpha} = (\alpha_t)_{0 \le t \le T} \in \mathbb{H}^2$, and $\boldsymbol{\beta} = (\beta_t)_{0 \le t \le T}$ and $\boldsymbol{\gamma} = (\gamma_t)_{0 \le t \le T}$ in $\mathbb{B}$. Since $\boldsymbol{\gamma}$ is bounded, we can define the probability measure $\mathbb{Q}$ by its density
$$\frac{d\mathbb{Q}}{d\mathbb{P}} = \exp\left[\int_0^T \gamma_t dW_t - \frac{1}{2}\int_0^T |\gamma_t|^2 dt\right],$$
and by Girsanov's theorem, $\tilde{W}_t = W_t - \int_0^t \gamma_s ds$ is a $\mathbb{Q}$-Wiener process, and $BSDE(\Psi, \xi)$
$$Y_t = \xi + \int_t^T (\alpha_s + \beta_s Y_s + \gamma_s Z_s) ds - \int_t^T Z_s dW_s$$
under $\mathbb{P}$ becomes
$$Y_t = \xi + \int_t^T (\alpha_s + \beta_s Y_s) ds - \int_t^T Z_s d\tilde{W}_s$$
under $\mathbb{Q}$. In other words, this simple Girsanov transform reduces the problem to the case $\gamma \equiv 0$ when the driver does not depend upon $\mathbf{Z}$. Now if we set
$$\tilde{Y}_t = Y_t e^{\int_0^t \beta_s ds},$$
then
$$d\tilde{Y}_t = e^{\int_0^t \beta_s ds} dY_t + \beta_t e^{\int_0^t \beta_s ds} Y_t dt$$
$$= -\alpha_t e^{\int_0^t \beta_s ds} dt + e^{\int_0^t \beta_s ds} Z_t d\tilde{W}_t,$$
which is a BSDE with a driver independent of $\mathbf{Y}$ and $\mathbf{Z}$, and which can be solved by a further reduction to a BSDE with a zero driver. Indeed, setting
$$\overline{Y}_t = \tilde{Y}_t + \int_0^t \alpha_s e^{\int_0^s \beta_r dr} ds,$$
we see that
$$d\overline{Y}_t = e^{\int_0^t \beta_s ds} Z_t d\tilde{W}_t,$$
which is a BSDE with a vanishing driver, which we can solve by the MRT. Reversing the steps, we find that the solution to our original affine BSDE is given by
$$Y_t = \mathbb{E}\left[\Gamma_t^T \xi + \int_t^T \Gamma_t^u \alpha_u du \big| \mathcal{F}_t\right] \tag{2.1}$$
with
$$\Gamma_t^u = \exp\left[\int_t^u \beta_s ds - \frac{1}{2}\int_t^u |\gamma_s|^2 ds + \int_t^u \gamma_s dW_s\right].$$

2.1.2 ▪ Existence and Comparison Results

The main existence result is the following theorem.

Theorem 2.2 (Pardoux and Peng (1990)). *If the driver Ψ is Lipschitz in (y,z) uniformly in $(t,\omega) \in [0,T] \times \Omega$, and if the process $\{\Psi(t,0,0)\}_{0 \le t \le T}$ is in $\mathbb{H}^{2,p}$, then the $BSDE(\Psi,\xi)$ has a unique solution.*

We say that the driver Ψ is Lipschitz in (y,z) uniformly in $(t,\omega) \in [0,T] \times \Omega$ if there exists a constant $C_\Psi > 0$ such that

$$\forall y_1, y_2 \in \mathbb{R}^p, \forall z_1, z_2 \in \mathbb{R}^{pm}, \ |\Psi(t,\omega,y_1,z_1) - \Psi(t,\omega,y_2,z_2)| \le C_\Psi(|y_1 - y_2| + |z_1 - z_2|)$$

for every $t \in [0,T]$ and $\omega \in \Omega$.

Proof. In full analogy with the proof of existence and uniqueness of solutions to ordinary SDEs, for each $\alpha > 0$ (to be chosen later) we define the norm

$$\|(\mathbf{Y},\mathbf{Z})\|_\alpha = \left(\mathbb{E} \int_0^T e^{\alpha t} (|Y_t|^2 + |Z_t|^2) dt \right)^{1/2}$$

for couples $(\mathbf{Y},\mathbf{Z})$ of progressively measurable processes with values in $\mathbb{R}^p$ and $\mathbb{R}^{pm}$ respectively. Next, we define the operator U by

$$\mathbb{H}^{2,p} \times \mathbb{H}^{2,pm} \ni (\mathbf{Y},\mathbf{Z}) \hookrightarrow U(\mathbf{Y},\mathbf{Z}) = (\tilde{\mathbf{Y}}, \tilde{\mathbf{Z}}),$$

where

$$\tilde{Y}_t = \xi + \int_t^T \Psi(s, Y_s, Z_s) ds - \int_t^T \tilde{Z}_s dW_s, \qquad t \le T.$$

This definition is nonambiguous because, since

$$|\Psi(s, Y_s, Z_s)| \le |\Psi(s,0,0)| + c(|Y_s| + |Z_s|),$$

the process $(\Psi(t, Y_t, Z_t))_{0 \le t \le T}$ is in $\mathbb{H}^{2,p}$ and $(\tilde{\mathbf{Y}}, \tilde{\mathbf{Z}})$ can be constructed by martingale representation. Indeed, if we apply the MRT to the martingale

$$M_t = \mathbb{E}\left[\xi + \int_0^T \Psi(s, Y_s, Z_s) ds \Big| \mathcal{F}_t \right],$$

we get the existence of $\tilde{\mathbf{Z}} \in \mathbb{H}^{2,pm}$ such that

$$M_t = M_0 + \int_0^t \tilde{Z}_s dW_s,$$

and if we define $\tilde{\mathbf{Y}}$ by

$$\tilde{Y}_t = \mathbb{E}\left[\xi + \int_t^T \Psi(s, Y_s, Z_s) ds \Big| \mathcal{F}_t \right],$$

we see that $\tilde{Y}_T = \xi$ and

$$d\tilde{Y}_t = dM_t - \Psi(t, Y_t, Z_t)dt = -\Psi(t, Y_t, Z_t)dt + \tilde{Z}_t dW_t.$$

Notice that

$$\mathbb{E}\sup_{0\leq t\leq T}|\tilde{Y}_t|^2 \leq c\left(\mathbb{E}|\xi|^2 + T\mathbb{E}\int_0^T |\Psi(t, Y_t, Z_t)|^2 dt \right.$$
$$\left. + \mathbb{E}\sup_{0\leq t\leq T}\left|\int_0^t Z_s dW_s\right|^2 + \mathbb{E}\left|\int_0^T Z_s dW_s\right|^2\right)$$
$$\leq c\left(\mathbb{E}|\xi|^2 + T\mathbb{E}\int_0^T |\Psi(t, 0, 0)|^2 dt + \mathbb{E}\int_0^T |Y_t|^2 dt + \mathbb{E}\int_0^T |Z_t|^2 dt\right)$$
$$< \infty,$$

which shows that $\tilde{\mathbf{Y}} \in \mathbb{S}^{2,p}$ and that the range of U is contained in $\mathbb{S}^{2,p} \times \mathbb{H}^{2,pm}$. To conclude the proof of existence and uniqueness of a solution to $BSDE(\Psi, \xi)$ we prove that U is a strict contraction. We pick $(\mathbf{Y}^1, \mathbf{Z}^1)$ and $(\mathbf{Y}^2, \mathbf{Z}^2)$ arbitrarily in $\mathbb{H}^{2,p} \times \mathbb{H}^{2,pd}$, and we apply Itô's formula to $e^{\alpha t}|\tilde{Y}_t^2 - \tilde{Y}_t^1|^2$. We get

$$e^{\alpha t}|\tilde{Y}_t^2 - \tilde{Y}_t^1|^2$$
$$= -\int_t^T e^{\alpha u}\left(\alpha|\tilde{Y}_u^2 - \tilde{Y}_u^1|^2 + 2(Y_u^2 - Y_u^1)d(\tilde{Y}_u^2 - \tilde{Y}_u^1) + |\tilde{Z}_u^2 - \tilde{Z}_u^1|^2 du\right)$$
$$= -\int_t^T e^{\alpha u}\left(\alpha|\tilde{Y}_u^2 - \tilde{Y}_u^1|^2 - 2(\tilde{Y}_u^2 - \tilde{Y}_u^1)[\Psi(u, Y_u^2, Z_u^2) - \Psi(u, Y_u^1, Z_u^1)]\right) du$$
$$- \int_t^T e^{\alpha u}|\tilde{Z}_u^2 - \tilde{Z}_u^1|^2 du - 2\int_t^T e^{\alpha u}(\tilde{Y}_u^2 - \tilde{Y}_u^1)(\tilde{Z}_u^2 - \tilde{Z}_u^1)dW^u.$$

We intend to take expectations on both sides and we expect that the stochastic integral will disappear. This is indeed the case as this local martingale is in fact uniformly integrable. This is a consequence of the Burkholder–Davis–Gundy inequality because if we set

$$N_t = \int_0^t e^{\alpha u}(\tilde{Y}_u^2 - \tilde{Y}_u^1)(\tilde{Z}_u^2 - \tilde{Z}_u^1)dW_u,$$

then we must have $\mathbb{E}\sup_{0\leq t\leq T}|N_t| \leq c\mathbb{E}[N, N]^{1/2}$ so that

$$\mathbb{E}\sup_{0\leq t\leq T}|N_t| \leq c\mathbb{E}\left(\int_0^T e^{2\alpha u}|\tilde{Y}_u^2 - \tilde{Y}_u^1|^2|\tilde{Z}_u^2 - \tilde{Z}_u^1|^2 du\right)^{1/2}$$
$$\leq c\mathbb{E}\left[\sup_{0\leq t\leq T}|\tilde{Y}_t^2 - \tilde{Y}_t^1|\left(\int_0^T |\tilde{Z}_t^2 - \tilde{Z}_t^1|^2 dt\right)^{1/2}\right]$$
$$\leq c\mathbb{E}\sup_{0\leq t\leq T}|\tilde{Y}_t^2 - \tilde{Y}_t^1|^2 + \mathbb{E}\int_0^T |\tilde{Z}_t^2 - \tilde{Z}_t^1|^2 dt$$
$$< \infty.$$

2.1. Introduction and First Definitions

Now for any $\epsilon > 0$ we have

$$\mathbb{E}\left(e^{\alpha t}|\tilde{Y}_t^2 - \tilde{Y}_t^1|^2 + \int_t^T e^{\alpha u}|\tilde{Z}_u^2 - \tilde{Z}_u^1|^2 du\right)$$

$$= \mathbb{E}\left[\int_t^T e^{\alpha u}\left(2(\tilde{Y}_u^2 - \tilde{Y}_u^1)[\Psi(u, Y_u^2, Z_u^2) - \Psi(u, Y_u^1, Z_u^1)] - \alpha|\tilde{Y}_u^2 - \tilde{Y}_u^1|^2\right) du\right]$$

$$\leq \mathbb{E}\int_t^T e^{\alpha u}\left(-\alpha|\tilde{Y}_u^2 - \tilde{Y}_u^1|^2 + |\tilde{Y}_u^2 - \tilde{Y}_u^1|c(|Y_u^2 - Y_u^1| + |Z_u^2 - Z_u^1|)\right) du$$

$$\leq \mathbb{E}\int_t^T e^{\alpha u}\left(-\alpha|\tilde{Y}_u^2 - \tilde{Y}_u^1|^2 + c(\epsilon^2|\tilde{Y}_u^2 - \tilde{Y}_u^1|^2 + \epsilon^{-2}(|Y_u^2 - Y_u^1| + |Z_u^2 - Z_u^1|)^2\right) du,$$

and if we choose $\epsilon = \sqrt{\alpha/c}$, we get

$$\leq \frac{c^2}{\alpha}\mathbb{E}\int_t^T e^{\alpha u}(|Y_u^2 - Y_u^1| + |Z_u^2 - Z_u^1|)^2 du$$

$$\leq 2\frac{c^2}{\alpha}\|(\mathbf{Y}^2, \mathbf{Z}^2) - (\mathbf{Y}^1, \mathbf{Z}^1)\|_\alpha^2.$$

Since

$$\|U(\mathbf{Y}^2, \mathbf{Z}^2) - U(\mathbf{Y}^1, \mathbf{Z}^1)\|_\alpha^2 = \mathbb{E}\int_0^T e^{\alpha t}(|\tilde{Y}_t^2 - \tilde{Y}_t^1| + |\tilde{Z}_t^2 - \tilde{Z}_t^1|)^2 dt$$

$$\leq 2\frac{c^2}{\alpha}(T+1)\|(\mathbf{Y}^2, \mathbf{Z}^2) - (\mathbf{Y}^1, \mathbf{Z}^1)\|_\alpha^2,$$

choosing $\alpha > 2c^2(1+T)$ proves that U is a strict contraction. $\square$

Remark 2.3. *As usual, the strict contraction property gives a construction of the solution by a Picard iteration. Indeed, if we start with $(\mathbf{Y}^0, \mathbf{Z}^0) = (0, 0)$ and construct by induction*

$$Y_t^{k+1} = \xi + \int_t^T \Psi(u, Y_u^k, Z_u^k) du + \int_t^T Z_u^{k+1} dW_u$$

by the martingale representation argument given in the above proof, then the sequence $\{(\mathbf{Y}^k, \mathbf{Z}^k)\}_{k \geq 1}$ is a Cauchy sequence in $\mathbb{H}^{2,p} \times \mathbb{H}^{2,pm}$ which converges to the unique solution $(\mathbf{Y}, \mathbf{Z})$ of $BSDE(\Psi, \xi)$. In fact, because the rate of convergence is geometric, one can use Borel–Cantelli's lemma to show that one can extract a subsequence for which the convergence is also almost sure.

2.1.3 ▪ Comparison Results

Many comparison results have been proven for BSDEs under different conditions. Most of their proofs are based on the same argument, which we give below in the proof of one specific comparison result, which we chose for the purpose of illustration.

Theorem 2.4. *Let us assume that $p = 1$ and let $(\mathbf{Y}^i, \mathbf{Z}^i)$ be the unique solutions of the $BSDE(\Psi^i, \xi^i)$ for $i = 1, 2$, where the conditions of the existence and uniqueness of Theorem 2.2 are satisfied and we assume furthermore that*

$$\xi^1 \leq \xi^2 \quad \text{and} \quad \Psi^1(t, Y_t^1, Z_t^1) \leq \Psi^2(t, Y_t^1, Z_t^1), \quad dt \otimes d\mathbb{P}\text{-}a.s. \tag{2.2}$$

Then for all $t \in [0, T]$ we have

$$Y_t^1 \leq Y_t^2, \quad \mathbb{P}\text{-a.s.} \tag{2.3}$$

Proof. We first remark that $\mathbf{Y}^2 - \mathbf{Y}^1$ is the solution of an affine BSDE. Indeed

$$d(Y_t^2 - Y_t^1) = -[\alpha_t(Y_t^2 - Y_t^1) + \beta_t(Z_t^2 - Z_t^1) + \gamma_t]dt + (Z_t^2 - Z_t^1)dW_t$$

with

$$\gamma_t = \Psi^2(t, Y_t^1, Z_t^1) - \Psi^1(t, Y_t^1, Z_t^1),$$

$$\alpha_t = \frac{\Psi^2(t, Y_t^2, Z_t^2) - \Psi^2(t, Y_t^1, Z_t^2)}{Y_t^2 - Y_t^1} \mathbf{1}_{\{Y_y^2 \neq Y_t^1\}},$$

and for $j = 0, \cdots, m-1$,

$$\beta_t^j = \frac{\Psi^2(t, Y_t^1, (Z_t^{1,1}, \cdots, Z_t^{1,j}, Z_t^{2,j+1}, \cdots, Z_t^{2,m})) - \Psi^2(t, Y_t^1, (Z_t^{1,1}, \cdots, Z_t^{1,j+1}, Z_t^{2,j+2}, \cdots, Z_t^{2,m}))}{Z_t^{2,j+1} - Z_t^{1,j+1}}.$$

Indeed, using a telescoping sum we have

$$\begin{aligned}
\Psi^2(t, Y_t^2, Z_t^2) - \Psi^1(t, Y_t^1, Z_t^1) &= \Psi^2(t, Y_t^2, Z_t^2) - \Psi^2(t, Y_t^1, Z_t^2) \\
&\quad + \Psi^2(t, Y_t^1, (Z_t^{2,1}, \cdots, Z_t^{2,m})) - \Psi^2(t, Y_t^1, (Z_t^{1,1}, Z_t^{2,2}, \cdots, Z_t^{2,m})) \\
&\quad + \Psi^2(t, Y_t^1, (Z_t^{1,1}, Z_t^{2,2}, \cdots, Z_t^{2,m})) - \Psi^2(t, Y_t^1, (Z_t^{1,1}, \cdots, Z_t^{1,2}, Z_t^{2,3}, \cdots, Z_t^{2,m})) \\
&\quad + \cdots \\
&\quad + \Psi^2(t, Y_t^1, (Z_t^{1,1}, \cdots, Z_t^{1,m-1}, Z_t^{2,m})) - \Psi^2(t, Y_t^1, (Z_t^{1,1}, \cdots, Z_t^{1,m})).
\end{aligned}$$

The processes $(\alpha_t)_{0 \leq t \leq T}$ and $(\beta_t)_{0 \leq t \leq T}$ are bounded because of the uniform Lipschitz assumption on the drivers Ψ^1 and Ψ^2. Since $Y_T^2 - Y_T^1 = \xi^2 - \xi^1 \geq 0$, we conclude using the explicit form (2.1) of the solution of affine BSDEs. $\square$

2.1.4 ▪ Pricing European Options

In this section, we present what is most likely the first application of the theory of BSDEs to financial mathematics.

The Market Model

We consider the standard model for a financial market with one nonrisky asset $\mathbf{P}^0 = (P_t^0)_{t \geq 0}$ and d risky assets $\mathbf{P}^i = (P_t^i)_{t \geq 0}$ for $i = 1, \cdots, d$, whose dynamics are given by

$$\begin{cases} dP_t^0 = P_t^0 r_t dt, \\ dP_t^i = P_t^i[\mu_t^i dt + \sigma_t^i dW_t], \quad i = 1, \cdots, d, \end{cases} \tag{2.4}$$

for $0 \leq t \leq T$ with $P_0^i > 0$ for $i = 0, 1, \cdots, d$. The processes $\mathbf{r}$, $\boldsymbol{\mu}^i$, and $\boldsymbol{\sigma}^i$ are supposed to be progressively measurable and bounded. Here r_t and μ_t^i are scalar, while σ_t^i and W_t are m-dimensional. We denote by $\boldsymbol{\pi} = (\pi_t^0, \pi_t^1, \cdots, \pi_t^d)_{0 \leq t \leq T}$ a small investor portfolio where π_t^0 and π_t^i represent the amounts invested in the bond (nonrisky asset) and the ith risky asset, respectively. The value at time t of such a portfolio is denoted by

$$V_t = V_t^\pi = \pi_t^0 + \pi_t^1 + \cdots + \pi_t^d.$$

2.1. Introduction and First Definitions

We denote by $\mathbb{A}$ the set of self-financing admissible portfolios. A portfolio π is said to be admissible if its components are progressively measurable and

$$\int_0^T |\pi_t^0| dt < \infty, \qquad \int_0^T |\pi_t^i \sigma_t^i|^2 dt < \infty, \ i = 1, \cdots, d, \quad \text{and} \quad V_t^\pi \geq 0, \ 0 \leq t \leq T$$

almost surely. A portfolio π is said to be self-financing if

$$dV_t^\pi = \pi_t^0 \frac{dP_t^0}{P_t^0} + \pi_t^1 \frac{dP_t^1}{P_t^1} + \cdots + \pi_t^d \frac{dP_t^d}{P_t^d}, \tag{2.5}$$

in which case, using (2.4)

$$\begin{aligned} dV_t^\pi &= r_t V_t dt + \sum_{i=1}^m [\pi_t^i(\mu_t^i - r_t)dt + \pi_t^i \sigma_t^i dW_t] \\ &= r_t V_t dt + \pi_t^{-0}(\mu_t - r_t \mathbf{1}_m)dt + \pi_t^{-0} \sigma_t dW_t, \end{aligned} \tag{2.6}$$

where we use obvious vector notations, π_t^{-0} being the row $(\pi_t^1, \cdots, \pi_t^d)$ and $\mathbf{1}_d$ the d-dimensional vector of ones. Notice that, except possibly for integrability properties, equality (2.6) looks very much like a BSDE.

The Replicating Portfolio

We now consider a nonnegative $\mathcal{F}_T$-measurable random variable ξ (a European contingent claim with maturity T), $\xi = (P_T^1 - K)^+$ being a typical example, and we look for the initial endowment x_0 a small investor should make in order to be able to replicate the contingent claim payoff ξ by investing in a self-financing (admissible) portfolio. Formally, we try to determine

$$x_0 = \inf\{x \in \mathbb{R}; V_0^\pi = x \text{ and } V_T^\pi = \xi \text{ for some } \pi \in \mathbb{A}\}.$$

Theorem 2.5. *Let us assume that $d = m$, the matrix $\sigma_t = [\sigma_t^1, \cdots, \sigma_t^d]$ is invertible, and the risk premium vector $\theta_t = [\sigma_t]^{-1}(\mu_t - r_t \mathbf{1}_d)$ is bounded. Then, for any nonnegative random variable $\xi \in L^2(\mathcal{F}_T)$, we have $x_0 = Y_0$, where $(Y_t, Z_t)_{0 \leq t \leq T}$ is the unique solution of the BSDE*

$$\begin{cases} dY_t = (r_t Y_t + Z_t \theta_t)dt + Z_t dW_t, \\ Y_T = \xi, \end{cases} \tag{2.7}$$

and the corresponding replicating self-financing portfolio π is given by $\pi_t^{-0} = Z_t[\sigma_t]^{-1}$.

Proof. Let $(\mathbf{Y}, \mathbf{Z})$ be the solution of the $BSDE(\Psi, \xi)$ for the linear driver $\Psi(t, y, z) = -r_t y - z\theta_t$, the BSDE for which existence and uniqueness are given by Theorem 2.2. Notice that Y_0 is a deterministic number since it is $\mathcal{F}_0$-measurable, which is trivial by the zero–one law. Notice also that the comparison results imply that $Y_t \geq 0$, since $\xi \geq 0$, and Y_t is the value of a self-financing portfolio. This portfolio π_t is obtained by setting $\pi_t = Z_t \sigma_t^{-1}$ since in that case we have $\int_0^T |\pi_t \sigma_t|^2 dt < \infty$ almost surely and hence the couple $(\mathbf{Y}, \pi\sigma)$ satisfies the BSDE (2.7).

Assumptions on the matrix σ_t imply that the market model is complete, and there exists a unique equivalent martingale measure, say $\mathbb{Q}$, and its density with respect to $\mathbb{P}$ on $\mathcal{F}_T$ is given by the Doléans exponential $\mathcal{E}(\int_0^\cdot \theta_t dW_t)$. Now, if we denote by $\Gamma_{t,T}^{r,\theta}$ the product of this density and the discount factor $D_{s,T}$ defined by $D_{t,u} = \exp[-\int_t^u r_s ds]$ for $t \leq u \leq T$, then $\Gamma_{t,T}^{r,\theta}$ appears as the solution of the SDE

$$d\Gamma_{t,s}^{r,\theta} = -\Gamma_{t,s}^{r,\theta}(r_s ds + \theta_s dW_s), \qquad t \leq s \leq T,$$

with initial condition $\Gamma_{t,t}^{r,\theta} = 1$. $\Gamma_{t,T}^{r,\theta}$ is called the state price density. Using the formula for the conditional expectation under an absolute continuous change of measure and the boundedness assumption on r_t and θ_t, it is plain to check that $(\Gamma_{0,t}^{r,\theta} Y_t)_{0 \leq t \leq T}$ is a uniformly integrable martingale (for the original probability measure $\mathbb{P}$). Now, if $\tilde{\pi}$ is any admissible replicating portfolio (i.e., an admissible portfolio such that $V_T^{\tilde{\pi}} = \xi$), using (2.7) for $V_t^{\tilde{\pi}}$ and Itô's formula, we see that $(\Gamma_{0,t}^{r,\theta} V_t^{\tilde{\pi}})_{0 \leq t \leq T}$ is a nonnegative local martingale (hence a supermartingale) with terminal value $\Gamma_{0,T}^{r,\theta} \xi = \Gamma_{0,T}^{r,\theta} Y_T$, which implies that for any prior time $t \leq T$,

$$\Gamma_{0,t}^{r,\theta} V_t^{\tilde{\pi}} \geq \mathbb{E}[\Gamma_{0,T}^{r,\theta} \xi | \mathcal{F}_t] = \mathbb{E}[\Gamma_{0,T}^{r,\theta} Y_T | \mathcal{F}_t] = \Gamma_{0,t}^{r,\theta} Y_t,$$

proving that Y_t is also the price at time t of the contingent claim ξ. $\square$

Remark 2.6. *The above argument shows that Y_t is in fact the unique no-arbitrage price at time t of the contingent claim ξ and that*

$$Y_t = \mathbb{E}[\Gamma_{t,T}^{r,\theta} \xi | \mathcal{F}_t] = \mathbb{E}^{\mathbb{Q}}[D_{t,T} \xi | \mathcal{F}_T],$$

2.1.5 ▪ Quadratic BSDEs

Solving BSDEs without the linear growth assumption is notoriously difficult, if not impossible. However, a little bit more can be said for the one-dimensional case $p = 1$. It is indeed possible to go beyond the case of linear growth under an extra quadratic growth assumption in z on the driver. More precisely, we say that assumption **(A)** is satisfied if

$$\textbf{(A)} \begin{cases} \forall (t, \omega) \in [0, T] \times \Omega, \ (y, z) \mapsto \Psi(t, \omega, y, z) \text{ is continuous}, \\ \forall (y, z) \in \mathbb{R} \times \mathbb{R}^d, \ \{\Psi(t, y, z)\}_{0 \leq t \leq T} \text{ is } \mathcal{P}\text{-measurable}, \\ \exists C_\Psi > 0, \forall (t, y, z) \in [0, T] \times \mathbb{R} \times \mathbb{R}^d, |\Psi(t, y, z)| \leq C_\Psi(1 + |y| + |z|^2), \quad \mathbb{P}\text{-a.s.} \end{cases}$$
(2.8)

The main existence result for quadratic growth drivers is the following.

Theorem 2.7 (Kobylanski [74]). *If $p = 1$, $\xi \in L^\infty(\Omega, \mathcal{F}, \mathbb{P})$, and if the driver Ψ satisfies hypothesis **(A)**, there exists at least one solution $(\mathbf{Y}, \mathbf{Z})$. Moreover, there exists a maximal solution $(\mathbf{Y}^*, \mathbf{Z}^*)$ in the sense that for any other solution $(\mathbf{Y}, \mathbf{Z})$, for all $t \in [0, T]$, $Y_t \leq Y_t^*$ a.s.*

We do not give the proof of this result as it would take us too far afield. We merely mention that the construction of the maximal solution is based on the following comparison result of independent interest.

Theorem 2.8 (Lepeltier and San Martin [82]). *If $p = 1$, $\xi \in L^2(\Omega, \mathcal{F}, \mathbb{P})$, if the drivers $\Psi^{(1)}$ and $\Psi^{(2)}$ satisfy hypothesis **(A)**, and*

$$\forall (t, y, z) \in [0, T] \times \mathbb{R} \times \mathbb{R}^d, \qquad \Psi^{(1)}(t, y, z) \leq \Psi^{(2)}(t, y, z), \quad a.s.,$$

then if $(\mathbf{Y}^{(1)}, \mathbf{Z}^{(1)})$ is a solution of the $BSDE(\Psi^{(1)}, \xi)$ and $(\mathbf{Y}^{(2)}, \mathbf{Z}^{(2)})$ is the maximal solution of the $BSDE(\Psi^{(2)}, \xi)$, then for all $t \in [0, T]$, $Y_t^{(1)} \leq Y_t^{(2)}$ a.s.

2.2 ▪ Mean-Field BSDEs

This section is devoted to an existence and uniqueness result for a special type of BSDE, which we will need in our analysis of the control of McKean–Vlasov dynamics- and mean-field games.

We consider a function ψ defined on $[0, T] \times \Omega \times \Omega \times \mathbb{R}^p \times \mathbb{R}^{p \times m} \times \mathbb{R}^p \times \mathbb{R}^{p \times m}$ with values in $\mathbb{R}^p$ which satisfies the following assumptions.

(MFA1) For each $\tilde{y}, y$ in $\mathbb{R}^p$ and $\tilde{z}, z$ in $\mathbb{R}^{p \times m}$, the process $\{\psi(t, \tilde{\omega}, \omega, \tilde{y}, \tilde{z}, y, z); 0 \le t \le T, (\tilde{\omega}, \omega) \in \Omega^2\}$ is progressively measurable for the filtration $\mathbb{F}^2 = \{\mathcal{F} \otimes \mathcal{F}_t; 0 \le t \le T\}$.

(MFA2) There exists a constant $c > 0$ such that for all $(\tilde{\omega}, \omega) \in \Omega^2, t \in [0, T], \tilde{y}_1, \tilde{y}_2, y_1, y_2$ in $\mathbb{R}^p$ and $\tilde{z}_1, \tilde{z}_2, z_1, z_2$ in $\mathbb{R}^{p \times m}$ we have

$$|\psi(t, \tilde{\omega}, \omega, \tilde{y}_2, \tilde{z}_2, y_2, z_2) - \psi(t, \tilde{\omega}, \omega, \tilde{y}_1, \tilde{z}_1, y_1, z_1)|$$
$$\le c(|\tilde{y}_2 - \tilde{y}_1| + |\tilde{z}_2 - \tilde{z}_1| + |y_2 - y_1| + |z_2 - z_1|).$$

(MFA3) $\psi(\,\cdot\,, 0, 0, 0, 0) \in \mathbb{H}^{2,p}_{\mathbb{F}^2}$.

In order to better understand the significance of these assumptions, let $\boldsymbol{\xi} = (\xi_t)_{0 \le t \le T}$ and $\boldsymbol{\zeta} = (\zeta_t)_{0 \le t \le T}$ be measurable square integrable processes on $(\Omega, \mathcal{F}, \mathbb{P})$ and let us define the function $\Psi^{\boldsymbol{\xi}, \boldsymbol{\zeta}}$ on $[0, T] \times \Omega \times \mathbb{R}^p \times \mathbb{R}^{p \times m}$ by

$$\Psi^{\boldsymbol{\xi}, \boldsymbol{\zeta}}(t, \omega, y, z) = \tilde{\mathbb{E}}[\psi(t, \,\cdot\,, \omega, \tilde{\xi}_t, \tilde{\zeta}_t, y, z)] = \int_\Omega \psi(t, \tilde{\omega}, \omega, \xi_t(\tilde{\omega}), \zeta_t(\tilde{\omega}), y, z) \mathbb{P}(d\tilde{\omega}).$$

Here and in what follows, we use the following convention. When a random variable X (or a random process) is defined on $(\Omega, \mathcal{F}, \mathbb{P})$, we use the same notation for its natural extension to Ω^2 defined by $X(\tilde{\omega}, \omega) = X(\omega)$, and we use the same letter with a tilde for its extension $\tilde{X}$ defined on $\Omega^2 = \Omega \times \Omega$ by $\tilde{X}(\tilde{\omega}, \omega) = X(\tilde{\omega})$. In this way, $\tilde{X}$ is an *independent copy* of X on $(\Omega^2, \mathcal{F}^2 = \mathcal{F} \otimes \mathcal{F}, \mathbb{P}^2 = \mathbb{P} \otimes \mathbb{P})$.

Notice that for each $y \in \mathbb{R}^p$ and $z \in \mathbb{R}^{p \times m}$ the process $(\Psi^{\boldsymbol{\xi}, \boldsymbol{\zeta}}(t, \,\cdot\,, y, z))_{0 \le t \le T}$ is progressively measurable. Moreover,

$$|\Psi^{\boldsymbol{\xi}, \boldsymbol{\zeta}}(t, \omega, y_2, z_2) - \Psi^{\boldsymbol{\xi}, \boldsymbol{\zeta}}(t, \omega, y_1, z_1)|$$
$$\le \int_\Omega |\psi(t, \tilde{\omega}, \omega, \xi_t(\tilde{\omega}), \zeta_t(\tilde{\omega}), y_2, z_2) - \psi(t, \tilde{\omega}, \omega, \xi_t(\tilde{\omega}), \zeta_t(\tilde{\omega}), y_1, z_1)|\mathbb{P}(d\tilde{\omega})$$
$$\le c(|y_2 - y_1| + |z_2 - z_1|)$$

because of assumption **(MFA2)**. Finally, using Hölder's inequality, the Lipschitz property of ψ, and the square integrability of $\boldsymbol{\xi}, \boldsymbol{\zeta}$, and $\psi(\,\cdot\,, \,\cdot\,, 0, 0, 0, 0)$, we get

$$\mathbb{E} \int_0^T |\Psi^{\boldsymbol{\xi}, \boldsymbol{\zeta}}(t, 0, 0)|^2 \, dt \le \tilde{\mathbb{E}} \mathbb{E} \int_0^T |\psi(t, \tilde{\xi}_t, \tilde{\zeta}_t, 0, 0)|^2 dt$$
$$\le 2 \mathbb{E} \int_0^T |\psi(t, 0, 0, 0, 0)|^2 dt + 2c \mathbb{E} \int_0^T |\xi_t|^2 dt + 2c \mathbb{E} \int_0^T |\zeta_t|^2 dt$$
$$< \infty.$$

In other words, $\Psi^{\boldsymbol{\xi}, \boldsymbol{\zeta}}$ satisfies the assumptions required by a driver in order to be able to apply the existence and uniqueness result for solutions of BSDEs proven in Theorem 2.2. We shall use this remark in the proof of the following theorem, which is the main result of this section.

Theorem 2.9. *Under the assumptions* **(MFA1)–(MFA3)**, *for any* $\xi \in L^2(\Omega, \mathcal{F}_T, \mathbb{P})$ *there exists a unique adapted solution* $(\mathbf{Y}, \mathbf{Z})$ *of*

$$dY_t = -\tilde{\mathbb{E}}[\psi(t, \tilde{Y}_t, \tilde{Z}_t, Y_t, Z_t)]\,dt + Z_t dW_t, \qquad Y_T = \xi, \tag{2.9}$$

in the space $\mathbb{S}^{2,p} \times \mathbb{H}^{2,pm}$.

Proof. Recall that the exact definition of the dt-term in the above BSDE is

$$\tilde{\mathbb{E}}[\psi(t, \tilde{Y}_t, \tilde{Z}_t, Y_t, Z_t)] = \int_\Omega \psi(t, \tilde{\omega}, \omega, Y_t(\tilde{\omega}), Z_t(\tilde{\omega}), Y_t(\omega), Z_t(\omega))\,\mathbb{P}(d\tilde{\omega}).$$

As before, we use the norm

$$\|(\mathbf{Y}, \mathbf{Z})\|_\alpha^2 = \mathbb{E}\int_0^T e^{\alpha t}(|Y_t|^2 + |Z_t|^2)\,dt$$

with a positive constant α to be chosen later in the proof. For any $(\mathbf{Y}', \mathbf{Z}') \in \mathbb{H}^{2,p+pm}$, we denote by $(\mathbf{Y}, \mathbf{Z})$ the unique solution of the BSDE

$$dY_t = -\tilde{\mathbb{E}}[\psi(t, \tilde{Y}'_t, \tilde{Z}'_t, Y_t, Z_t)]\,dt + Z_t dW_t, \qquad Y_T = \xi,$$

as given by Theorem 2.2. This defines a map U by $(\mathbf{Y}', \mathbf{Z}') \hookrightarrow (\mathbf{Y}, \mathbf{Z}) = U(\mathbf{Y}', \mathbf{Z}')$ from $\mathbb{H}^{2,p+pm}$ into itself. Note however that $\mathbf{Y} \in \mathbb{S}^{2,p}$. As before in the case of Theorem 2.2, the proof consists in showing that one can choose α so that the mapping U is a strict contraction, its unique fixed point giving the desired solution to the mean-field BSDE (2.9). Let us choose $(\mathbf{Y}'^1, \mathbf{Z}'^1)$ and $(\mathbf{Y}'^2, \mathbf{Z}'^2)$ in $\mathbb{H}^{2,p+pm}$ and let us set $(\mathbf{Y}^1, \mathbf{Z}^1) = U(\mathbf{Y}'^1, \mathbf{Z}'^1)$, $(\mathbf{Y}^2, \mathbf{Z}^2) = U(\mathbf{Y}'^2, \mathbf{Z}'^2)$, $(\hat{\mathbf{Y}}', \hat{\mathbf{Z}}') = (\mathbf{Y}'^2 - \mathbf{Y}'^1, \mathbf{Z}'^2 - \mathbf{Z}'^1)$, and $(\hat{\mathbf{Y}}, \hat{\mathbf{Z}}) = (\mathbf{Y}^2 - \mathbf{Y}^1, \mathbf{Z}^2 - \mathbf{Z}^1)$. Applying Itô's formula to $e^{\alpha t}|\hat{Y}_t|^2$, we get for any $t \in [0, T]$

$$|\hat{Y}_t|^2 + \mathbb{E}\left[\int_t^T \alpha e^{\alpha(r-t)}|\hat{Y}_r|^2 dr\Big|\mathcal{F}_t\right] + \mathbb{E}\left[\int_t^T e^{\alpha(r-t)}|\hat{Z}_r|^2 dr\Big|\mathcal{F}_t\right]$$
$$= \mathbb{E}\left[\int_t^T e^{\alpha(r-t)} 2\hat{Y}_r[\Psi^{\mathbf{Y}'^2, \mathbf{Z}'^2}(r, Y_r^2, Z_r^2) - \Psi^{\mathbf{Y}'^1, \mathbf{Z}'^1}(r, Y_r^1, Z_r^1)]dr\Big|\mathcal{F}_t\right].$$

From the uniform Lipschitz assumption **(MFA2)** and the integrability assumption **(MFA3)** we deduce that

$$\left(\frac{\alpha}{2} - 2c + 2c^2\right)\mathbb{E}\int_0^T e^{\alpha r}|\hat{Y}_r|^2 dr + \frac{1}{2}\mathbb{E}\int_0^T e^{\alpha r}|\hat{Z}_r|^2 dr$$
$$\leq \frac{4c^2}{\alpha}\left[\mathbb{E}\int_0^T e^{\alpha r}|\hat{Y}'_r|^2 dr + \mathbb{E}\int_0^T e^{\alpha r}|\hat{Z}'_r|^2 dr\right],$$

which gives

$$\mathbb{E}\int_0^T e^{\alpha t}(|\hat{Y}_t|^2 + |\hat{Z}_t|^2)\,dt \leq \frac{1}{2}\mathbb{E}\int_0^T e^{\alpha t}(|\hat{Y}'_t|^2 + |\hat{Z}'_t|^2)\,dt,$$

or equivalently $\|(\hat{\mathbf{Y}}, \hat{\mathbf{Z}})\|_\alpha \leq 2^{-1/2}\|(\hat{\mathbf{Y}}', \hat{\mathbf{Z}}')\|_\alpha$ provided we choose $\alpha = 16c^2 + 4c + 1$. This completes the proof. $\square$

2.3 • Reflected Backward Stochastic Differential Equations (RBSDEs)

In this section we assume that the backward component Y_t is unidimensional so that $p = 1$, and we try to compare Y_t to a *barrier* and make sure that Y_t remains above this barrier. We assume that the driver Ψ and the terminal condition ξ are as before, and we add to these data an adapted stochastic process $\mathbf{S} = (S_t)_{0 \le t \le T}$, which stands for the obstacle acting as a barrier for the backward component of the potential solution of the problem. A solution for the reflected BSDE with data $(\Psi, \xi, \mathbf{S})$ is a triple of stochastic processes $(\mathbf{Y}, \mathbf{Z}, \mathbf{K}) = (Y_t, Z_t, K_t)_{0 \le t \le T}$, where the process $\mathbf{K} = (K_t)_{0 \le t \le T}$ is continuous and nondecreasing, satisfying

$$\begin{cases} Y_t = \xi + \int_t^T \Psi(s, \omega, Y_s, Z_s) ds + K_T - K_t - \int_t^T Z_s dW_s, \\ Y_t \ge S_t, \quad 0 \le t \le T, \\ \int_0^T (Y_t - S_t) dK_t = 0. \end{cases} \quad (2.10)$$

The role of K_t is to push Y_t up in order to keep it above the barrier S_t, the third condition implying that K_t increases only when $Y_t = S_t$, forcing $\mathbf{K}$ to be minimal in some sense.

Theorem 2.10. *The reflected BSDE (2.10) associated with the data $(\Psi, \xi, \mathbf{K})$ has a unique solution.*

Proof. The existence proof is rather technical, and presenting it would take us far beyond the intended scope of this book. We refer the interested reader to the Notes & Complements section at the end of the chapter for precise references. We give only the details of the proof of uniqueness which is simpler. Let us assume that the triples of stochastic processes $(Y_t^1, Z_t^1, K_t^1)_{0 \le t \le T}$ and $(Y_t^2, Z_t^2, K_t^2)_{0 \le t \le T}$ are both solutions of the reflected BSDE (2.10). Since K_t^1 increases only when $Y_t^1 = S_t$ and $Y_t^2 \ge S_t$, we know that $\mathbf{1}_{\{Y_t^1 > Y_t^2\}} dK_t^1 = 0$. Similarly, it holds that $\mathbf{1}_{\{Y_t^2 > Y_t^1\}} dK_t^2 = 0$. Consequently, we have

$$[\mathbf{1}_{\{Y_t^1 > Y_t^2\}} - \mathbf{1}_{\{Y_t^2 > Y_t^1\}}](dK_t^1 - dK_t^2) \le 0 \quad (2.11)$$

for all $t \le T$. Since

$$d(Y_t^1 - Y_t^2) = -[\Psi(t, Y_t^1, Z_t^1) - \Psi(t, Y_t^2, Z_t^2)]dt - d(K_t^1 - K_t^2) + (Z_t^1 - Z_t^2)dW_t,$$

using Tanaka's formula we get

$$d|Y_t^1 - Y_t^2| = \text{sgn}(Y_t^1 - Y_t^2)\left[-[\Psi(t, Y_t^1, Z_t^1) - \Psi(t, Y_t^2, Z_t^2)]dt - d(K_t^1 - K_t^2) + (Z_t^1 - Z_t^2)dW_t\right] + dL_t,$$

where the local time L_t increases only when $Y_t^1 = Y_t^2$. So consequently, if we use (2.11) we obtain

$$|Y_t^1 - Y_t^2| \le \int_t^T \text{sgn}(Y_s^1 - Y_s^2)[\Psi(t, Y_s^1, Z_s^1) - \Psi(t, Y_s^2, Z_s^2)]ds$$

$$- \int_t^T \text{sgn}(Y_s^1 - Y_s^2)(Z_s^1 - Z_s^2)dW_s.$$

The Lipschitz assumption on the driver Ψ gives the existence of two bounded processes $(C_t^Y)_{0 \le t \le T}$ and $(C_t^Z)_{0 \le t \le T}$ such that

$$\Psi(t, Y_t^1, Z_t^1) - \Psi(t, Y_t^2, Z_t^2) = C_t^Y (Y_t^1 - Y_t^2) + C_t^Z (Z_t^1 - Z_t^2).$$

This implies that

$$|Y_t^1 - Y_t^2| \leq \int_t^T C_s^Y |Y_s^1 - Y_s^2| + C_s^Z \text{sgn}(Y_s^1 - Y_s^2)(Z_s^1 - Z_s^2) ds$$
$$- \int_t^T \text{sgn}(Y_s^1 - Y_s^2)(Z_s^1 - Z_s^2) dW_s.$$

We can define the process $(\tilde{W}_t)_{0 \leq t \leq T}$ by $\tilde{W}_s = W_s + \int_t^s C_u^Z du$, and using Girsanov's theorem, we can change from probability $\mathbb{P}$ to an equivalent probability $\tilde{\mathbb{P}}$, turning $\tilde{W}_t$ into a Brownian motion so that the above inequality becomes

$$|Y_t^1 - Y_t^2| \leq C \int_t^T |Y_s^1 - Y_s^2| ds - \int_t^T \text{sgn}(Y_s^1 - Y_s^2)(Z_s^1 - Z_s^2) d\tilde{W}_s.$$

Now, taking $\tilde{\mathbb{P}}$-expectations of both sides, we get

$$\tilde{\mathbb{E}} |Y_t^1 - Y_t^2| \leq C \int_t^T \tilde{\mathbb{E}} |Y_s^1 - Y_s^2| ds$$

for all $t \leq T$, so that using Gronwall's inequality we can conclude that $\tilde{\mathbb{E}}[|Y_t^1 - Y_t^2|] = 0$ for all $t \leq T$. This implies that $\mathbf{Y}^1 = \mathbf{Y}^2$ almost surely for $\mathbb{P}$ as well, since $\mathbb{P}$ and $\tilde{\mathbb{P}}$ are equivalent. From this we conclude that $\mathbf{Z}^1 = \mathbf{Z}^2$, since $(Z_t^1 - Z_t^2) dW_t$ is the martingale part of the continuous Itô semimartingale $Y_t^1 - Y_t^2$, and a fortiori $\mathbf{K}^1 = \mathbf{K}^2$ as well. $\square$

2.3.1 • Itô's Obstacles

In this subsection, we assume that the obstacle $\mathbf{S} = (S_t)_{t \geq 0}$ is an Itô process of the form

$$S_t = S_0 + \int_0^t U_s ds + \int_0^t V_s dW_s, \qquad t \geq 0, \tag{2.12}$$

where $\mathbf{U} = (U_t)_{t \geq 0}$ and $\mathbf{V} = (V_t)_{t \geq 0}$ are progressively measurable processes in $\mathbb{R}$ and $\mathbb{R}^m$, respectively, and satisfy

$$\int_0^t [|U_s| + |V_s|^2] ds < \infty \qquad \text{for all } t \geq 0 \tag{2.13}$$

almost surely. In this case, we can identify the solution of the reflected BSDE.

Proposition 2.11. *Assuming (2.12) and (2.13), the solution $(\mathbf{Y}, \mathbf{Z}, \mathbf{K})$ of $RBSDE(\Psi, \xi, S)$ satisfies*
(i) $Z_t = V_t$, $d\mathbb{P} \times dt$-a.e. whenever $Y_t = S_t$;
(ii) $0 \leq dK_t \leq \mathbf{1}_{\{Y_t = S_t\}} [\Psi(t, S_t, V_t) + U_t]^- dt$; and
(iii) *there exists a predictable process $\alpha = (\alpha_t)_{t \geq 0}$ such that $0 \leq \alpha_t \leq 1$ and*

$$dK_t = \alpha_t \mathbf{1}_{\{Y_t = S_t\}} [\Psi(t, S_t, V_t) + U_t]^- dt.$$

Proof. Putting together (2.10) and (2.12), we get

$$d(Y_t - S_t) = -[\Psi(t, Y_t, Z_t) + U_t] dt - dK_t + (Z_t - V_t) dW_t,$$

2.3. Reflected Backward Stochastic Differential Equations (RBSDEs)

and using Tanaka's formula for the positive part of the continuous semimartingale $Y_t - S_t$, we get

$$d(Y_t - S_t)^+ = -\mathbf{1}_{\{Y_t > S_t\}}[\Psi(t, Y_t, Z_t) + U_t]dt + \mathbf{1}_{\{Y_t > S_t\}}(Z_t - V_t)dW_t + \frac{1}{2}dL_t,$$

where L_t is now the local time of the semimartingale $(Y_t - S_t)_{0 \le t \le T}$ at 0, and where we have used the fact that $dK_t = 0$ whenever $Y_t > S_t$. Now, since S_t is the barrier and Y_t is a solution of the $RBSDE(\Psi, \xi, \mathbf{S})$, we have $Y_t \ge S_t$ and hence $(Y_t - S_t)^+ = Y_t - S_t$. So we have two Itô decompositions for the same continuous semimartingale, so we can identify the bounded variation parts and the martingale parts separately. In particular, we have $\mathbf{1}_{\{Y_t = S_t\}}(Z_t - V_t)dW_t = 0$, proving the first statement. Moreover, from this identification we also get

$$dK_t + \frac{1}{2}dL_t = -\mathbf{1}_{\{Y_t = S_t\}}[\Psi(t, Y_t, Z_t) + U_t]dt$$
$$= \mathbf{1}_{\{Y_t = S_t\}}[\Psi(t, Y_t, Z_t) + U_t]^- dt.$$

We can now conclude, using the fact that K_t is nondecreasing in t and $K_0 = 0$. □

2.3.2 ▪ Pricing and Hedging American Contingent Claims

Let $\mathcal{M}$ be a standard complete financial market model (d risky assets and a riskless bond), like the one introduced in Subsection 2.1.4. Given a self-financing portfolio π, the wealth process $\mathbf{X}$ satisfies

$$dX_t = b(t, X_t, \pi_t)dt + \pi_t \sigma_t dW_t$$

for some possibly random function b, which is Lipschitz and convex in (x, π) uniformly in (t, ω). Recall that the classical case presented in Subsection 2.1.4 corresponds to

$$b(t, x, \pi) = r_t x + \pi \sigma_t \theta_t, \quad (2.14)$$

where r_t is the short interest rate and θ_t is the risk premium vector. In this subsection, we consider slightly more general wealth dynamics as given by (2.14), and we assume furthermore that $d = m$, the process $(b(t,0,0))_{0 \le t \le T}$ is in $\mathbb{H}^{2,1}$, and the volatility matrix $(\sigma_t)_{0 \le t \le T}$ of the d risky assets is invertible and its inverse $(\sigma_t)^{-1}$ is bounded. Assuming that the terminal payoff (i.e., the scrap value at maturity T) is given by an $\mathcal{F}_T$-measurable square integrable random variable ξ, and the running payoff by an adapted process $\mathbf{S} = (S_t)_{0 \le t \le T}$ in $\mathcal{H}^{2,1}$, the actual payoff of the contingent claim is given by

$$\tilde{S}_\tau = S_\tau \mathbf{1}_{\{\tau < T\}} + \xi \mathbf{1}_{\{\tau = T\}} \quad (2.15)$$

if the time of exercise is chosen to be the stopping time τ.

From the discussion in Subsection 2.1.4, we learned that for each fixed time $t \in [0, T]$, and for each stopping time τ with values in $[t, T]$, we can replicate the contingent claim with payoff $\tilde{S}_\tau$ by constructing the pair $(Y_s^\tau, Z_s^\tau)_{t \le s \le \tau}$ satisfying

$$dY_s^\tau = b(s, Y_s^\tau, Z_s^\tau)ds + Z_s^\tau dW_s, \quad t \le s \le \tau,$$

and $Y_\tau^\tau = \tilde{S}_\tau$. This was done in the particular case (2.14) and for a deterministic time τ, but the proof applies mutatis mutandis to the more general case considered here. Therefore, a simple arbitrage argument implies that the price at time t of the American contingent claim should be given by

$$Y_t = \operatorname{ess\,sup}_{\tau \in \mathcal{S}_{t,T}} Y_t^\tau.$$

We now show that this price process can be characterized in terms of the solution of an RBSDE.

Proposition 2.12. *Assuming that $\xi \geq S_T$, there exist a hedging portfolio process $\pi \in \mathbb{H}^{2,d}$ and a nondecreasing continuous process $\mathbf{K}$ such that*

$$\begin{cases} Y_t = \xi - \int_t^T b(s, Y_s, \pi_s)ds + K_T - K_t - \int_t^T \pi_s dW_s, \\ Y_t \geq S_t, \quad 0 \leq t \leq T, \\ \int_0^T (Y_t - S_t) dK_t = 0. \end{cases} \quad (2.16)$$

Furthermore, the stopping time

$$\tau^* = \inf\{s \geq t;\ Y_s = S_s\} \wedge T$$

is an optimal exercise time after t and $Y_t = Y_t^{\tau^*}$.

Proof. The existence of a solution $(\mathbf{Y}, \pi, \mathbf{K})$ of (2.16) is given by Theorem 2.10. Now, if τ is any stopping time in $[t, T]$, then we have

$$dY_t^\tau = b(t, Y_t^\tau, \pi_t^\tau)dt + \pi_t^\tau dW_t$$

for $s \leq \tau$ and $Y_\tau^\tau = \tilde{S}_\tau$. But since

$$dY_t = b(t, Y_t, \pi_t)dt - dK_t + \pi_t dW_t$$

and $Y_\tau \geq \tilde{S}_\tau$, we have $Y_t^\tau \leq Y_t$ by the comparison result given in Theorem 2.4. Consequently, we have

$$Y_t \geq \underset{\tau \in \mathcal{S}_{t,T}}{\text{ess sup}}\ Y_t^\tau.$$

On the other hand, we have

$$dY_s = b(s, Y_s, \pi_s)dt + \pi_s\, dW_t, \qquad t \leq s \leq \tau^*,$$

with $Y_{\tau^*} = \tilde{S}_{\tau^*}$ proving that $Y_t = Y_t^{\tau^*}$ and $Y_t \leq \text{ess sup}_{\tau \in \mathcal{S}_{t,T}}\ Y_t^\tau$. □

Remark 2.13. *It is possible to extend our presentation of reflected BSDEs to include a second barrier. This generalization is helpful for the formalization and the solution of Dynkin games of timing. The latter are tailor-made to the valuation of convertible bonds which offer a convincing application of this theory. We shall refrain from presenting this extension here. The interested reader may consult the Notes & Complements section of Chapter 5 for references.*

2.4 ▪ Forward-Backward Stochastic Differential Equations (FBSDEs)

We now consider the case of BSDEs, where the randomness in the driver comes from the presence of a stochastic process which satisfies a standard (forward) SDE. In other words, we now consider systems of stochastic differential equations of mixed type, some components being forward (starting from a prescribed initial condition) and others being backward (satisfying a specific random terminal condition). When the coefficients of the forward equations do not depend upon the solution of the backward equation, the forward

2.4. Forward-Backward Stochastic Differential Equations (FBSDEs)

equation can be solved first, and its solution can be injected into the backward equation, which can then be solved as a BSDE with random coefficients by techniques introduced in the previous section. These examples are degenerate in some sense, and we will need to develop tools to handle the important cases when the coupling between forward and backward equations is nontrivial. These results will be crucial to our discussions of stochastic control problems and stochastic differential games.

2.4.1 ▪ Statement of the Problem

In its most general form, the problem of fully coupled FBSDEs is to construct solutions $(\mathbf{X}, \mathbf{Y}, \mathbf{Z})$ of the system

$$\begin{cases} dX_t = b(t, X_t, Y_t, Z_t)dt + \sigma(t, X_t, Y_t, Z_t)dW_t, \\ dY_t = -f(t, X_t, Y_t, Z_t)dt + Z_t dW_t, \\ X_0 = x, \ Y_T = g(X_T), \end{cases} \quad (2.17)$$

where the coefficients b, σ, f, and g,

$$(b, \sigma, f) : [0, T] \times \Omega \times \mathbb{R}^d \times \mathbb{R}^p \times \mathbb{R}^{pm} \hookrightarrow \mathbb{R}^d \times \mathbb{R}^{dm} \times \mathbb{R}^p,$$
$$g : \Omega \times \mathbb{R}^d \hookrightarrow \mathbb{R}^p$$

satisfy the following assumptions.

(A0) *Measurability condition.* For each fixed $(x, y, z) \in \mathbb{R}^d \times \mathbb{R}^p \times \mathbb{R}^{pm}$, the processes $(b(t, x, y, z))_{0 \le t \le T}$, $(\sigma(t, x, y, z))_{0 \le t \le T}$, and $(f(t, x, y, z))_{0 \le t \le T}$ are in $\mathbb{H}^{2,d}$, $\mathbb{H}^{2,dm}$, and $\mathbb{H}^{2,p}$, respectively. Furthermore, for each $x \in \mathbb{R}^d$, $g(x) \in L^2(\Omega, \mathcal{F}, \mathbb{P}; \mathbb{R}^p)$.

(A1) *Growth condition.* There exists a constant $c > 0$ such that for each $(t, \omega, x, y, z) \in [0, T] \times \Omega \times \mathbb{R}^d \times \mathbb{R}^p \times \mathbb{R}^{pm}$, we have

$$|(b, \sigma, f)(t, \omega, x, y, z)| + |g(\omega, x)| \le c(1 + |x| + |y| + |z|).$$

(A2) *Lipschitz condition.* There exists a constant $c > 0$ such that $\forall t \in [0, T]$, $\forall \omega \in \Omega$, $\forall x, x' \in \mathbb{R}^d$, $\forall y, y' \in \mathbb{R}^p$, and $\forall z, z' \in \mathbb{R}^{pm}$,

$$|(b, \sigma, f)(t, \omega, x, y, z) - (b, \sigma, f)(t, \omega, x', y', z')| + |g(\omega, x) - g(\omega, x')|$$
$$\le c(|x - x'| + |y - y'| + |z - z'|).$$

Notice that assumption **(A1)** is not needed as it is implied by **(A2)** whenever the processes $\{b(t, 0, 0, 0)\}_{0 \le t \le T}$, $\{\sigma(t, 0, 0, 0)\}_{0 \le t \le T}$, and $\{f(t, 0, 0, 0)\}_{0 \le t \le T}$ are bounded and $g(0)$ is a bounded random variable in $\mathbb{R}^d$. Because of the limitations and extreme technicality of the existence theory in the general case, even when we try to work in the most general setting, we will assume throughout that the forward volatility σ is independent on Z_t. As a curiosity, we notice that even in the innocent-looking simple case $p = m$ and $\sigma(t, \omega, x, y, z) = z$, things can go wrong. Indeed, if b and f are taken to be identically zero and $g(x) = x$, then the FBSDE becomes

$$\begin{cases} X_t = x + \int_0^t Z_s dW_s, \\ Y_t = X_T - \int_t^T Z_s dW_s, \end{cases}$$

which has infinitely many solutions, one for each choice of $\mathbf{Z} \in \mathbb{H}^{2,dm}$!

2.4.2 • Connection with PDEs: The Decoupled Case

As a warm-up, we first consider the case where the coefficients of the forward equation are deterministic and do not depend upon the solution of the backward equation, namely when

$$(b, \sigma)(t, \omega, x, y, z) \equiv (b, \sigma)(t, x),$$

and we use the decoupling that this form of the coefficients provides to highlight the connection between FBSDEs and partial differential equations (PDEs).

The linear growth and Lipschitz assumptions imply existence and uniqueness, for each initial condition $X_0 \in L^2$ independent of the Wiener process, of a solution of the SDE

$$dX_t = b(t, X_t)dt + \sigma(t, X_t)dW_t, \qquad 0 \leq t \leq T. \tag{2.18}$$

As we explained in Chapter 1, solutions of this equation are closely connected to the linear partial differential operator $\mathscr{L}_t$ defined by

$$[\mathscr{L}_t\varphi](t, x) = \frac{1}{2}\mathrm{trace}[\sigma(t, x)\sigma(t, x)^\dagger \partial^2_{x,x}\varphi(t, x)] + b(t, x)\partial_x\varphi(t, x),$$

where we use a superscript † to denote the transpose of a vector or a matrix. If we also assume that the coefficients of the backward equation are independent of $\omega \in \Omega$,

$$f(t, \omega, x, y, z) \equiv f(t, x), \qquad \text{and} \qquad g(\omega, x) \equiv g(x),$$

the following verification result shows that classical solutions of the semilinear parabolic equations based on this linear operator provide solutions of $BSDE(\Psi, \xi)$ with

$$\Psi(t, \omega, y, z) = f(t, X_t(\omega), y, z) \qquad \text{and} \qquad \xi = g(X_T). \tag{2.19}$$

Proposition 2.14. *Let us assume that the function $(t, x) \hookrightarrow u(t, x)$ is jointly continuous on $[0, T] \times \mathbb{R}^d$, continuously differentiable in t, and twice continuously differentiable in x, the first derivative $\partial_x u(t, x)$ being of polynomial growth in x. Let us also assume that this function u is a classical solution of the PDE*

$$\begin{cases} \partial_t u(t, x) + [\mathscr{L}_t u](t, x) + f(t, x, u(t, x), \partial_x u(t, x)\sigma(t, x)) = 0, & (t, x) \in [0, T) \times \mathbb{R}^d, \\ u(T, x) = g(x), & x \in \mathbb{R}^d, \end{cases} \tag{2.20}$$

Then $(\mathbf{Y}, \mathbf{Z}) = (Y_t, Z_t)_{t \in [0, T]}$, defined by

$$Y_t = u(t, X_t), \qquad Z_t = \partial_x u(t, X_t)\sigma(t, X_t), \qquad 0 \leq t \leq T, \tag{2.21}$$

is the unique solution of the BSDE

$$dY_t = -f(t, X_t, Y_t, Z_t)dt + Z_t dW_t, \qquad 0 \leq t \leq T, \quad Y_T = g(X_T), \tag{2.22}$$

for the filtration of $\mathbf{W}$.

As we shall see later in Chapter 3, the nonlinear PDE (2.20) is the epitome of the Hamilton–Jacobi–Bellman (HJB) equations satisfied by the value functions of optimal control problems for diffusion processes. For this reason, a function u providing such a representation (2.21) of the solution Y_t in terms of time t and X_t is often called the value function of the FBSDE. However, we will also use the more appropriate terminology of the decoupling field of the FBSDE.

Proof. The proof is an immediate consequence of the computation of dY_t using Itô's formula and the fact that u is a solution of (2.20). □

We now explain that, conversely, solving a nonlinear PDE can be done by solving FBSDEs. For each fixed $t \in [0,T]$ and $x \in \mathbb{R}^d$, the linear growth and Lipschitz assumptions imply the existence and uniqueness of the $\mathbf{X}^{t,x} = (X_s^{t,x})_{s\in[t,T]}$ solution of the forward stochastic differential equation

$$dX_s = b(s, X_s)ds + \sigma(s, X_s)dW_s, \quad t \leq s \leq T, \ X_t = x. \tag{2.23}$$

Using the finiteness of the moments of $\mathbf{X}^{t,x}$ proved in Theorem 1.2, the assumptions on the coefficients f and g implying that the terminal condition $\xi = g(X_T)$ and the driver Ψ defined by $\Psi(t,\omega,y,z) = f(t, X_t^{t,x}(\omega), y, z)$ satisfy the assumptions of Theorem 2.2, existence and uniqueness of solutions for these BSDEs hold. Now, if we denote by $Y_s^{t,x} = Y_s$ the solution of the backward stochastic differential equation

$$dY_s = -f(s, X_s, Y_s, Z_s)ds + Z_s dW_s, \quad t \leq s \leq T, \ Y_T = g(X_T)$$

for the filtration generated by $\mathbf{W}$, the zero-one law applied to $\{W_s - W_t\}_{t \leq s \leq T}$ implies that the random variable $Y_t^{t,x}$ is deterministic and its value defines a function u through the formula

$$u(t,x) = Y_t^{t,x}. \tag{2.24}$$

While this function is not always twice continuously differentiable—in which case we cannot expect that it will provide a classical solution of (2.20)—it is a solution of (2.20) in the sense of viscosity solutions.

Proposition 2.15. *The function u defined by (2.24) is a continuous viscosity solution of (2.20).*

We do not prove this result as we will not use it. But the reader interested in the notion of viscosity solutions to PDEs is referred to Chapter 3, in which we discuss this notion rather extensively.

2.4.3 ▪ Markovian Reflected FBSDEs

We continue to work in the same Markovian setup: the coefficients of the forward equation (2.23) are Lipschitz, the driver and the terminal condition are given by equation (2.19), f is jointly continuous with linear growth and Lipschitz in (y,z) uniformly in (t,x), and g is measurable and with linear growth. We now consider a *Markovian barrier* of the form $S_t = h(X_t)$ for some function h which we assume to be continuous, with linear growth, and satisfying $h \leq g$. Using the same argument as before, the Markov property of the forward diffusion $\mathbf{X}$ and the existence and uniqueness result for solutions of reflected BSDEs can be used to show that the solution of the reflected BSDE

$$\begin{cases} Y_t = g(X_T) + \int_t^T f(s, X_s, Y_s, Z_s)ds + K_T - K_t - \int_t^T Z_s dW_s, \\ Y_t \geq h(X_t), \quad 0 \leq t \leq T, \\ \int_0^T (Y_t - h(X_t))dK_t = 0 \end{cases} \tag{2.25}$$

satisfies $Y_t = u(t, X_t)$ for $0 \leq t \leq T$ for some deterministic function $(t,x) \hookrightarrow u(t,x)$, which is defined as $u(t,x) = Y_t^{t,x}$, where for each $(t,x) \in [0,T] \times \mathbb{R}^d$, the triple of stochastic processes $(X_s^{t,x}, Y_s^{t,x}, Z_s^{t,x})_{t \leq s \leq T}$ is the unique solution of the reflected BSDE (2.25) over the interval $[t, T]$, with forward component $(X_s^{t,x})_{t \leq s \leq T}$ in lieu of $(X_s)_{t \leq s \leq T}$. The following verification result is the analogue of Proposition 2.14 in the reflected case.

Proposition 2.16. *Let us assume that the function $(t, x) \hookrightarrow u(t, x)$ is jointly continuous on $[0, T] \times \mathbb{R}^d$, continuously differentiable in t, and twice continuously differentiable in x, the first derivative $\partial_x u(t, x)$ being of polynomial growth in x. Let us also assume that this function u is a classical solution of the quasi-variational inequality (QVI):*

$$\begin{cases} \min\Big[\partial_t u(t,x) + [\mathscr{L}_t u](t,x) + f(t,x,u(t,x), \partial_x u(t,x)\sigma(t,x)), u(t,x) - h(x)\Big] = 0, (t,x) \in [0,T) \times \mathbb{R}^d, \\ u(T, x) = g(x), \quad x \in \mathbb{R}^d, \end{cases} \quad (2.26)$$

Then the triple $(\mathbf{Y}, \mathbf{Z}, \mathbf{K}) = (Y_t, Z_t, K_t)_{t \in [0,T]}$ of processes defined by

$$Y_t = u(t, X_t), \qquad Z_t = \partial_x u(t, X_t)\sigma(t, X_t), \qquad 0 \leq t \leq T, \quad (2.27)$$

$$K_t = \int_0^t \Big[\partial_t u(s, X_s) + [\mathscr{L}_s u](s, X_s) + f(s, X_s, u(s, X_s), \partial_x u(s, X_s)\sigma(s, X_s))\Big] ds$$

is the unique solution of the reflected BSDE (2.25).

Proof. As before, the fact that the first equation of the reflected BSDE (2.25) is satisfied is merely a consequence of the definitions (2.27) of Y_t and Z_t, Itô's formula applied to $u(t, X_t)$, and the terminal condition $Y_T = g(X_T)$. Moreover, the growth assumptions on u and its gradient in x, together with the existence of moments for X_t (in the case of the second moment; recall (1.19) for example) imply that $\mathbf{Y} \in \mathbb{S}^{2,p}$ and $\mathbf{Z} \in \mathbb{H}^{2,pm}$. Furthermore, $Y_t = u(t, X_t) \geq h(X_t)$ because of the form (2.26) of the QVI. Finally, the QVI also implies both that K_t is continuous and nondecreasing (since the integrand in the definition (2.27) of K_t is nonnegative), and the minimality condition because the minimum of the two terms appearing in the QVI has to be identically equal to 0. □

As before, after proving that a classical solution of the QVI (2.26) provided a solution of the reflected BSDE (2.25), we now state without proof the fact that the existence and uniqueness of solutions of the reflected BSDE (2.25) over all the intervals $[t, T]$ and initial conditions $x \in \mathbb{R}^p$ provide a solution of the QVI (2.26) in the viscosity sense.

Proposition 2.17. *The deterministic function $(t, x) \hookrightarrow u(t, x)$ defined as $u(t, x) = Y_t^{t,x}$, where for each $(t, x) \in [0, T] \times \mathbb{R}^d$, the triple $(Y_s^{t,x}, Z_s^{t,x}, K_s^{t,x})_{t \leq s \leq T}$ is the unique solution of the reflected BSDE (2.25) over the interval $[t, T]$ with forward component $(X_s^{t,x})_{t \leq s \leq T}$ in lieu of $(X_s)_{t \leq s \leq T}$, is jointly continuous on $[0, T] \times \mathbb{R}^d$ and is a solution of the QVI (2.26) in the viscosity sense.*

2.4.4 ▪ Monte Carlo Simulations of Decoupled FBSDEs

We will not discuss the generation of Monte Carlo scenarios of solutions of general FBSDEs. This is a very challenging problem which does not have a satisfactory solution in general. We shall limit ourselves to the case of decoupled equations identified in the previous subsection, and for which we know that the solution Y_t of the backward equation is a function of the forward component X_t.

For the purpose of Monte Carlo simulations, we first discretize time, replacing the interval $[0, T]$ by the finite time grid $\pi = \{t_0, t_1, \cdots, t_N\}$, where $t_i = i\Delta t$ for $i = 0, 1, \cdots, N$

2.4. Forward-Backward Stochastic Differential Equations (FBSDEs)

with $\Delta t = T/N$. We emphasize that in this subsection, we use the notation π for a subdivision of the interval $[0, T]$, and not for a portfolio as we did earlier in our discussion of the pricing and hedging of contingent claims. This should not be a source of confusion.

The forward component $\mathbf{X}$ can be simulated by a plain Euler scheme $\mathbf{X}^\pi$ on the grid π. In other words we can choose

$$X_{t_{i+1}}^\pi = X_{t_i}^\pi + b(t_i, X_{t_i}^\pi)\Delta t + \sigma(t_i, X_{t_i}^\pi)(W_{t_{i+1}} - W_{t_i}).$$

The backward Euler scheme $(\mathbf{Y}^\pi, \mathbf{Z}^\pi)$ for $(\mathbf{Y}, \mathbf{Z})$ starting from $Y_{t_N}^\pi = g(X_{t_N}^\pi)$ is defined as

$$Y_{t_i}^\pi = Y_{t_{i+1}}^\pi + \Psi(t_i, X_{t_i}^\pi, Y_{t_{i+1}}^\pi, Z_{t_i}^\pi)\Delta t - Z_{t_i}^\pi(W_{t_{i+1}} - W_{t_i}) \qquad (2.28)$$

so that, taking conditional expectations, we get

$$Y_{t_i}^\pi = \mathbb{E}\left[Y_{t_{i+1}}^\pi + \Psi(t_i, X_{t_i}^\pi, Y_{t_{i+1}}^\pi, Z_{t_i}^\pi)\Delta t \,\Big|\, X_{t_i}^\pi\right], \qquad (2.29)$$

and in order to get the Z-component providing the integrand, we multiply (2.28) by $W_{t_{i+1}} - W_{t_i}$ and take conditional expectations. We obtain

$$Z_{t_i}^\pi = \frac{1}{\Delta t}\mathbb{E}\left[Y_{t_{i+1}}^\pi (W_{t_{i+1}} - W_{t_i}) \,\Big|\, X_{t_i}^\pi\right]. \qquad (2.30)$$

Equations (2.29) and (2.30) point to a clear strategy for the simulation of Monte Carlo scenarios, and reduce the simulation problem to the computation of the above conditional expectations. Several methods have been proposed and implemented. We list a few.

- The basis function least squares regression method is the most popular in the financial industry. It relies on the choice of a finite set of feature functions $\psi_1, \cdots, \psi_q$, usually referred to as basis functions, and the approximation of the conditional expectations by linear combinations of these functions, the coefficients being chosen by minimization of a least squares criterion. To be more specific, we approximate

$$\mathbb{E}\left[Y_{t_{i+1}}^\pi (W_{t_{i+1}} - W_{t_i}) \,\Big|\, X_{t_i}^\pi\right] \quad \text{by} \quad \sum_{k=1}^q \alpha_k \psi_k(X_{t_i}^\pi)$$

and choose the coefficients α_k in

$$\arg\inf_{\alpha \in \mathbb{R}^q} \mathbf{E}\left|Y_{t_{i+1}}^\pi(W_{t_{i+1}} - W_{t_i}) - \sum_{k=1}^q \alpha_k \psi_k(X_{t_i}^\pi)\right|^2,$$

where $\mathbf{E}$ denotes the empirical mean based on the (forward Euler) Monte Carlo simulations of $X_{t_i}^\pi$, $X_{t_{i+1}}^\pi$, and $W_{t_{i+1}} - W_{t_i}$. The same procedure is applied to compute the conditional expectation giving $Y_{t_i}^\pi$. It has been demonstrated empirically that it is preferable to use the same set of Monte Carlo sample paths to compute *all* the conditional expectations needed.

- An alternative method based on a specific application of the Malliavin calculus has been proposed but it did not rival the method described above in popularity because of a less intuitive and more mathematically involved rationale.

- Quantization methods, based on the quantization of the probability distributions needed to evaluate the conditional expectation over specific tesselations of the space, have also been proposed, but their appeal remained limited.

- Finally, we mention the recent attempts to use cubature and ideas from the theory of rough paths to compute these conditional expectations and simulate solutions of special FBSDEs.

2.5 • Existence and Uniqueness of Solutions

While the set of assumptions required for existence and uniqueness of solutions for BSDEs was not too different from the classical case of standard (forward) stochastic differential equations, the case of FBSDEs is an order of magnitude more complicated because of the two conflicting directions in which the time variable has to evolve. We address this delicate issue in this section, starting with an illuminating example of some of the challenges we shall have to overcome, even in the absence of random shocks.

2.5.1 • A Sobering Counterexample

As we already saw at the end of the previous section, and are about to demonstrate, solving FBSDEs is a very tricky business, and before we attempt to solve them in any kind of generality, it is important to be aware of the fact that existence does not hold without strong assumptions. The following example shows that even when σ is bounded, or even 0, i.e., $\sigma \equiv 0$, existence of a solution may not hold. Indeed, if we consider the case

$$b(t,x,y,z) = b_{1,1}x + b_{1,2}y, \quad f(t,x,y,z) = b_{2,1}x + b_{2,2}y, \quad \text{and} \quad g(x) = cx,$$

and if $(\mathbf{X}, \mathbf{Y}, \mathbf{Z})$ is a solution of (2.17), then by taking expectations of all sides, we see that $(x(t) = \mathbb{E}X_t, y(t) = \mathbb{E}Y_t)$ is a solution of the deterministic first-order differential equation

$$\frac{d}{dt}\begin{bmatrix} x(t) \\ y(t) \end{bmatrix} = \begin{bmatrix} b_{1,1} & b_{1,2} \\ -b_{2,1} & -b_{2,2} \end{bmatrix} \begin{bmatrix} x(t) \\ y(t) \end{bmatrix}$$

with $x(0) = x$ and $y(T) = cx(T)$. We shall often denote the derivative with respect to time of deterministic functions by a dot. This ordinary differential equation (ODE) does not always have a solution. Indeed, if we take $b_{1,1} = b_{2,2} = 0$, $b_{2,1} = b_{1,2} = 1$, $c = 1$, and $T = k\pi + \pi/4$ for some integer k, then the ODE becomes

$$\begin{cases} \dot{x}(t) &= y(t), \\ \dot{y}(t) &= -x(t), \end{cases}$$

whose solution has to be of the form $x(t) = \alpha \sin t + \beta \cos t$ so that $\beta = x(0) = x$, and since $y(T) = x(T)$ is equivalent to $\alpha(\cos T - \sin T) = \beta(\cos T + \sin T)$, our choice of T implies that there are infinitely many solutions if $x = 0$, and no solution if $x \neq 0$.

2.5.2 • Existence by Monotonicity Arguments

Here, we work with the assumptions **(A0)**, **(A1)**, and **(A2)** stated above, which we collectively call **(H1)**, and we add a set of technical monotonicity conditions which we now describe. Throughout this subsection, we use the notation $\xi \in \mathbb{R}^{d+p+pm}$ for the vector $\xi = [x, y, z]^\dagger$, and for a fixed $p \times d$ full rank matrix G (to be chosen later), we set

$$A(t, \xi) = \begin{bmatrix} -G^\dagger f \\ Gb \\ G\sigma \end{bmatrix}(t, x, y, z),$$

which has the same dimension and the same structure as ξ. The monotonicity assumption used in this section can be stated as follows: for all $\xi = [x, y, z]^\dagger$ and $\tilde{\xi} = [\tilde{x}, \tilde{y}, \tilde{z}]^\dagger$ we have

(H2) $\begin{cases} [A(t,\xi) - A(t,\tilde{\xi})] \cdot (\xi - \tilde{\xi}) \leq -\beta_1 |G(x - \tilde{x})|^2 - \beta_2 |G^\dagger(y - \tilde{y})|^2, \\ (g(x) - g(\tilde{x})) \cdot G(x - \tilde{x}) \geq 0, \end{cases}$

2.5. Existence and Uniqueness of Solutions

where $\beta_1 \geq 0$, $\beta_2 \geq 0$ are such that $\beta_1 + \beta_2 > 0$, and $\beta_1 > 0$ (resp., $\beta_2 > 0$) when $p > d$ (resp., $d > p$). The results of this section are borrowed from [105]. We provide a complete proof for the existence and uniqueness result presented in Theorem 2.20 only for the sake of completeness, as we will not use these results in the text.

Theorem 2.18. *Under* **(H1)** *and* **(H2)**, *the FBSDE* (2.17) *has at most one solution.*

Under assumptions **(H1)** and **(H2)**, existence and uniqueness can only be proven under an extra assumption on the terminal condition of a rather restrictive nature.

Theorem 2.19. *Under* **(H1)** *and* **(H2)**, *if $g(x)$ is independent of $x \in \mathbb{R}^d$, then the FBSDE* (2.17) *has a unique solution.*

Finally, in order to prove existence and uniqueness for a terminal condition of the form $Y_T = g(X_T)$, we need to strengthen the monotonicity assumption **(H2)** to

$$\textbf{(H3)} \begin{cases} [A(t,\xi) - A(t,\tilde{\xi})] \cdot (\xi - \tilde{\xi}) \leq -\beta_1 |G(x - \tilde{x})|^2 - \beta_2 |G^\dagger(y - \tilde{y})|^2 + |G(z - \tilde{z})|^2, \\ (g(x) - g(\tilde{x})) \cdot G(x - \tilde{x}) \geq \mu_1 |G(x - \tilde{x})|^2 \end{cases}$$

for all $\xi = [x, y, z]^\dagger$ and $\tilde{\xi} = [\tilde{x}, \tilde{y}, \tilde{z}]^\dagger$, where $\beta_1 \geq 0$, $\beta_2 \geq 0$, and $\mu_1 \geq 0$ are such that $\beta_1 + \beta_2 > 0$ and $\mu_1 + \beta_2 > 0$. Furthermore, $\beta_1 > 0$ and $\mu_1 > 0$ (resp., $\beta_2 > 0$) when $p > d$ (resp., $d > p$). Then we can prove the following.

Theorem 2.20. *Under* **(H1)** *and* **(H3)**, *the FBSDE* (2.17) *has a unique solution.*

The proof is broken down into two technical lemmas.

Lemma 2.21. *For any positive constant $\lambda \geq 0$, the FBSDE*

$$\textbf{(FB1)} \begin{cases} dX_t = [-\beta_2 G^\dagger Y_t + \phi_t] dt + [-\beta_2 G^\dagger Z_t + \psi_t] dW_t, \\ dY_t = -[\beta_1 G X_t + \gamma_t] dt + Z_t dW_t, \\ X_0 = x, \quad Y_T = \lambda G X_T + \xi \end{cases}$$

has a unique solution.

Proof. (1) We first assume that $d \leq p$. G being assumed to have full rank, the matrix $G^\dagger G$ is a $d \times d$ symmetric matrix with strictly positive eigenvalues. We set

$$\begin{bmatrix} x' \\ y' \\ z' \end{bmatrix} = \begin{bmatrix} x \\ G^\dagger y \\ G^\dagger z \end{bmatrix} \quad \text{and} \quad \begin{bmatrix} y'' \\ z'' \end{bmatrix} = \begin{bmatrix} [I_p - (G(G^\dagger G)^{-1} G^\dagger)] y \\ [I_p - (G(G^\dagger G)^{-1} G^\dagger)] z \end{bmatrix}.$$

Multiplying both sides of the backward part of (FB1) by $G^\dagger$ we get

$$\textbf{(FB2)} \begin{cases} dX'_t = [-\beta_2 Y'_t + \phi_t] dt + [-\beta_2 Z'_t + \psi_t] dW_t, \\ dY'_t = -[\beta_1 G^\dagger G X'_t + G^\dagger \gamma_t] dt + Z'_t dW_t, \\ X'_0 = x, \quad Y'_T = \lambda G^\dagger G X'_T + G^\dagger \xi. \end{cases}$$

Multiplying by $I_p - G(G^\dagger G)^{-1} G^\dagger$ instead, we get

$$\begin{cases} dY''_t = -[I_p - G(G^\dagger G)^{-1} G^\dagger] \gamma_t dt + Z''_t dW_t, \\ Y''_T = [I_p - G(G^\dagger G)^{-1} G^\dagger] \xi, \end{cases}$$

which is a simple BSDE for which Theorem 2.2 provides existence and uniqueness of a solution. Uniqueness of a solution for (FB2) follows from Theorem 2.18. In order to prove existence, we consider the following Riccati equation for $d \times d$ matrices:

$$\begin{cases} -\dot{K}(t) = -\beta_2 K(t)^2 + \beta_1 G^\dagger G, \\ K(T) = \lambda G^\dagger G, \end{cases}$$

which is known to have a solution (see Subsection 2.6.2). Next, we consider the solution $(\mathbf{P}, \mathbf{Q})$ of the linear FBSDE

$$\begin{cases} dP_t = -[-\beta_2 K(t) P_t + K(t)\phi_t + G^\dagger \gamma_t]dt + [-K(t)\psi_t + (I_p + \beta_2 K(t))]Q_t dW_t, \\ P_T = G^\dagger \xi. \end{cases}$$

We now let $\mathbf{X}'$ be the solution of the SDE

$$dX'_t = [-\beta_2(K(t)X'_t + P_t) + \phi_t]dt + [\gamma_t - \beta_2 Q_t]dW_t$$

with initial condition $X'_0 = x$. From this, one can easily check that $(\mathbf{X}', \mathbf{Y}', \mathbf{Z}')$, where $\mathbf{Y}'$ and $\mathbf{Z}'$ are defined by $Y'_t = K(t)X'_t + P_t$ and $Z'_t = Q_t$, is a solution of (FB2). Finally, one recovers $(\mathbf{X}, \mathbf{Y}, \mathbf{Z})$ from $(\mathbf{X}', \mathbf{Y}', \mathbf{Z}')$ and $(\mathbf{Y}'', \mathbf{Z}'')$ via the formula

$$\begin{bmatrix} x \\ y \\ z \end{bmatrix} = \begin{bmatrix} x' \\ G(G^\dagger G)^{-1} y' + y'' \\ G(G^\dagger G)^{-1} z' + z'' \end{bmatrix},$$

and it is easy to check that $(\mathbf{X}, \mathbf{Y}, \mathbf{Z})$ solves (FB1).

(2) We now assume that $p < d$. In this case, the matrix $GG^\dagger$ is a $p \times p$ matrix with full rank. We set

$$\begin{bmatrix} x' \\ x'' \\ y' \\ z' \end{bmatrix} = \begin{bmatrix} Gx \\ [I_d - G^\dagger(GG^\dagger)^{-1}G]x \\ y \\ z \end{bmatrix},$$

and we construct $\mathbf{X}''$ as the solution of the SDE

$$\begin{cases} dX''_t = [I_d - G^\dagger(GG^\dagger)^{-1}G]\phi_t dt + [I_d - G^\dagger(GG^\dagger)^{-1}G]\psi_t dW_t, \\ X''_0 = [I_d - G^\dagger(GG^\dagger)^{-1}G]x, \end{cases}$$

and $(\mathbf{X}', \mathbf{Y}', \mathbf{Z}')$ as the solution of the FBSDE

$$\text{(FB3)} \begin{cases} dX'_t = [-\beta_2 GG^\dagger Y'_t + G\phi_t]dt + [G\psi_t - \beta_2 GG^\dagger Z'_t]dW_t, \\ dY'_t = -[\beta_1 X'_t + \gamma_t]dt + Z'_t dW_t, \\ X'_0 = Gx, \quad Y'_T = \lambda X'_T + \xi. \end{cases}$$

In order to solve this FBSDE, we first construct the solution $t \mapsto K(t)$ of the Riccati equation for $p \times p$ matrices

$$\begin{cases} -\dot{K}(t) = -\beta_2 K(t) GG^\dagger K(t) + \beta_1 I_d, \\ K(T) = \lambda I_d, \end{cases}$$

and then the unique solution $(\mathbf{P}, \mathbf{Q})$ of the FBSDE

$$\begin{cases} dP_t = -[-\beta_2 K(t) GG^\dagger P_t + K(t) G\phi_t + \gamma_t]dt \\ \qquad\quad + [-K(t) G\psi_t + (I_d + \beta_2 K(t)) GG^\dagger]Q_t dW_t, \\ P_T = \xi, \end{cases}$$

2.5. Existence and Uniqueness of Solutions

from which we solve the (forward) SDE

$$dX'_t = [-\beta_2 GG^\dagger(K(t)X'_t + P_t) + G\phi_t]dt + [G\gamma_t - \beta_2 GG^\dagger Q_t]dW_t$$

with initial condition $X'_0 = Gx$. Completing the definition of $(\mathbf{X'}, \mathbf{Y'}, \mathbf{Z'})$ by $Y'_t = K(t)X'_t + P_t$ and $Z'_t = Q_t$, we see that it solves (FB3) and that, defining $(\mathbf{X}, \mathbf{Y}, \mathbf{Z})$ via

$$\begin{bmatrix} x \\ y \\ z \end{bmatrix} = \begin{bmatrix} G^\dagger(GG^\dagger)^{-1}x' + x'' \\ y' \\ z' \end{bmatrix},$$

it is easy to check that $(\mathbf{X}, \mathbf{Y}, \mathbf{Z})$ so defined solves (FB1). As before, uniqueness follows from Theorem 2.18. □

While several *variations on the same theme* exist, the result of the following lemma is the epitome of the continuation methods used to prove well-posedness of FBSDEs.

Lemma 2.22. *Under* **(H1)** *and* **(H3)**, *there exists* $\delta_0 > 0$ *such that if for some* $\alpha_0 \in [0, 1)$ *there exists a solution* $\Xi^{\alpha_0} = (\mathbf{X}^{\alpha_0}, \mathbf{Y}^{\alpha_0}, \mathbf{Z}^{\alpha_0})$ *of the FBSDE (*) below, then for each* $\delta \in [0, \delta_0]$ *there exists a solution* $\Xi^{\alpha_0+\delta} = (\mathbf{X}^{\alpha_0+\delta}, \mathbf{Y}^{\alpha_0+\delta}, \mathbf{Z}^{\alpha_0+\delta})$ *of (*) with* $\alpha = \alpha_0 + \delta$.

Proof. Let us assume that, for each $\tilde{X}_T \in L^2(\Omega, \mathcal{F}_T, \mathbb{P}; \mathbb{R}^d)$ and $\tilde{\Xi} = (\tilde{\mathbf{X}}, \tilde{\mathbf{Y}}, \tilde{\mathbf{Z}}) \in \mathbb{H}^{2,d+p+dm}$, there exists a unique solution $\Xi = (\mathbf{X}, \mathbf{Y}, \mathbf{Z}) \in \mathbb{H}^{2,d+p+dm}$ of the FBSDE

$$\begin{cases} dX_t = [(1-\alpha_0)\beta_2(-G^\dagger Y_t) + \alpha_0 b(t, \Xi_t) + \delta[\beta_2 G^\dagger \tilde{Y}_t + b(t, \tilde{\Xi}_t)] + \phi_t]dt \\ \qquad + [(1-\alpha_0)\beta_2(-G^\dagger Z_t) + \alpha_0 \sigma(t, \Xi_t) + \delta[\beta_2 G^\dagger \tilde{Z}_t + \sigma(t, \tilde{\Xi}_t)] + \psi_t]dW_t, \\ dY_t = -[(1-\alpha_0)\beta_1 G X_t + \alpha_0 f(t, \Xi_t) + \delta[-\beta_1 G \tilde{X}_t + f(t, \tilde{\Xi}_t)] + \gamma_t]dt + Z_t^\alpha dW_t, \\ X_0 = x, \quad Y_T = \alpha_0 g(X_T) + (1-\alpha_0)G X_T + \delta[g(X_T) - G\tilde{X}_T] + \xi. \end{cases}$$

Now, for $\delta > 0$ we consider the mapping $I_{\alpha_0+\delta}$ from $\mathbb{H}^{2,d+p+dm} \times L^2(\Omega, \mathcal{F}_T, \mathbb{P}; \mathbb{R}^d)$ into itself defined by

$$(\tilde{\Xi}, \tilde{X}_T) \hookrightarrow I_{\alpha_0+\delta}(\tilde{\Xi}, \tilde{X}_T) = (\Xi, X_T)$$

and prove that it is a strict contraction for $\delta > 0$ small enough. Clearly, this will complete the proof. Let us pick $(\tilde{\Xi}', \tilde{X}'_T) \in \mathbb{H}^{2,d+p+dm} \times L^2(\Omega, \mathcal{F}_T, \mathbb{P}; \mathbb{R}^d)$, let us set $(\Xi', X'_T) = I_{\alpha_0+\delta}(\tilde{\Xi}', \tilde{X}'_T)$,

$$\Delta\tilde{\Xi} = (\Delta\tilde{\mathbf{X}}, \Delta\tilde{\mathbf{Y}}, \Delta\tilde{\mathbf{Z}}) = (\tilde{\mathbf{X}}' - \tilde{\mathbf{X}}, \tilde{\mathbf{Y}}' - \tilde{\mathbf{Y}}, \tilde{\mathbf{Z}}' - \tilde{\mathbf{Z}}),$$

and

$$\Delta\Xi = (\Delta\mathbf{X}, \Delta\mathbf{Y}, \Delta\mathbf{Z}) = (\mathbf{X}' - \mathbf{X}, \mathbf{Y}' - \mathbf{Y}, \mathbf{Z}' - \mathbf{Z}).$$

X and X' being solutions of standard Itô SDEs, usual estimates of the difference ΔX give

$$\sup_{0 \le t \le T} \mathbb{E}|\Delta X_t|^2 \le C\delta\mathbb{E}\int_0^T |\Delta\tilde{\Xi}_t|^2 dt + C\mathbb{E}\int_0^T \left(|\Delta Y_t|^2 + |\Delta Z_t|^2\right)dt, \qquad (2.31)$$

and consequently

$$\mathbb{E}\int_0^T |\Delta X_t|^2 dt \leq CT\delta\, \mathbb{E}\int_0^T |\Delta\tilde{\Xi}_t|^2 dt + CT\mathbb{E}\int_0^T \left(|\Delta Y_t|^2 + |\Delta Z_t|^2\right) dt. \quad (2.32)$$

Similarly, using the same standard technique already used in this chapter to control the stability of solutions of BSDEs, we estimate the difference $(\Delta\mathbf{Y}, \Delta\mathbf{Z})$ by

$$\mathbb{E}\int_0^T \left(|\Delta Y_t|^2 + |\Delta Z_t|^2\right) dt \leq C\delta\mathbb{E}\int_0^T |\Delta\tilde{\Xi}_t|^2 dt + C\mathbb{E}\int_0^T |\Delta X_t|^2 dt. \quad (2.33)$$

Above, and throughout the remainder of the proof, the constant $C > 0$ depends upon the Lipschitz constant c, G, β_1, β_2, and T. Applying Itô's formula to $|\Delta Y_t|^2$ and using the BSDEs solved by $\mathbf{Y}$ and $\mathbf{Y}'$, we get

$$|\Delta Y_0|^2 + \mathbb{E}\int_0^T |\Delta Z_t|^2 dt \quad (2.34)$$

$$= \mathbb{E}\int_0^T 2\Delta Y_t\left(\alpha_0 \Delta f_t + (1-\alpha_0)\beta_1 G\Delta X_t + \delta\Delta f_t - \delta\beta_1 G\Delta\tilde{X}_t\right) dt$$

$$\leq \frac{1}{4}\mathbb{E}\int_0^T |\Delta Z_t|^2 dt + \frac{1}{4C''}\mathbb{E}\int_0^T |\Delta|X_t|^2 dt$$
$$+ C'\mathbb{E}\int_0^T |\Delta Y_t|^2 dt + C'\delta\mathbb{E}\int_0^T |\Delta\tilde{\Xi}_t|^2 dt,$$

where we used the notations Δf_t for $f(t, \Xi_t') - f(t, \Xi_t)$ and $C'' = \max(CT, 1)$, and C' is large enough, again depending upon the same constants as C. Applying Itô's formula to $G\Delta X_t \cdot \Delta Y_t$ and taking expectations, we get

$$0 = \mathbb{E}\int_0^T \alpha_0[A(t, \Xi_t') - A(t, \Xi_t)] \cdot \Delta\Xi_t \, dt$$

$$- (1-\alpha_0)\mathbb{E}\int_0^T \left(\beta_1 |G\Delta X_t|^2 + \beta_2 |G^\dagger\Delta Y_t|^2 + \beta_2 |G^\dagger\Delta Z_t|^2\right) dt$$

$$+ \delta\mathbb{E}\int_0^T \left(\beta_1 G\Delta X_t \cdot G\Delta\tilde{X}_t + \beta_2 G^\dagger\Delta Y_t \cdot G^\dagger\Delta\tilde{Y}_t + \beta_2 G^\dagger\Delta Z_t \cdot G^\dagger\Delta\tilde{Z}_t\right.$$

$$\left. - \Delta X_t \cdot G^\dagger\Delta f_t + G^\dagger\Delta Y_t \cdot \Delta b_t + \Delta Z_t \cdot G\Delta\sigma_t\right) dt,$$

where we used the notation Δb_t and $\Delta\sigma_t$ for $b(t, \Xi_t') - b(t, ,\Xi_t)$ and $\sigma(t, \Xi_t') - \sigma(t, ,\Xi_t)$, respectively. Finally, using assumptions **(H1)** and **(H3)**, we get

$$\mathbb{E}\int_0^T \left(\beta_1 |G\Delta X_t|^2 + \beta_2 |G^\dagger\Delta Y_t|^2\right) dt \leq C\delta\mathbb{E}\int_0^T \left(|\Delta\tilde{\Xi}_t|^2 + |\Delta\Xi_t|^2\right) dt. \quad (2.35)$$

Combining the estimates (2.31)–(2.35) we get, for a new constant C'''

$$\mathbb{E}\int_0^T |\Delta\Xi_t|^2 \leq C'''\delta\mathbb{E}\int_0^T |\Delta\tilde{\Xi}_t|^2 dt$$

as long as ($\beta_1 > 0$ and $\beta_2 \geq 0$) or ($\beta_1 \geq 0$ and $\beta_2 > 0$). We can now choose δ_0 so that $\delta_0 C''' < 1$ to conclude the proof. □

2.5. Existence and Uniqueness of Solutions

Proof of Theorem 2.20. For each $\alpha \in [0,1]$, $\phi = (\phi_t)_{0 \le t \le T} \in \mathbb{H}^{2,d}$, $\psi = (\psi_t)_{0 \le t \le T} \in \mathbb{H}^{2,dm}$, $\gamma = (\gamma_t)_{0 \le t \le T} \in \mathbb{H}^{2,p}$, and $\xi \in L^2(\Omega, \mathcal{F}, \mathbb{P}; \mathbb{R}^p)$, let us consider the FBSDE

$$(*) \begin{cases} dX_t^\alpha = [(1-\alpha)\beta_2(-G^\dagger Y_t^\alpha) + \alpha b(t, \Xi_t^\alpha) + \phi_t]dt \\ \qquad\qquad + [(1-\alpha)\beta_2(-G^\dagger Z_t^\alpha) + \alpha \sigma(t, \Xi_t^\alpha) + \psi_t]dW_t, \\ dY_t^\alpha = -[(1-\alpha)\beta_1 G X_t^\alpha + \alpha f(t, \Xi_t^\alpha) + \gamma_t]dt + Z_t^\alpha dW_t, \\ X_0^\alpha = x, \quad Y_T^\alpha = \alpha g(X_T^\alpha) + (1-\alpha)G X_T^\alpha + \xi. \end{cases}$$

Clearly, our goal is to prove existence and uniqueness for $\alpha = 1$ and $\xi = 0$. Choosing $\lambda = 1$ and $\xi = 0$ in Lemma 2.21, we see that (*) has a unique solution for $\alpha = 0$. Next, applying Lemma 2.22 N times for N chosen such that $1 \le N\delta_0 \le 1 + \delta_0$, we conclude that (*) has a unique solution for $\alpha = 1$. □

2.5.3 ▪ Existence and Uniqueness for Short Time

As before, we consider the FBSDE

$$\begin{cases} dX_t = b(t, X_t, Y_t, Z_t)dt + \sigma(t, X_t, Y_t)dW_t, \\ dY_t = -f(t, X_t, Y_t, Z_t)dt + Z_t dW_t \end{cases} \quad (2.36)$$

with the initial condition $X_0 = x$ and terminal condition $Y_T = g(X_T)$. Notice that Z_t does not appear in the volatility of the forward component X_t. Despite the drastic limitations due to the difficulties in checking the monotonicity condition in practical situations, the existence and uniqueness results proven above can be very useful when the coefficients are random. In this section, we limit ourselves to the case of coefficients b, σ, f, and g given by deterministic functions $(t, x, y, z) \hookrightarrow b(t, x, y, z)$, $(t, x, y) \hookrightarrow \sigma(t, x, y)$, $(t, x, y, z) \hookrightarrow f(t, x, y, z)$, and $x \hookrightarrow g(x)$. We will use the following classical Lipschitz continuity assumption together with the usual linear growth condition. They are just reformulations of the assumptions (**A1**) and (**A2**) which we used earlier in the chapter.

(i) *Linear growth condition.* There exists a constant $c > 0$ such that

$$|b(t,x,y,z)| + |f(t,x,y,z)| \le c(1 + |x| + |y| + |z|),$$
$$|\sigma(t,x,y)| \le c(1 + |x| + |y|),$$
$$|g(x)| \le c(1 + |x|)$$

for all $t \in [0,T]$, $x \in \mathbb{R}^d$, $y \in \mathbb{R}^p$, and $z \in \mathbb{R}^{pm}$.

(ii) *Lipschitz condition.* There exists a constant $K > 0$ such that

$$|b(t,x,y,z) - b(t,x',y',z')| + |f(t,x,y,z) - f(t,x',y',z')| \quad (2.37)$$
$$+ |\sigma(t,x,y) - \sigma(t,x',y')| + |g(x) - g(x')| \le c(|x-x'| + |y-y'| + |z-z'|)$$

for all $t \in [0,T]$, $x, x' \in \mathbb{R}^d$, $y, y' \in \mathbb{R}^p$, and $z, z' \in \mathbb{R}^{pd}$. Under these conditions we have existence and uniqueness of a solution to the FBSDE (2.36) if T is small enough.

Theorem 2.23. *Under assumptions* (i) *and* (ii) *above, there exists $\delta > 0$ such that if $T < \delta$, then for any $x \in \mathbb{R}^d$, existence and uniqueness hold for* (2.36).

We prove this result for a filtration $\mathbb{F}$, which is not necessarily the filtration generated by $\mathbf{W}$ as long as $\mathbf{W}$ remains an $\mathbb{F}$-Brownian motion in the sense that it is adapted to $\mathbb{F}$ and for each $s \le t$, $W_t - W_s$ is independent of $\mathcal{F}_s$.

Proof. Let $x \in \mathbb{R}^d$ be fixed. If $\mathbf{X} = (X_t)_{0 \leq t \leq T} \in \mathbb{S}^{2,d}$ is given such that $X_0 = x$, we denote by $(\mathbf{Y}, \mathbf{Z})$ the unique solution of $BSDE(\Psi, \xi)$, where $\Psi(t, \omega, y, z) = f(t, X_t(\omega), y, z)$ and $\xi = g(X_T)$, whose existence and uniqueness are given by Theorem 2.2, and we let $\mathbf{X}' = (X'_t)_{0 \leq t \leq T}$ be the unique solution of the SDE

$$dX'_t = b(t, X'_t, Y_t, Z_t)dt + \sigma(t, X'_t, Y_t)dW_t, \qquad X'_0 = x,$$

whose existence is given by Theorem 1.2. In this way, we have defined a map

$$\mathbb{S}^{2,d} \ni \mathbf{X} \hookrightarrow \Phi(\mathbf{X}) = \mathbf{X}' \in \mathbb{S}^{2,d},$$

and we proceed to show that Φ is a strict contraction for T small enough. If $\mathbf{X}' = \Phi(\mathbf{X})$ and $\tilde{\mathbf{X}}' = \Phi(\tilde{\mathbf{X}})$ with $X_0 = \tilde{X}_0 = x$, then using the obvious notation $(\tilde{\mathbf{Y}}, \tilde{\mathbf{Z}})$ for the solution of the BSDE with the driver depending upon $\tilde{X}_t$, Itô's formula gives

$$|X'_t - \tilde{X}'_t|^2 = 2 \int_0^t (X'_s - \tilde{X}'_s) \cdot [b(s, X'_s, Y_s, Z_s) - b(s, \tilde{X}'_s, \tilde{Y}_s, \tilde{Z}_s)]ds$$
$$+ 2 \int_0^t (X'_s - \tilde{X}'_s) \cdot [\sigma(s, X'_s, Y_s) - \sigma(s, \tilde{X}'_s, \tilde{Y}_s)]dW_s$$
$$+ \int_0^t \text{trace}[(\sigma(s, X'_s, Y_s) - \sigma(s, \tilde{X}'_s, \tilde{Y}_s))(\sigma(s, X'_s, Y_s) - \sigma(s, \tilde{X}'_s, \tilde{Y}_s))^\dagger]ds,$$

and, using the Lipschitz property of b and σ,

$$\mathbb{E} \sup_{0 \leq t \leq T} |X'_t - \tilde{X}'_t|^2 \leq c\mathbb{E} \int_0^T |X'_t - \tilde{X}'_t|(|X'_t - \tilde{X}'_t| + |Y_t - \tilde{Y}_t| + |Z_t - \tilde{Z}_t|)dt$$
$$+ c\mathbb{E} \sup_{0 \leq t \leq T} \left| \int_0^t (X'_t - \tilde{X}'_t) \cdot [\sigma(s, X'_s, Y_s) - \sigma(s, \tilde{X}'_s, \tilde{Y}_s)]dW_s \right|$$
$$+ c\mathbb{E} \int_0^T (|X'_t - \tilde{X}'_t|^2 + |Y_t - \tilde{Y}_t|^2)dt.$$

Using the Burkholder–Davis–Gundy Inequality (see for example (2.43)), we get

$$\mathbb{E} \sup_{0 \leq t \leq T} |X'_t - \tilde{X}'_t|^2 \leq c\mathbb{E} \int_0^T |X'_t - \tilde{X}'_t|(|X'_t - \tilde{X}'_t| + |Y_t - \tilde{Y}_t| + |Z_t - \tilde{Z}_t|)dt$$
$$+ c\mathbb{E} \left(\int_0^T |X'_t - \tilde{X}'_t|^2 (|X'_t - \tilde{X}'_t|^2 + |Y_t - \tilde{Y}_t|^2) ds \right)^{1/2}$$
$$+ c\mathbb{E} \int_0^T (|X'_t - \tilde{X}'_t|^2 + |Y_t - \tilde{Y}_t|^2)dt.$$

Limiting ourselves to values of T which are bounded from above, this gives

$$\mathbb{E} \sup_{0 \leq t \leq T} |X'_t - \tilde{X}'_t|^2$$
$$\leq cT^{1/2} \left(\mathbb{E} \sup_{0 \leq t \leq T} |X'_t - \tilde{X}'_t|^2 + \mathbb{E} \sup_{0 \leq t \leq T} |Y_t - \tilde{Y}_t|^2 + \mathbb{E} \int_0^T |Z_t - \tilde{Z}_t|^2 dt \right),$$

and, consequently,

$$(1 - cT^{1/2})\mathbb{E} \sup_{0 \leq t \leq T} |X'_t - \tilde{X}'_t|^2 \leq cT^{1/2} \left(\mathbb{E} \sup_{0 \leq t \leq T} |Y_t - \tilde{Y}_t|^2 + \mathbb{E} \int_0^T |Z_t - \tilde{Z}_t|^2 dt \right). \tag{2.38}$$

2.5. Existence and Uniqueness of Solutions

We now estimate the right-hand side in terms of $\mathbb{E}\sup_{0\leq t\leq T}|X_t - \tilde{X}_t|^2$, proceeding very much in the same way as in the proof of Theorem 2.2. Using Itô's formula again, we get

$$|Y_t - \tilde{Y}_t|^2 = |g(X_T) - g(\tilde{X}_T)|^2 + 2\int_t^T (Y_s - \tilde{Y}_s)$$
$$\cdot [f(s, X_s, Y_s, Z_s) - f(s, \tilde{X}_s, \tilde{Y}_s, \tilde{Z}_s)]ds$$
$$- 2\int_t^T (Y_t - \tilde{Y}_t) \cdot [Z_s - \tilde{Z}_s]dW_s - \int_t^T |Z_s - \tilde{Z}_s|^2 ds \quad (2.39)$$

so that

$$\int_0^T |Z_s - \tilde{Z}_s|^2 ds \leq |g(X_T) - g(\tilde{X}_T)|^2$$
$$+ 2\int_0^T |Y_s - \tilde{Y}_s||f(s, X_s, Y_s, Z_s) - f(s, \tilde{X}_s, \tilde{Y}_s, \tilde{Z}_s)|ds$$
$$- 2\int_0^T (Y_t - \tilde{Y}_t) \cdot [Z_s - \tilde{Z}_s]dW_s. \quad (2.40)$$

The expectation of the stochastic integral is zero because

$$\mathbb{E}\left(\int_0^T |Y_s - \tilde{Y}_s|^2|Z_s - \tilde{Z}_s|^2 ds\right)^{1/2} \leq \frac{1}{2}\mathbb{E}\left(\sup_{0\leq s\leq T}|Y_s - \tilde{Y}_s|^2 + \int_0^T |Z'_s - \tilde{Z}'_s|^2 ds\right),$$

which is finite by our definition of a solution of a BSDE. So, taking expectations on both sides of (2.40) and using the Lipschitz property of f and g, we find the existence of a constant $c > 0$ such that

$$\mathbb{E}\int_0^T |Z_s - \tilde{Z}_s|^2 ds$$
$$\leq c\left(\mathbb{E}|X_T - \tilde{X}_T|^2 + \mathbb{E}\int_0^T |Y_t - \tilde{Y}_t|(|X_t - \tilde{X}_t| + |Y_t - \tilde{Y}_t| + |Z_t - \tilde{Z}_t|)dt\right)$$
$$\leq c\left((1+T)\mathbb{E}\sup_{0\leq t\leq T}|X_t - \tilde{X}_t|^2 + T\mathbb{E}\sup_{0\leq t\leq T}|Y_t - \tilde{Y}_t|^2 + \frac{1}{2}\mathbb{E}\int_0^T |Z_t - \tilde{Z}_t|^2 dt\right)$$

so that

$$\mathbb{E}\int_0^T |Z_s - \tilde{Z}_s|^2 ds \leq c\left((1+T)\mathbb{E}\sup_{0\leq t\leq T}|X_t - \tilde{X}_t|^2 + T\mathbb{E}\sup_{0\leq t\leq T}|Y_t - \tilde{Y}_t|^2\right). \quad (2.41)$$

Returning to Itô's formula (2.39), and using once more the Burkholder–Davis–Gundy Inequality and the Lipschitz property of f and g, we get

$$\mathbb{E}\sup_{0\leq t\leq T}|Y_s - \tilde{Y}_s|^2 ds \leq c\left(\mathbb{E}|X_T - \tilde{X}_T|^2 + \mathbb{E}\left(\int_0^T |Y_t - \tilde{T}_t|^2|Z_t - \tilde{Z}_t|^2 dt\right)^{1/2}\right)$$
$$+ \mathbb{E}\int_0^T |Y_t - \tilde{Y}_t|(|X_t - \tilde{X}_t| + |Y_t - \tilde{Y}_t| + |Z_t - \tilde{Z}_t|)dt.$$

Using (2.41), this gives

$$\mathbb{E} \sup_{0 \leq t \leq T} |Y_t - \tilde{Y}_t|^2$$
$$\leq c\left((1+T)\mathbb{E} \sup_{0 \leq t \leq T} |X_t - \tilde{X}_t|^2 + T\mathbb{E} \sup_{0 \leq t \leq T} |Y_t - \tilde{Y}_t|^2\right) + \frac{1}{2}\mathbb{E} \sup_{0 \leq t \leq T} |Y_t - \tilde{Y}_t|^2,$$

and, consequently,

$$(1 - cT)\mathbb{E} \int_0^T |Z_s - \tilde{Z}_s|^2 ds \leq c(1+T)\mathbb{E} \sup_{0 \leq t \leq T} |X_t - \tilde{X}_t|^2. \qquad (2.42)$$

Finally, plugging (2.41) and (2.42) into (2.38) shows that Φ is indeed a strict contraction if T is small enough.

Remark 2.24. *Burkhölder–Davis–Gundy Inequality. Since we shall use this inequality repeatedly, we state it for future reference. For any $p \geq 1$, there exist positive constants c_p and C_p such that, for all local martingale, $\mathbf{X}$ with $X_0 = 0$ and stopping times τ, the following inequalities hold:*

$$c_p \mathbb{E}[X, X]_\tau^{p/2} \leq \mathbb{E} \sup_{0 \leq t \leq \tau} |X_t|^p \leq C_p \mathbb{E}[X, X]_\tau^{p/2}, \qquad (2.43)$$

and the statement holds for all $0 < p < \infty$ for continuous local martingales.

Once we have existence and uniqueness for a given $T > 0$, however small, and assuming now that the filtration is generated by the Wiener process $\mathbf{W}$, we can repeat the construction of the FBSDE value function u on $[0, T] \times \mathbb{R}^d$ which we gave in the decoupled case. We first notice that for each $(t, x) \in [0, T] \times \mathbb{R}^d$ fixed, if we denote by $(\mathbf{X}^{t,x}, \mathbf{Y}^{t,x}, \mathbf{Z}^{t,x})$ the solution of the FBSDE (2.36) over the interval $[t, T]$ with initial condition $X_t^{t,x} = x$ and terminal condition $Y_T^{t,x} = g(X_T^{t,x})$, then $Y_t^{t,x}$ is deterministic and we denote its value by $u(t, x)$. Indeed, the randomness comes only from the increments $\{W_r - W_t\}_{t \leq r \leq T}$, and the 0–1 law forces $Y_t^{t,x}$ to be deterministic. This is due to the fact that we are actually working with the filtration $\mathcal{F}_s^t = \sigma\{W_r - W_t; t \leq r \leq T\}$. We close this section with the proofs, still for small time T, of two important properties of the FBSDE value function. While we cannot prove that u is differentiable, we can prove that it is Lipschitz for T small enough.

Lemma 2.25. *For T small enough, there exists a constant $c > 0$ such that for all x and x' in $\mathbb{R}^d$, we have*

$$|u(t, x) - u(t, x')| \leq c|x - x'|, \qquad 0 \leq t \leq T.$$

Proof. Starting from the definition of the decoupling field, we get

$$|u(t, x) - u(t, x')| = |Y_t^{t,x} - Y_t^{t,x'}|$$
$$= |\mathbb{E}[Y_t^{t,x} - Y_t^{t,x'}]|$$
$$= \left|\mathbb{E}[g(X_T^{t,x}) - g(X_T^{t,x'})] + \mathbb{E}\int_t^T [f(s, X_s^{t,x}, Y_s^{t,x}, Z_s^{t,x}) - f(s, X_s^{t,x'}, Y_s^{t,x'}, Z_s^{t,x'})]ds\right|$$
$$\leq c\mathbb{E}|X_T^{t,x} - X_T^{t,x'}| + c\mathbb{E}\int_t^T (|X_s^{t,x} - X_s^{t,x'}| + |Y_s^{t,x} - Y_s^{t,x'}| + |Z_s^{t,x} - Z_s^{t,x'}|)ds,$$

and we conclude using Gronwall's inequality. $\square$

2.5. Existence and Uniqueness of Solutions

Finally, we show that the value function gives the solution of the backward equation at all times.

Lemma 2.26. *For T small enough, if $(\mathbf{X}, \mathbf{Y}, \mathbf{Z})$ is the unique solution of the FBSDE (2.36), then for all $t \in [0, T]$ we have*

$$Y_t = u(t, X_t) \quad a.s.$$

Proof. For a given fixed $t \in [0, T]$, we have $u(t, x) = Y_t^{t,x}$ for all $x \in \mathbb{R}^d$, and by uniqueness (and a few subtle measure-theoretical arguments), $Y_t = Y_t^{t, X_t}$, which is the desired result. □

2.5.4 ▪ Existence and Uniqueness for Fully Coupled FBSDEs

As before, we consider the FBSDE

$$\begin{cases} dX_t = b(t, X_t, Y_t, Z_t)dt + \sigma(t, X_t, Y_t)dW_t, \\ dY_t = -f(t, X_t, Y_t, Z_t)dt + Z_t dW_t \end{cases} \quad (2.44)$$

with initial condition $X_0 = x$ and terminal condition $Y_T = g(X_T)$. Notice that, as in the derivation of existence for small time T, Z_t does not appear in the volatility of the forward component X. As before, we limit ourselves to the case of coefficients b, σ, f, and g given by deterministic functions $(t, x, y, z) \hookrightarrow b(t, x, y, z)$, $(t, x, y) \hookrightarrow \sigma(t, x, y)$, $(t, x, y, z) \hookrightarrow f(t, x, y, z)$, and $x \hookrightarrow g(x)$. In addition to the linear growth and Lipschitz assumptions (i) and (ii) of the previous subsection we add the following.

(iii) *Nondegeneracy condition.* There exists a constant $\lambda > 0$ such that

$$\sigma(t, x, y)\sigma(t, x, y)^\dagger \geq \lambda \quad (2.45)$$

for all $t \in [0, T]$, $x \in \mathbb{R}^d$, and $y \in \mathbb{R}^m$. Inequality (2.45) has to be understood in the sense of quadratic forms. It means that the spectrum of the symmetric matrix $\sigma(t, x, y)\sigma(t, x, y)^\dagger$ is bounded from below by λ. This *uniform ellipticity* condition will guarantee that the noise driving the forward equation will be transmitted to the backward equation.

From what we have seen earlier in the analysis of both the decoupled case considered in Proposition 2.14 and the short time case considered in Theorem 2.23, it is reasonable to expect that $\mathbf{Y}$ and $\mathbf{Z}$ are given by deterministic functions of $\mathbf{X}$ in the sense that

$$Y_t = u(t, X_t) \quad \text{and} \quad Z_t = v(t, X_t) \quad (2.46)$$

for all $t \in [0, T]$ for some deterministic functions $(t, x) \hookrightarrow u(t, x)$ and $(t, x) \hookrightarrow v(t, x)$. Recall that we use the terminology *FBSDE value function* or *decoupling field* for u. Accordingly, we also expect that the function v is given by

$$v(t, x) = \partial_x u(t, x)\sigma(t, x, u(t, x)). \quad (2.47)$$

Remark 2.27. *Clearly, things would change dramatically and become more involved if σ also depended upon z, namely if $\sigma = \sigma(t, x, y, z)$—hence our decision not to discuss this more general case here. Indeed, in this case the relationship (2.47) between u and v would be*

$$v(t, x) = \partial_x u(t, x)\sigma(t, x, u(t, x), v(t, x))$$

which is an implicit equation in v.

Applying Itô's formula (at least formally, since we do not know if the functions u and v are sufficiently differentiable) to $dY_t = du(t, X_t)$, using (2.46) and (2.47), we find that for the backward equation to be satisfied, the function u needs to satisfy

$$\begin{cases} \partial_t u + \frac{1}{2}a(t,x,u(t,x))\partial_{xx}^2 u + b(t,x,u(t,x),\partial_x u(t,x)\sigma(t,x,u(t,x)))\partial_x u \\ \qquad + f(t,x,u(t,x),\partial_x u(t,x)\sigma(t,x,u(t,x))) = 0, \quad 0 \leq t \leq T, \\ u(T,x) = g(x), \end{cases} \quad (2.48)$$

at least for $x = X_t$, where as usual, the $d \times d$ symmetric matrix $a(t,x,y)$ is defined as $a(t,x,y) = \sigma(t,x,y)\sigma(t,x,y)^\dagger$. This PDE is highly nonlinear and existence and uniqueness results could be problematic in general. In 1994, Ma, Protter, and Yong [90] assumed that (2.48) is solvable in the classical sense (i.e., they assumed that there exists a function $u : [0,T] \times \mathbb{R}^d \hookrightarrow \mathbb{R}^p$ for which $\partial_t u$, $\partial_x u$, and $\partial_{xx}^2 u$ exist as functions, are continuous and bounded on the whole space, and satisfy (2.48)) and considered the SDE

$$\begin{aligned} dX_t &= b(t, X_t, u(t, X_t), \partial_x u(t, X_t)\sigma(t, X_t, u(t, X_t)))dt \\ &\quad + \sigma(t, X_t, u(t, X_t), \sigma(t, X_t u(t, X_t)))dW_t, \end{aligned} \quad (2.49)$$

which has uniformly Lipschitz coefficients and hence is uniquely solvable. Without worrying about the issue of uniqueness for the PDE (2.48), they considered (2.44) solved. The following verification theorem formalizes this fact: existence and uniqueness for the FBSDE (2.44) follows from the existence (not necessarily uniqueness) of a classical solution for the PDE (2.48). Unfortunately, the main shortcoming of this result is due to the fact that the existence of classical solutions of the nonlinear PDE (2.48) is known to hold under very restrictive conditions. This discussion is very similar to the discussion we had when we proved Proposition 2.14 in the decoupled case. One of the major differences is that the current nonlinear PDE (2.48) is more involved because of the coupling. Indeed, the solution function u did not appear in σ then, but it does now!

Theorem 2.28. Verification Theorem. *Let us assume that there exists a function $u : [0,T] \times \mathbb{R}^d \hookrightarrow \mathbb{R}^p$ which is $C^{1,2}$ and such that $\partial_t u$, $\partial_x u$, and $\partial_{xx}^2 u$ are continuous and bounded on the whole space, and which satisfies the PDE (2.48). Then $(\mathbf{Y}, \mathbf{Z})$, defined by $Y_t = u(t, X_t)$ and $Z_t = \partial_x u(t, X_t)\sigma(t, X_t, u(t, X_t))$, where X_t is the unique solution of the SDE (2.49) and v is defined by (2.47), is the unique solution of the FBSDE (2.44).*

Proof. While the result is true in the generality of the statement, we give a proof in the case where, like the volatility σ, the coefficients b and f do not depend upon z. Also, for the sake of notation, we assume that $d = p = m = 1$. The fact that $\mathbf{Y}$ defined in this way satisfies the backward equation follows the computation of $dY_t = du(t, X_t)$ using Itô's formula. Uniqueness is more of a challenge, especially since we did not assume that there was uniqueness for the PDE (2.48). Let us assume that (X'_t, Y'_t, Z'_t) is another solution of the FBSDE (2.44). We first prove that for every $t \in [0,T]$, $Y'_t = u(t, X'_t)$ almost surely. Itô's formula gives

$$\begin{aligned} d[u(t, X'_t)] &= \left[\partial_t u(t, X'_t) + \partial_x u(t, X'_t)b(t, X'_t, Y'_t, Z'_t) + \frac{1}{2}a(t, X'_t, Y'_t)\partial_{xx}u(t, X_t)\right]dt \\ &\quad + \partial_x u(t, X'_t)\sigma(t, X'_t, Y'_t)dW_t. \end{aligned}$$

2.5. Existence and Uniqueness of Solutions

So if we set $\Delta_t = Y'_t - u(t, X'_t)$, then $\Delta_T = Y'_T - g(X'_T) = 0$, and using (2.48)

$$\begin{aligned}d\Delta_t &= [-f(t, X'_t, Y'_t, Z'_t) - \partial_t u(t, X'_t) - \partial_x u(t, X'_t) b(t, X'_t, Y'_t) \\ &\quad - \frac{1}{2} a(t, X'_t, Y'_t) \partial^2_{xx} u(t, X'_t)] dt + [Z'_t - \partial_x u(t, X'_t) \sigma(t, X'_t, Y'_t)] dW_t \\ &= [f(t, X'_t, u(t, X'_t)(\partial_x u(t, X'_t) \sigma(t, X'_t, u(t, X'_t)))) \\ &\quad - f(t, X'_t, Y'_t, Z'_t) + [b(t, X'_t, u(t, X'_t)) \\ &\quad - b(t, X'_t, Y'_t, Z'_t)] \partial_x u(t, X'_t) + \frac{1}{2} [a(t, X'_t, u(t, X'_t)) - a(t, X'_t, Y'_t)] \partial^2_{xx} u(t, X'_t)] dt \\ &\quad + [Z'_t - \partial_x u(t, X'_t) \sigma(t, X'_t, Y'_t)] dW_t.\end{aligned}$$

Using the Lipschitz property of the coefficients and the boundedness of the derivatives of u, one sees that there exists a constant $c > 0$ such that the coefficient of dt is bounded from above by $c|Y'_t - u(t, X'_t)| = c|\Delta_t|$. Now, notice that $\boldsymbol{\Delta}' = (\Delta'_t)_{0 \le t \le T}$ with $\Delta'_t \equiv 0$ is the solution of the BSDE

$$d\Delta'_t = -c|\Delta'_t| dt + Z'_t dW_t, \qquad \Delta'_T = 0.$$

So the drivers of the BSDEs satisfied by $\boldsymbol{\Delta}$ and $\boldsymbol{\Delta}'$ compare, and since they have the same terminal values, we conclude that $\Delta_t \le \Delta'_t = 0$ almost surely for all $t \in [0, T]$. Going through the same argument for $-\Delta_t$ instead of Δ_t, we prove that $\Delta_t \ge 0$ almost surely, which proves that $\Delta_t = 0$ and hence that $Y'_t = u(t, X'_t)$. Plugging this expression into the forward SDE satisfied by X'_t, we get

$$dX'_t = b(t, X'_t, u(t, X'_t)) dt + \sigma(t, X'_t, u(t, X'_t)) dW_t,$$

which is exactly the same equation as the equation satisfied by X_t. By uniqueness of the solutions of SDEs with Lipschitz coefficients (recall Theorem 1.2), we conclude that $X'_t = X_t$ almost surely for each $t \in [0, T]$, and consequently that $Y'_t = Y_t$ as well. Finally, we conclude that $Z'_t = Z_t$ by uniqueness of the martingale representation. $\square$

The main result of this section is the following. It is due to F. Delarue [49]. Its proof is highly technical, so we skip most of the details and only describe its strategy.

Theorem 2.29. *Under the above three assumptions* (i), (ii), *and* (iii), *for any* $x \in \mathbb{R}^d$, *existence and uniqueness hold for* (2.36).

Proof. As before, for the sake of notation, we restrict ourselves to the particular case $d = p = m = 1$. The proof relies on an induction argument based on the existence and uniqueness results for small time. Let T be arbitrary. We proved in Theorem 2.23 the existence of a constant $\delta > 0$ depending only upon the Lipschitz constant of the coefficients such that existence and uniqueness for the FBSDE (2.44) hold on the interval $[T - \delta, T]$. In other words, for each $(t, x) \in [T - \delta, T] \times \mathbb{R}^d$, there is a unique solution $(\mathbf{X}^{t,x}, \mathbf{Y}^{t,x}, \mathbf{Z}^{t,x})$ over the time interval $[t, T]$ where the solution of the forward SDE starts from x at time t. The construction of the FBSDE value function shows that one can define a function $u : [T - \delta, T] \times \mathbb{R}^d \hookrightarrow \mathbb{R}^d$ such that $Y^{t,x}_s = u(s, X^{t,x}_s)$ for all $t \le s \le T$. The goal is to extend the definition of the function u for times smaller than $T - \delta$. If we assume that $(\tilde{\mathbf{X}}, \tilde{\mathbf{Y}}, \tilde{\mathbf{Z}})$ is a solution of (2.44) over $[0, T] \times \mathbb{R}^d$, then it is a solution over the interval $[T - \delta, T]$ with initial condition $\tilde{X}_{T-\delta}$. Consequently,

$$\tilde{Y}_{T-\delta} = u(T - \delta, \tilde{X}_{T-\delta}),$$

and $(\tilde{X}_t, \tilde{Y}_t, \tilde{Z}_t)$ is a solution of (2.44) over $[0, T-\delta] \times \mathbb{R}^d$ with the terminal condition given by the function $u(T-\delta, \cdot)$. We proved in Lemma 2.26 that this function is Lipschitz, so that we can find $\delta' > 0$ and extend u over the interval $[T - \delta - \delta', T]$. The larger the Lipschitz constant of u, the smaller the δ'. If we can prove that the Lipschitz constants of u are uniformly bounded, then the δ' will be bounded from below by a strictly positive constant, and a finite number of iterations of this procedure will provide a function u defined on the entire $[0, T] \times \mathbb{R}^d$. The conclusion of the proof relies on a uniform bound on the gradient of u. The existence of such a bound is due to the nondegeneracy condition (iii), but the technicalities required by the proof are beyond the scope of this book, and we refer the interested reader to Delarue's original proof in [49]. □

2.6 ▪ The Affine Case

While our intention is not to consider the most general case, we discuss a reasonably general case of an affine FBSDE which can be solved in many instances. In any case, this exercise will be good enough for the applications we consider later on in this book. The FBSDE of interest in this section is

$$\begin{cases} dX_t = [A_t X_t + B_t Y_t + \beta_t]dt + [C_t X_t + D_t Y_t + \sigma_t]dW_t, & X_0 = x, \\ dY_t = -[\hat{A}_t X_t + \hat{B}_t Y_t + \hat{\beta}_t]dt + Z_t dW_t, & Y_T = GX_T + g, \end{cases} \quad (2.50)$$

where the coefficients $\mathbf{A}$, $\mathbf{B}$, $\boldsymbol{\beta}$, $\mathbf{C}$, $\mathbf{D}$, $\boldsymbol{\sigma}$, $\hat{\mathbf{A}}$, $\hat{\mathbf{B}}$, and $\hat{\boldsymbol{\beta}}$ are bounded progressively measurable stochastic processes with values in Euclidean spaces $\mathbb{R}^n$ of appropriate dimensions n, and G and g are $\mathcal{F}_T$-measurable bounded random variables with values in $\mathbb{R}^{pd}$ and $\mathbb{R}^d$, respectively. Notice that the forward and backward equations are coupled, but that the coupling is not of the strongest type since, here again, Z_t does not appear in the forward equation, or even in the drift of the backward equation.

Except for possibly in the case $C_t = D_t = 0$ and $\sigma_t \equiv \sigma > 0$, and despite the innocent-looking form of the coefficients of FBSDE (2.50), none of the existence theorems which we *slaved* to prove in the previous sections apply to this simple particular case. Clearly, the monotonicity assumptions can only be satisfied in very special cases. Moreover, being affine, the diffusion coefficient of the forward equation is not bounded below, so the results of the previous section do not apply either.

2.6.1 ▪ Guessing the Form of the Solution

Motivated by the discussion of the previous sections and the role of the FBSDE value function, we search for a solution in the form $Y_t = u(t, X_t)$, and we make an ansatz on the form of the value function. It appears that this approach is merely a form of the so-called *four step scheme*, for which we refer the reader to [90]. Given the specific form (2.50) of the FBSDE, we search for a solution in the form

$$Y_t = P_t X_t + p_t \quad \text{with} \quad P_T = G \quad \text{and} \quad p_T = g \quad (2.51)$$

for some adapted processes $\mathbf{P}$ and $\mathbf{p}$ with values in $\mathbb{R}^{pd}$ and $\mathbb{R}^p$, respectively, and which do not depend upon $\mathbf{X}$ and $\mathbf{Y}$. The strategy is to use the forward and backward equations of (2.50) to derive equations that the processes $\mathbf{P}$ and $\mathbf{p}$ have to satisfy in order for $(\mathbf{X}, \mathbf{Y})$ to be a solution of (2.50), and then to solve for $\mathbf{P}$ and $\mathbf{p}$. Since the latter have natural terminal conditions given by G and g, it is natural to assume that they are solutions of BSDEs of the form

$$dP_t = -\Gamma(t, P_t, \Lambda_t)dt + \Lambda_t dW_t, \quad P_T = G, \quad (2.52)$$

and

$$dp_t = -\gamma(t, p_t, \lambda_t)dt + \lambda_t dW_t, \quad p_T = g, \quad (2.53)$$

2.6. The Affine Case

for appropriate drivers Γ and γ satisfying the conditions for existence and uniqueness of Theorem 2.2. Later in the section, we consider cases where we can choose **P** and **p** as progressively measurable processes with bounded variations, and even deterministic functions of time. Computing dY_t from (2.51) and using (2.52) and (2.53) with the short notation Γ_t and γ_t, respectively, for their drifts, we get

$$\begin{aligned}
dY_t &= dP_t\, X_t + P_t\, dX_t + d[P, X]_t + dp_t \\
&= \big[-\Gamma_t X_t + P_t A_t X_t + P_t B_t(P_t X_t + p_t) + P_t \beta_t + \Lambda_t C_t X_t + \Lambda_t D_t(P_t X_t + p_t) \\
&\quad + \Lambda_t \sigma_t - \gamma_t\big]dt + \big[\Lambda_t X_t + P_t C_t X_t + P_t D_t(P_t X_t + p_t) + P_t \sigma_t + \lambda_t\big]dW_t \\
&= \big[(-\Gamma_t + P_t A_t + P_t B_t P_t + \Lambda_t C_t + \Lambda_t D_t P_t) X_t \\
&\quad + (P_t \beta_t + \Lambda_t D_t p_t + P_t B_t p_t + \Lambda_t \sigma_t - \gamma_t)\big]dt \\
&\quad + \big[(\Lambda_t + P_t C_t + P_t D_t P_t) X_t + (P_t D_t p_t + P_t \sigma_t + \lambda_t)\big]dW_t.
\end{aligned}$$

Our goal is to identify the affine coefficients term by term with

$$\begin{aligned}
dY_t &= -(\hat{A}_t X_t + \hat{B}_t (P_t X_t + p_t) + \hat{\beta}_t)dt + Z_t dW_t \\
&= \big[-(\hat{A}_t + \hat{B}_t P_t) X_t - (\hat{B}_t p_t + \hat{\beta}_t)\big]dt + Z_t dW_t.
\end{aligned}$$

Notice that identifying the diffusion terms gives

$$Z_t = [\Lambda_t + P_t C_t + P_t D_t P_t] X_t + P_t D_t p_t + P_t \sigma_t + \lambda_t. \tag{2.54}$$

Identifying the coefficients of X_t in the drift terms gives

$$\Gamma_t = \hat{A}_t + \Lambda_t C_t + P_t A_t + (\hat{B}_t + \Lambda_t D_t) P_t + P_t B_t P_t \tag{2.55}$$

so that the process **P** can be found if we can solve the matrix-valued BSDE

$$dP_t = -\big[\hat{A}_t + \Lambda_t C_t + P_t A_t + (\hat{B}_t + \Lambda_t D_t) P_t + P_t B_t P_t\big]dt + \Lambda_t dW_t, \qquad P_T = G, \tag{2.56}$$

whose driver can be written as

$$\Psi(t, \omega, y, z) = \hat{A}_t(\omega) + \hat{B}_t(\omega) y + y A_t(\omega) + z C_t(\omega) + z D_t(\omega) y + y B_t(\omega) y,$$

which is not Lipschitz and which is quadratic in the variables y and z. Even in the particular case $d = p = m = 1$, which we consider in greater detail later, we cannot use known results to prove existence. Indeed, using corresponding lowercase symbols for the coefficients which are now scalars instead of matrices, and using the fact that the product of scalars is commutative, the driver rewrites as

$$\Psi(t, \omega, y, z) = \hat{a}_t(\omega) + [\hat{b}_t(\omega) + a_t(\omega)]y + c_t(\omega)z + d_t(\omega)yz + b_t(\omega)y^2,$$

and the term in y^2 shows that this case is not covered by Kobylanski's result in Theorem 2.7, nor can (2.56) be solved by the mutidimensional extensions of these results which appeared recently. See, for example, [71, 60, 11] and the Notes & Complements section at the end of the chapter for further references.

Identifying the coefficients independent of X_t in the drift terms gives

$$\gamma_t = [\hat{B}_t + \Lambda_t D_t + P_t B_t] p_t + \Lambda_t \sigma_t + P_t \beta_t + \hat{\beta}_t, \tag{2.57}$$

so assuming that a solution $(\mathbf{P}, \boldsymbol{\lambda})$ to (2.56) can be found, the process **p** can be obtained via the solution of the BSDE

$$dp_t = -\big[(\hat{B}_t + \Lambda_t D_t + P_t B_t) p_t + \Lambda_t \sigma_t + P_t \beta_t + \hat{\beta}_t\big]dt + \lambda_t dW_t, \qquad p_T = g, \tag{2.58}$$

whose driver can be written as

$$\Psi(t,\omega,y,z) = \left[\hat{B}_t(\omega)+\Lambda_t(\omega)D_t(\omega)+P_t(\omega)B_t(\omega)\right]y+\Lambda_t(\omega)\sigma_t(\omega)+P_t(\omega)\beta_t(\omega)+\hat{\beta}_t(\omega).$$

This driver satisfies the Lipschitz condition of the existence result in Theorem 2.2 if $\mathbf{P}$ and $\boldsymbol{\lambda}$ are bounded. In fact, the discussion of the affine example we gave before this existence theorem shows that if $\mathbf{P}$ and $\boldsymbol{\lambda}$ have exponential moments in the sense that $\mathbb{E}\exp\int_0^t |P_u|du < \infty$, then the existence result still holds. In any case, if these two BSDEs can be solved and $\mathbf{P}$ and $\mathbf{p}$ can be found, then if the SDE

$$dX_t = \left[(A_t+B_tP_t)X_t+B_tp_t+\beta_t\right]dt + \left[(C_t+D_tP_t)X_t+D_tp_t+\sigma_t\right]dW_t, \qquad X_0 = x, \tag{2.59}$$

can be solved (again, this will be the case if $\mathbf{P}$ and $\mathbf{p}$ are bounded processes but also in some slightly more general situations), then the triple $(\mathbf{X},\mathbf{Y},\mathbf{Z})$ where $\mathbf{Y}$ is defined by (2.51) and $\mathbf{Z}$ by (2.54) is a solution of the affine FBSDE (2.50).

So the strategy we propose in order to solve the affine FBSDE (2.50) is the following:

1. Solve the BSDE of Riccati type (2.56).
2. Substitute the value found in the first step for P_t and solve the BSDE (2.58) with the linear driver.
3. Substitute the values of P_t and p_t found in the first two steps, and solve the affine SDE (2.59).
4. The triple $(\mathbf{X},\mathbf{Y},\mathbf{Z})$, where $\mathbf{Y}$ is defined by (2.51) and $\mathbf{Z}$ by (2.54), is a solution of the affine FBSDE (2.50).

Of the above four steps, the first one is clearly the most challenging. Proving existence of a solution of the BSDE of Riccati type is difficult because of the quadratic and multidimensional characters of these equations. And if a solution exists, it has to be *regular* enough for the next two steps to be possible.

We conclude this section with the analysis of a couple of particular cases for which we can answer these questions positively.

2.6.2 • The Case of Deterministic Coefficients

In this subsection, we assume that the coefficients $\mathbf{A}$, $\mathbf{B}$, $\boldsymbol{\beta}$, $\mathbf{C}$, $\mathbf{D}$, $\boldsymbol{\sigma}$, $\hat{\mathbf{A}}$, $\hat{\mathbf{B}}$, and $\hat{\boldsymbol{\beta}}$ are deterministic, and for the sake of simplicity we also assume that they are continuous functions on $[0,T]$ with values in Euclidean spaces of appropriate dimensions. Similarly, we assume that G and g are deterministic. We are still searching for a solution $\mathbf{Y}$ in the form (2.51), but we now require that $\mathbf{P}$ and $\mathbf{p}$ are also deterministic, and in other words, that the diffusion coefficients $\boldsymbol{\Lambda}$ and $\boldsymbol{\lambda}$ are identically zero.

We now proceed to the search for a solution. Since $\Gamma_t = \dot{P}_t$ is the derivative of $\mathbf{P}$ with respect to time, the identification (2.55) of Γ_t which led to the BSDE (2.56) for P_t now reduces to a matrix-valued Riccati equation for $\mathbf{P}$. It reads

$$\dot{P}_t + \hat{A}_t + P_tA_t + \hat{B}_tP_t + P_tB_tP_t = 0, \qquad P_T = G. \tag{2.60}$$

If and when this equation can be solved over the interval $[0,T]$, and if the solution happens to be a continuous function of time, the solution P_t can be injected into the equation, giving $\mathbf{p}$, which then becomes a linear ODE:

$$\dot{p}_t = -(\hat{B}_t + P_tB_t)p_t - P_t\beta_t - \hat{\beta}_t, \qquad p_T = g. \tag{2.61}$$

2.6. The Affine Case

It is interesting to notice that, in the case of deterministic coefficients, the Riccati equation (2.60) and the linear ODE (2.61) do not depend upon the coefficients of the volatility of the forward component since neither C_t, nor D_t, nor σ_t appear in these equations. Notice also that the linear ODE (2.61) is always solvable. Indeed, its solution is given by

$$p_t = U(t,T)g + \int_t^T U(t,s)[P_s\beta_s + \hat{\beta}_s]ds,$$

where for each $s \in [0,T]$ and $t \in [0,s]$, $U(t,s)$ is the $p \times p$ matrix solving the matrix ODE

$$\frac{d}{dt}U(t,s) = -(\hat{B}_t + P_tB_t)U(t,s), \qquad U(s,s) = I_p.$$

So the major difficulty in this approach is the solution of the matrix Riccati equation (2.60), and to this effect we propose the following procedure. We describe the details of this procedure in the purely deterministic case, but the same idea can be extended (with appropriate modifications) to handle some of the stochastic Riccati equations derived in the previous subsection. We denote by $\mathbf{A}_t$ the $(d+p) \times (d+p)$ matrix

$$\mathbf{A}_t = \begin{bmatrix} A_t & B_t \\ \hat{A}_t & \hat{B}_t \end{bmatrix}$$

and by $[\Psi(t,s)]_{0 \leq t,s \leq T}$ the propagator of the ODE $\dot{X}_t = \mathbf{A}_t X_t$. In other words, for each $s \in [0,T]$, the $(d+p) \times (d+p)$ matrix-valued function $[0,T] \ni t \hookrightarrow \Psi(t,s)$ is the solution of the matrix-value ODE

$$\frac{d}{dt}\Psi(t,s) = \mathbf{A}_t\Psi(t,s), \qquad \Psi(s,s) = I_{d+p}.$$

Notice that the matrices $\Psi(t,s)$ are invertible and that $\Psi(t,s) = \Psi(t,0)\Psi(s,0)^{-1}$. Also, as a function of its second variable, Ψ is the solution of the adjoint equation

$$\frac{d}{dt}\Psi(s,t) = -\Psi(s,t)\mathbf{A}_t, \qquad \Psi(s,s) = I_{d+p}.$$

Next, we define the $(d+p) \times (d+p)$ matrix $\mathbf{G}$ by

$$\mathbf{G} = \begin{bmatrix} I_d & 0 \\ -G & I_p \end{bmatrix},$$

and for each $t \in [0,T]$ the $p \times p$ matrix Θ_t by $\Theta_t = [0, I_p]\mathbf{G}\Psi(T,t)\begin{bmatrix} 0 \\ I_p \end{bmatrix}$. Notice that

$$\Theta_T = [0, I_p]\mathbf{G}\begin{bmatrix} 0 \\ I_p \end{bmatrix} = I_p.$$

Moreover, this function of time is differentiable and its derivative satisfies

$$\dot{\Theta}_t = [0, I_p]\mathbf{G}\frac{d}{dt}\Psi(T,t)\begin{bmatrix} 0 \\ I_p \end{bmatrix}$$

$$= -[0, I_p]\mathbf{G}\Psi(T,t)\mathbf{A}_t\begin{bmatrix} 0 \\ I_p \end{bmatrix}$$

$$= -[0, I_p]\mathbf{G}\Psi(T,t)\begin{bmatrix} B_t \\ \hat{B}_t \end{bmatrix}$$

$$= -[0, I_p]\mathbf{G}\Psi(T,t)\begin{bmatrix} I_d \\ 0 \end{bmatrix}B_t - \Theta_t\hat{B}_t.$$

Lemma 2.30. *If for each $t \in [0,T]$, the matrix Θ_t is invertible and the inverse is a continuous function of time, then the matrix-valued function*

$$P_t = -\Theta_t^{-1}[0, I_p]\mathbf{G}\Psi(T,t)\begin{bmatrix} I_d \\ 0 \end{bmatrix} \tag{2.62}$$

solves the Riccati equation (2.60).

Proof. Since

$$\frac{d}{dt}\Theta_t^{-1} = -\Theta_t^{-1}\dot{\Theta}_t\Theta_t^{-1},$$

by definition of P_t and Θ_t we have

$$\begin{aligned}\dot{P}_t &= \Theta_t^{-1}\dot{\Theta}_t\Theta_t^{-1}[0,I_p]\mathbf{G}\Psi(T,t)\begin{bmatrix}I_d\\0\end{bmatrix} + \Theta_t^{-1}[0,I_p]\mathbf{G}\Psi(T,t)\mathbf{A}_t\begin{bmatrix}I_d\\0\end{bmatrix}\\ &= -\Theta_t^{-1}\dot{\Theta}_t P_t + \Theta_t^{-1}[0,I_p]\mathbf{G}\Psi(T,t)\begin{bmatrix}A_t\\\hat{A}_t\end{bmatrix}\\ &= \Theta_t^{-1}[0,I_p]\mathbf{G}\Psi(T,t)\begin{bmatrix}I_d\\0\end{bmatrix}B_t P_t + \hat{B}_t P_t - \Theta_t \hat{B}_t P_t\\ &\quad + \Theta_t^{-1}[0,I_p]\mathbf{G}\Psi(T,t)\begin{bmatrix}I_d\\0\end{bmatrix}A_t + \Theta_t^{-1}[0,I_p]\mathbf{G}\Psi(T,t)\begin{bmatrix}0\\I_p\end{bmatrix}\hat{A}_t\\ &= -P_t B_t P_t + \hat{B}_t P_t - \Theta_t \hat{B}_t P_t - P_t A_t + \hat{A}_t,\end{aligned}$$

which proves that P_t defined by (2.62) solves the matrix Riccati equation (2.60) since it also satisfies the terminal condition $P_T = G$. □

Remark 2.31. *Later, we shall encounter stochastic control and stochastic games models for which the volatility is not controlled, and for which this volatility is assumed to be deterministic. In the present situation, this amounts to assuming that $C_t = D_t = 0$. So if we can find the proper deterministic functions $\mathbf{P}$ and $\mathbf{p}$, then the forward process $\mathbf{X}$ solves the SDE*

$$dX_t = \big[(A_t + B_t P_t)X_t + B_t p_t + \beta_t\big]dt + \sigma_t dW_t, \qquad X_0 = x, \tag{2.63}$$

and consequently, $\mathbf{X}$ is a Gaussian process whose mean and auto-covariance functions can be determined and computed explicitly. Moreover, since $Y_t = P_t X_t + p_t$ with P_t and p_t deterministic, $\mathbf{Y}$ is also a Gaussian process, which is consistent with the fact that $Z_t = P_t \sigma_t$ is also deterministic!

2.6.3 ▪ The One-Dimensional Case $d = p = m = 1$.

While the assumption of invertibility of Θ_t is rather innocent, it is not easy to check in practice, not even in the present one-dimensional case. Indeed the matrix $\mathbf{A}_t$ is 2×2 and the propagator $\Psi(t,s)$ cannot be computed explicitly, at least when the coefficients are actually time-dependent. When the coefficients are time-independent, $\Psi(t,s) = e^{(t-s)\mathbf{A}}$, and explicit formulas can be derived in some cases. In this subsection, we present a *pedestrian approach* based on a direct analysis of the ordinary differential equations at hand. First we reiterate the fact that once the Riccati equation is solved and provides a continuous function

$[0,T] \ni t \hookrightarrow P_t$, then **p** can be solved for explicitly

$$p_t = g e^{\int_t^T (\hat{B}_u + P_u B_u) du} + \int_t^T [P_s \beta_s + \hat{\beta}_s] e^{\int_t^s (\hat{B}_u + P_u B_u) du} ds.$$

For the remainder of the discussion, we assume that B_t does not vanish.

If $t \hookrightarrow P_t$ is continuous and solves the Riccati equation (2.60), then $t \hookrightarrow \theta_t = e^{-\int_t^T P_s B_s ds}$ is strictly positive (in particular, it does not vanish) and twice continuously differentiable and satisfies $\theta_T = 1$, $\dot{\theta}_T = GB_T$, and

$$B_t \ddot{\theta}_t + [B_t(A_t + \hat{B}_t) - \dot{B}_t]\dot{\theta}_t + \hat{A}_t B_t^2 \theta_t = 0, \qquad 0 \le t \le T. \tag{2.64}$$

Conversely, if $t \hookrightarrow \theta_t$ is twice continuously differentiable on $[0,T]$ and does not vanish, then the function

$$P_t = \frac{1}{B_t} \frac{\dot{\theta}_t}{\theta_t}$$

solves the Riccati equation (2.60).

The purpose of this final section was to show that in the one-dimensional case, the problem of the existence of a nice solution to the Riccati equation could be reduced to the positivity of the solution of a linear second-order ordinary differential equation, and to question which is easier to settle, especially when the coefficients are independent of the time variable.

2.7 • Notes & Complements

The theory of BSDEs is well established and can easily be found in book form. See, for example, the monographs by Touzi [114], Pham [106], Ma and Yong [91], the book by Yong and Zhou [117], and the more recent account by Pardoux and Răşcanu [103]. The result of Theorem 2.7 on quadratic BSDEs, stated without proof in the text, is due to Kobylanski. The reader is referred to the original work [74] for a proof. Recently, several multidimensional extensions of this result have been proposed. See, for example, [71, 60, 11]. However, their scopes are rather specialized and the conditions under which they apply remain contrived. As a result, we were not able to use them in the linear-quadratic control problems and games considered later in the text.

The reader interested in recent attempts at the use of cubature methods and ideas from the theory of rough paths in the simulation of BSDEs is referred to[45]. The existence and uniqueness result for mean-field BSDEs presented in Section 2.2 is borrowed from [23]. Reflected BSDEs were introduced in [54]. The existence part of Theorem 2.10 is proved either by means of a penalization argument, as in Pham's book [106], or through the use of Snell envelope arguments, as described by El Karoui, Hamadène, and Matoussi in [53]. Pricing and hedging American contingent claims were done in [55].

We have tried to present tools that will be useful in the solution of stochastic control problems and in the search for Nash equilibriums of stochastic differential games. For this reason, we did not limit ourselves to BSDEs, and devoted the major part of the chapter to the well-posedness of FBSDEs. These equations are more difficult to solve and the usual Lipschitz conditions are not enough to guarantee the existence and uniqueness of a solution on a given time interval. Indeed, Lipschitz coefficients guarantee existence and uniqueness on a small enough time interval. Extending the solution to a time interval of arbitrary length requires extra work and technical assumptions, and the proofs become highly involved. We presented the original result of Peng and Wu [105] based on monotonicity conditions (Theorem 2.18 in the text), as well as the approach developed by Delarue in [49] based on fine estimates on the gradient of the decoupling field of the equation.

As we shall see later in the book, linear-quadratic (LQ) FBSDEs appear naturally in the analysis of LQ stochastic control problems and LQ games. Because of these important applications, they have been studied for their own sake. The interested reader may want to consult Yong's paper [116] for an example.

Part II

Stochastic Control

Chapter 3
Continuous Time Stochastic Optimization and Control

This chapter introduces stochastic optimization for the control of continuous time (stochastic) dynamical systems. In the Markovian case, we review the dynamic programming principle and its connections with nonlinear partial differential equations and the Hamilton–Jacobi–Bellman equation. We give a verification theorem. We concentrate on the finite horizon problems and leave the discussion of the case of the infinite horizon to a set of remarks and references in the Notes & Complements section at the end of the chapter.

3.1 ▪ Optimization of Stochastic Dynamical Systems

The backbone of the study of stochastic optimization problems considered in this chapter is a stochastic dynamical system whose *state* at time $t \geq 0$ in scenario $\omega \in \Omega$ is an element $X_t(\omega)$ of the *state space* $\mathbb{R}^d$. We assume that the set Ω of scenarios is equipped with a σ-field $\mathcal{F}$ of events and a probability measure $\mathbb{P}$ *weighting the likelihoods of the scenarios.* As most often done in probability, we do not write the scenario ω explicitly, and we use the same notation X_t for the random variable defined on the probability space $(\Omega, \mathcal{F}, \mathbb{P})$ and its specific (random) realizations in $\mathbb{R}^d$. The dynamics of the state will be specified in continuous time by means of Itô's stochastic differentials.

In the cases of interest, the time evolution $t \hookrightarrow X_t$ of the state of the system depends upon a stochastic process $\alpha = (\alpha_t)_{t \geq 0}$. In this chapter, α_t should be thought of as a *control*, the choice of which is made at time t as a function of the information available at that time by the state controller.

3.1.1 ▪ Information Structures and Admissible Controls

For the sake of convenience, we assume that the actions or controls are chosen in a measurable space $(A, \mathcal{A})$. A could be a subset of a Polish space, $\mathcal{A}$ being the trace of the Borel σ-field, but most often A will be a subset of the Euclidean space $\mathbb{R}^k$. We will denote by $\mathbb{A}$ the set of all the admissible actions or controls. "Control" is the standard terminology in stochastic control, but because of our intention to discuss stochastic games as well, we also use the term "action," which is commonly used in game theory. Typically, $\mathbb{A}$ will be the set of processes $\alpha = (\alpha_t)_{t \geq 0}$ taking values in A which satisfy a set of admissibility conditions specific to the application at hand. For example, when A is not assumed to be a fixed compact subset of $\mathbb{R}^k$, the processes α in $\mathbb{A}$ are often required to be bounded, or at least to satisfy an integrability condition of the form

$$\mathbb{E}\int_0^T |\alpha_t|^2 dt < \infty. \tag{3.1}$$

In most applications, further restrictions are imposed on the controls deemed to be admissible. These restrictions have to do with the actual information the controller has at any given time $t \in [0,T]$ when he/she makes his/her decision, and chooses an action or control α_t. We shall often denote by $\mathcal{I}_t$ the σ-field to which the controls need to be adapted. In other words, $\mathcal{I}_t$ represents the information available to the controller at time t. Because optimal control problems are an integral part of research in many different fields of engineering, computer science, economics, and applied mathematics, the terminology used to characterize its elements is not consistent across some of these fields. For example, the following four classes of control processes are often used as sets of admissible controls. We introduce them for the sake of compatibility with the existing literature.

OL We say that the admissible controls are open loop (OL) when $\mathcal{I}_t \equiv \sigma\{X_0\}$ for all $t \in [0,T]$ and the control α_t is in fact a function of t and X_0.

CLPS We say that the admissible controls are closed loop perfect state (CLPS) when $\mathcal{I}_t = \sigma\{X_s : 0 \le s \le t\}$ for all $t \in [0,T]$ and the control α_t is in fact a function of t and state history $X_{[0,t]}$.

MPS We say that the admissible controls are memoryless perfect state (MPS) when $\mathcal{I}_t = \sigma\{X_0, X_t\}$ for all $t \in [0,T]$ and the control α_t is a function of t, X_0 and X_t.

FPS We say that the admissible controls are feedback perfect state (FPS) or Markovian when $\mathcal{I}_t = \sigma\{X_t\}$ for all $t \in [0,T]$ and the control α_t is a function of t and X_t.

Most of the time we will drop the term "perfect state" from our terminology. Moreover, we shall give a different meaning to the notion of open loop control, especially in the case of stochastic games. Throughout the text, we shall make sure that the reader is informed of the class of admissible controls used each time a specific problem is studied, or a specific result is stated and possibly proved.

3.1.2 ▪ Dynamics of the Controlled System

We assume that the dynamics of the controlled state are given by Itô processes. More precisely, we assume that the probability space is equipped with a right continuous filtration $\mathbb{F} = (\mathcal{F}_t)_{0 \le t \le T}$ such that $\mathcal{F}_0$ contains the $\mathbb{P}$-null sets, and we work with an m-dimensional $\mathbb{F}$-Wiener process $\mathbf{W} = (W_t)_{t \in [0,T]}$. We will sometimes assume that $\mathbb{F} = (\mathcal{F}_t)_{t \in [0,T]}$ is the filtration generated by $\mathbf{W}$, but for the time being, we only assume that $\mathbf{W}$ is adapted to $\mathbb{F}$ and that for each $0 \le s \le t$, $W_t - W_s$ is independent of $\mathcal{F}_s$. We denote by $\mathcal{P}$ the σ-field of $\mathbb{F}$-progressively measurable subsets of $[0,T] \times \Omega$ and by $\mathcal{B}(E)$ the Borel σ-field of E for any topological space E.

Remark 3.1. *As we already did when we discussed the Wasserstein distances on spaces of probability measures we will often work with the space $C([0,T];\mathbb{R}^n)$ of continuous functions on the interval $[0,T]$ with values in the Euclidean space $\mathbb{R}^n$, and for a generic value of n, we shall denote this space by $\mathcal{C}$. When dealing with this space, one major convenience of progressive measurability is the following. Whenever a function f defined on $[0,T] \times \mathcal{C}$ is progressively measurable (for the natural filtration $\mathbb{F}^x$ generated by the*

3.1. Optimization of Stochastic Dynamical Systems

coordinate process $x_t(w) = w(t)$ for $t \in [0,T]$ and $w \in \mathcal{C}$), then for any continuous and $\mathbb{F}$-adapted process $\mathbf{p} = (p_t)_{0 \le t \le T}$, the process $\{f(t, \mathbf{p})\}_{0 \le t \le T}$ defined by $f(t,p)(\omega) = f(t,p.(\omega))$ (where we replace the continuous function $w \in \mathcal{C}$ in $f(t,w)$ by the entire trajectory $p.(\omega)$) is $\mathcal{P}$-measurable. In fact, this property is essentially equivalent to the progressive measurability of f.

We assume that the state of the system at time $t \in [0,T]$ is given by a random vector $X_t \in \mathbb{R}^d$ so that $\mathbf{X} = (X_t)_{0 \le t \le T}$ is an Itô process satisfying

$$dX_t = b(t, X_t, \alpha_t)dt + \sigma(t, X_t, \alpha_t)dW_t, \tag{3.2}$$

where the drift coefficient b and the volatility σ are functions

$$(b, \sigma) : [0,T] \times \Omega \times \mathbb{R}^d \times A \hookrightarrow \mathbb{R}^d \times \mathbb{R}^{d \times m}$$

satisfying the following assumptions:

(S1) b and σ are measurable with respect to $\mathcal{P} \times \mathcal{B}(\mathbb{R}^d) \times \mathcal{A}$;

(S2) $\exists c > 0, \forall t \in [0,T], \forall \omega \in \Omega, \forall x, x' \in \mathbb{R}^d, \forall \alpha \in A,$

$$|(b, \sigma)(t, \omega, x, \alpha) - (b, \sigma)(t, \omega, x', \alpha)| \le c|x - x'|.$$

As always, we follow the standard practice of, whenever possible, not making the dependence upon $\omega \in \Omega$ explicit in the formulas. Given these assumptions on the drift and volatility of the state of the system, we make explicit the condition we shall require for the controls to be admissible. From now on, and unless stated otherwise, the set $\mathbb{A}$ of admissible controls will be the set of A-valued progressively measurable processes $\boldsymbol{\alpha} = (\alpha_t)_{0 \le t \le T}$ satisfying

$$\mathbb{E}\int_0^T (|b(t,0,\alpha_t)|^2 + |\sigma(t,0,\alpha_t)|^2) dt < \infty. \tag{3.3}$$

From this admissibility constraint, it follows that for each admissible control $\boldsymbol{\alpha} \in \mathbb{A}$, and for each initial condition $X_0 = \xi$ independent of $\mathbf{W}$, Theorem 1.2 says that there exists a unique strong solution $\mathbf{X}$ of the SDE (3.2). We shall denote this solution by $\mathbf{X}^{\boldsymbol{\alpha}}$. It is the state process controlled by $\boldsymbol{\alpha}$. In analogy with our discussion of Chapter 1, for each $(t,x) \in [0,T] \times \mathbb{R}^d$ we shall denote by $\mathbf{X}^{t,x,\boldsymbol{\alpha}} = (X_s^{t,x,\boldsymbol{\alpha}})_{t \le s \le T}$ the unique solution of (3.2) over the interval $[t,T]$ satisfying $X_t = x$.

3.1.3 ▪ The Optimization Problem

We assume that we are given an $\mathcal{F}_T$-measurable random variable representing the *terminal cost*. It is assumed to be square integrable. Most often, it will be of the form $g(X_T)$, where $g : \Omega \times \mathbb{R}^d \hookrightarrow \mathbb{R}$ is $\mathcal{F}_T \times \mathcal{B}(\mathbb{R}^d)$-measurable, and of polynomial growth in $x \in \mathbb{R}^d$ uniformly in $\omega \in \Omega$. We also assume that the cost includes a *running cost* given by a function $f : [0,T] \times \Omega \times \mathbb{R}^d \times A \hookrightarrow \mathbb{R}$ satisfying the same assumptions (S1) and (S2) as the drift b. Finally, we define the cost functional J by

$$J(\boldsymbol{\alpha}) = \mathbb{E}\left[\int_0^T f(s, X_s, \alpha_s)ds + g(X_T)\right], \qquad \boldsymbol{\alpha} \in \mathbb{A}. \tag{3.4}$$

As explained earlier, the goal of a stochastic control problem is to find an admissible control $\boldsymbol{\alpha} \in \mathbb{A}$ which minimizes the cost functional $J(\boldsymbol{\alpha})$. The cost functional J is often called the objective, or objective functional.

Remark 3.2. *Terminal Cost Only.* *For the purposes of some proofs, it is sometimes convenient to assume that the running cost is not present, namely that $f \equiv 0$, and that the objective function of the problem is of the form $J(\alpha) = \mathbb{E}g(X_T)$. This is indeed possible by defining a new controlled state process $\tilde{\mathbf{X}} = (\mathbf{X}, \mathbf{Y})$, where the process $\mathbf{Y}$ is defined as*

$$Y_t = \int_0^t f(s, X_s, \alpha_s) ds,$$

and by defining the new cost functional $\tilde{J}(\alpha) = \mathbb{E}\tilde{g}(\tilde{X}_T)$, where the function $\tilde{g}$ is defined by $\tilde{g}(\tilde{x}) = \tilde{g}(x, y) = y + g(x)$. Using the same set $\mathbb{A}$ of admissible controls, the fact that $\tilde{J}(\alpha) = J(\alpha)$ gives the desired equivalence. Notice that this apparent simplification of the optimization problem comes at a cost. The underlying state process lives in a higher dimensional space (which is always a concern for anyone who has had to deal with the curse of dimensionality at least once), and the dynamics of this controlled state process are degenerate in the sense that the y-component does not have a diffusion part: this may be the source of delicate problems down the road.

Remark 3.3. *Existence of an Optimum.* *This can be proven in some cases by purely functional analytic techniques. Indeed, it is sometimes possible to show that $\mathbb{A}$ is a convex subset of a topological vector space, and if the function J can be proven to be convex and lower semicontinuous with compact level sets, then existence of a minimum follows. Also, when trying to numerically compute approximations of an optimal control, gradient descent and other methods have been suggested and implemented. When the stochastic control problem is based on the dynamics of an Itô process of the kind considered here, the theory known as (optimal) stochastic control aims to take full advantage of the form of the stochastic dynamics of the controlled state, and the techniques brought to bear in the solution of the stochastic optimization problem are tailored to the form of the Itô dynamics. As a result, they depart from the general stochastic optimization algorithms which are more global in nature, and do not always try to take advantage of the time dynamics of the state.*

3.1.4 ▪ The Hamiltonian of the System

The Hamiltonian of the system is a function of time t, possibly the random scenario ω, the state variable x, an action $\alpha \in A$, and two new variables y and z (often called *dual variables* or *covariables*). These variables should belong to the dual spaces of the spaces in which the drift b and the volatility matrix σ take their values. Since we are only considering finite-dimensional systems, these spaces are Euclidean spaces, which can be identified with their dual spaces, the inner product providing the duality. The Hamiltonian of the system is the function H,

$$[0, T] \times \Omega \times \mathbb{R}^d \times \mathbb{R}^d \times \mathbb{R}^{d \times m} \times A \ni (t, \omega, x, y, z, \alpha) \hookrightarrow H(t, \omega, x, y, z, \alpha) \in \mathbb{R},$$

defined by

$$H(t,\omega,x,y,z,\alpha) = \underbrace{b(t,\omega,x,\alpha) \cdot y}_{\substack{\text{inner product of} \\ \text{state drift } b \text{ and} \\ \text{covariable } y}} + \underbrace{\sigma(t,\omega,x,\alpha) \cdot z}_{\substack{\text{inner product of} \\ \text{state volatility } \sigma \text{ and} \\ \text{covariable } z}} + \underbrace{f(t,\omega,x,\alpha)}_{\text{running cost}},$$

(3.5)

As usual, we shall not write ω unless its presence is necessary to emphasize the randomness of the expressions in which it occurs. We use the notation $\cdot$ for the scalar product in $\mathbb{R}^d$, and more generally in any Euclidean space. Notice that since σ is a matrix, z is a matrix with the same dimensions, and their inner product is given by the trace of a square matrix. To be more specific,

$$\sigma(t,x,\alpha) \cdot z = \text{trace}\,[\sigma(t,x,\alpha)^\dagger z] = \text{trace}\,[z^\dagger \sigma(t,x,\alpha)], \tag{3.6}$$

where we use the notation $\dagger$ to denote the transpose of a matrix.

Remark 3.4. *When the controller can only access the drift of the state (in other words, when the control variable α does not appear in the volatility σ of the state), we use a simpler form of the Hamiltonian function. In such a case, we use the* reduced Hamiltonian *$\tilde{H}$ defined by*

$$\tilde{H}(t,\omega,x,y,\alpha) = b(t,\omega,x,\alpha) \cdot y + f(t,\omega,x,\alpha), \tag{3.7}$$

which does not depend upon the dual variable z.

3.1.5 ▪ The Particular Case of Markovian/Diffusion Dynamics

In most applications for which actual numerical computations are possible, the coefficients of the state dynamics (3.2) depend only and deterministically on the *present value* X_t of the state. In this case, the dynamics are given by a diffusion-like equation

$$dX_t = b(t, X_t, \alpha_t)dt + \sigma(t, X_t, \alpha_t)dW_t, \qquad 0 \le t \le T, \tag{3.8}$$

with the same initial condition $X_0 = x$, for drift and diffusion coefficients given by deterministic functions $(b,\sigma) : [0,T] \times \mathbb{R}^d \times A \ni (t,x,\alpha) \hookrightarrow (b,\sigma)(t,x,\alpha) \in \mathbb{R}^d \times \mathbb{R}^{d \times m}$. In this case, we also assume that the running and terminal costs are of a Markovian type in the sense that they are given by deterministic functions f and g, $f : [0,T] \times \mathbb{R}^d \times A \ni (t,x,\alpha) \hookrightarrow f(t,x,\alpha) \in \mathbb{R}$ and a measurable function $g : \mathbb{R}^d \ni x \hookrightarrow g(x) \in \mathbb{R}$ such that $\mathbb{E}|g(X_T)|^2 < \infty$. In most cases, g will be assumed to be bounded so that this assumption will be automatically satisfied. In other words, the cost functional is of the form

$$J(\boldsymbol{\alpha}) = \mathbb{E}\left[\int_0^T f(s, X_s, \alpha_s)ds + g(X_T)\right], \qquad \boldsymbol{\alpha} \in \mathbb{A}. \tag{3.9}$$

The Hamiltonian is now deterministic:

$$(t,x,y,z,\alpha) \hookrightarrow H(t,x,y,z,\alpha) = b(t,x,\alpha) \cdot y + \sigma(t,x,\alpha) \cdot z + f(t,x,\alpha) \in \mathbb{R}. \tag{3.10}$$

In the analysis of control problems, the stochastic maximum principle (which we prove later on in this chapter) suggests the minimization of the Hamiltonian with respect to the control variable α. When the volatility is not controlled (in other words, when σ does not depend upon α), the dual variable z does not affect the search for a minimizer, and for this search it is convenient to use the *reduced Hamiltonian* $\tilde{H}$ already introduced in (3.7), which takes the form

$$[0,T] \times \mathbb{R}^d \times \mathbb{R}^d \times A \ni (t,x,y,\alpha) \hookrightarrow \tilde{H}(t,x,y,\alpha) = y \cdot b(t,x,\alpha) + f(t,x,\alpha) \in \mathbb{R} \tag{3.11}$$

in the present Markovian case. In our discussion of Chapter 1, we emphasized the connection between Markovian SDEs and partial differential equations, and the powerful tools the latter can bring to bear in the analysis of the properties of the solutions. In the present situation, these techniques cannot be applied if the controls α are merely adapted, and so depend upon the past. So, in order to take full advantage of the Markovian structure, the optimization is often restricted to closed loop controls in feedback form, namely, controls of the form $\alpha_t = \alpha(t, X_t)$ for some deterministic function $[0,T] \times \mathbb{R}^d \ni (t,x) \hookrightarrow \alpha(t,x) \in \mathbb{R}^k$. In this case, the dynamics of the state are given by the Markovian SDE

$$dX_t = b(t, X_t, \alpha(t, X_t))dt + \sigma(t, X_t, \alpha(t, X_t))dW_t, \qquad 0 \leq t \leq T, \qquad (3.12)$$

and the cost functional becomes

$$J(\boldsymbol{\alpha}) = \mathbb{E}\left[\int_0^T f(s, X_s, \alpha(s, X_s))ds + g(X_T)\right], \qquad \boldsymbol{\alpha} \in \mathbb{A}. \qquad (3.13)$$

Remark 3.5. *Strong and Weak Forms of the Optimal Control Problem.* *The above setup for a stochastic control problem is often called the* strong form *of the problem. In this setting the probability space* $(\Omega, \mathcal{F}, \mathbb{P})$, *the filtration* $\mathbb{F} = (\mathcal{F}_t)_{0 \leq t \leq T}$, *and the set* $\mathbb{A}$ *of admissible controls are given, and the optimization problem needs to be solved from these data. In particular, the optimal control needs to be a stochastic process on* $(\Omega, \mathcal{F}, \mathbb{P})$ *adapted to the filtration* $\mathbb{F} = (\mathcal{F}_t)_{0 \leq t \leq T}$. *In full analogy with the theory of stochastic differential equations, it is possible to define a weak form of the problem when the coefficients (drift function b, volatility σ, running cost function f, and terminal cost g) are deterministic. Indeed, they can be viewed as the core data of the problem, and a control can then be defined as a probability space* $(\Omega, \mathcal{F}, \mathbb{P})$, *a filtration* $\mathbb{F} = (\mathcal{F}_t)_{0 \leq t \leq T}$, *a Wiener process* $\mathbf{W} = (W_t)_{0 \leq t \leq T}$ *for this filtration, and a progressively measurable process* $\boldsymbol{\alpha}$ *such that, if* $\mathbf{X} = \mathbf{X}^{\boldsymbol{\alpha}}$ *is the unique solution of (3.2) on this particular probability space, for this filtration and this Wiener process, then the running cost* $f(\cdot, X., \alpha.)$ *is in* $L^1([0,T] \times \Omega, dt \otimes d\mathbb{P})$ *and the terminal cost* $g(X_T)$ *is in* $L^1(\Omega, \mathbb{P})$. *The notion of weak form still makes sense when the coefficients b, σ, f, and g are random, in which case changes of the probability space* $(\Omega, \mathcal{F}, \mathbb{P})$ *are limited to changes of the probability measure* $\mathbb{P}$ *on* $(\Omega, \mathcal{F})$, *and possibly the filtration* $\mathbb{F}$. *Most often, practical applications are meaningful for the strong form only. The usefulness of the weak form seems to be limited to its convenience when it comes to some technical mathematical arguments, to issues concerning the optimal value of the cost* $J(\boldsymbol{\alpha})$, *statistics of the optimal control and/or the optimally controlled state process, and models for which the search for optimal controls can be restricted to the search among Markovian controls in closed loop feedback form. Indeed the flexibility that it offers makes computations and some mathematical proofs possible, if not easier. See for example our discussion of the dynamic programming principle later in the chapter.*

3.1.6 ▪ Risk Sensitive Control

The models discussed in this section have been introduced for the purpose of accounting for the risk appetite/aversion of the controller. While we assume that the dynamics of the controlled state of the system are still given by the same Itô SDE

$$dX_t = b(t, X_t, \alpha_t)dt + \sigma(t, X_t)dW_t,$$

the criterion minimized by the controller is now given in the form

$$G(\boldsymbol{\alpha}) = \mathbb{E}\exp\left[\theta\left(\int_0^T f(t, X_t, \alpha_t)dt + \xi\right)\right], \qquad (3.14)$$

where the value of the parameter $\theta \in \mathbb{R}$ is expected to influence the kind of optimality reached by the controller. Setting

$$\hat{J}(\boldsymbol{\alpha}) = \int_0^T f(t, X_t, \alpha_t)dt + \xi \quad \text{and} \quad J_\theta(\boldsymbol{\alpha}) = \frac{1}{\theta}\log \mathbb{E}[\exp[\theta \hat{J}(\boldsymbol{\alpha})]], \qquad (3.15)$$

one can see that minimizing $G(\alpha)$ is essentially equivalent to minimizing $J_\theta(\alpha)$, and since

$$J_\theta(\alpha) \sim \mathbb{E}[\hat{J}(\alpha)] + \frac{\theta}{2}\text{var}\{\hat{J}(\alpha)\} \qquad (3.16)$$

whenever $\theta \text{var}\{\hat{J}(\alpha)\}$ is small, the problem looks very much like a *mean-variance* optimization problem. Here and in what follows we use the notation $\text{var}\{\eta\}$ for the variance of the random variable η. When $\theta < 0$, larger variations of $\hat{J}(\alpha)$—a situation which is perceived as corresponding to a greater risk—give a smaller criterion and help the minimization. On the other hand, when $\theta > 0$, larger variations of $\hat{J}(\alpha)$ go in the opposite direction of the minimization of J. For this reason, a controller trying to minimize the criterion G for a negative parameter θ will be considered as risk-seeking, while a controller minimizing G for a positive θ will be said to be risk-averse. Finally, since

$$\lim_{\theta \to 0} J_\theta(\alpha) = \mathbb{E}\hat{J}(\alpha),$$

the controller is said to be risk-neutral when $\theta = 0$.

3.1.7 ▪ The Case of Uncontrolled Volatility: Weak Formulation

Under some restrictive assumptions on the volatility (typically, that σ does not depend upon the control α and is invertible with a reasonable inverse), we present a specific construction of the stochastic evolution of the system which will come in handy in a few arguments later in this chapter. In these models, the controller can only affect the drift of the dynamics and, not surprisingly, we use Girsanov's theorem to approach the computation of the objective expected cost $J(\alpha)$. This fits in with the framework of the above remark discussing the weak formulation of control problems.

Assumptions on the Diffusion Coefficient of the State Dynamics

We assume that $m = d$ and that the volatility is given by a function $\sigma : [0,T] \times \Omega \times \mathbb{R}^d \hookrightarrow \mathbb{R}^{d \times d}$ satisfying the following.

(A.1) σ is progressively measurable ($\mathcal{P}$-measurable).

(A.2) There exists a constant $c > 0$ such that

(a) $\forall t \in [0,T], \forall \omega \in \Omega, \forall x, x' \in \mathbb{R}^d, |\sigma(t,\omega,x) - \sigma(t,\omega,x')| \leq c|x - x'|$;

(b) $\forall (t,\omega,x) \in [0,T] \times \Omega \times \mathbb{R}^d$, σ is invertible, and $\sigma(t,\omega,x)^{-1}$ satisfies

$$|\sigma(t,\omega,x)^{-1}| \leq K(1 + |x|^\delta)$$

for some $\delta > 0$ and $K > 0$ independent of t, ω, and x;

(c) $\forall t \in [0,T], \forall \omega \in \Omega, \forall x \in \mathbb{R}^d, |\sigma(t,\omega,x)| \leq c(1 + |x|)$.

Under these assumptions, Theorem 1.2 says that there exists a unique strong solution, say $\mathbf{X}^0 = (X_t^0)_{0 \leq t \leq T}$, of the SDE

$$dX_t = \sigma(t, X_t)dW_t, \qquad 0 \leq t \leq T, \qquad X_0 = x, \qquad (3.17)$$

and that this solution has uniform moments of all orders in the sense that

$$\forall p \geq 1, \quad \mathbb{E}\sup_{0 \leq t \leq T}|X_t^0|^p < \infty. \qquad (3.18)$$

Assumptions on the Drift of the State Dynamics

The drift term of the dynamics of the controlled state is given by a function $b : [0,T] \times \Omega \times \mathbb{R}^d \times A \hookrightarrow \mathbb{R}^d$ satisfying the following.

(A.3) For each $\alpha \in A$ and $x \in \mathbb{R}^d$, the process $(b(t,x,\alpha))_{0 \leq t \leq T}$ is $\mathcal{P}$-measurable and $\exists c > 0, \forall t \in [0,T], \forall \omega \in \Omega, \forall x \in \mathbb{R}^d, |b(t,\omega,x,\alpha)| \leq c(1+|x|)$.

Under these assumptions, for each admissible control $\alpha \in \mathbb{A}$, we can define the new probability measure $\mathbb{P}^\alpha$ by its density with respect to $\mathbb{P}$:

$$\frac{d\mathbb{P}^\alpha}{d\mathbb{P}} = \mathcal{E}\left(\int_0^\cdot \sigma(s,X_s^0)^{-1} b(s,X_s^0,\alpha_s) dW_s\right)_T, \tag{3.19}$$

where we use the notation $\mathcal{E}$ for the Doléans exponential of a martingale defined by

$$\mathcal{E}(M)_t = \exp\left[M_t - \frac{1}{2}\langle M \rangle_t\right], \qquad 0 \leq t \leq T. \tag{3.20}$$

Notice that, because of the assumption on the drift b and the growth of the inverse of the diffusion matrix, this Doléans exponential is a bona fide martingale and $\mathbb{P}^\alpha$ is a probability measure equivalent to $\mathbb{P}$.

Dynamics of the Controlled System

For each admissible control $\alpha \in \mathbb{A}$,

$$dX_t^0 = b(t,X_t^0,\alpha_t)dt + \sigma(t,X^0)dW_t^\alpha, \qquad 0 \leq t \leq T, \qquad X_0 = x, \tag{3.21}$$

where the process $\mathbf{W}^\alpha$ is defined by

$$W_t^\alpha = W_t - \int_0^t \sigma(s,X_s^0)^{-1} b(s,X_s^0,\alpha_s) ds, \qquad 0 \leq t \leq T, \tag{3.22}$$

and by Girsanov's theorem, is a Wiener process under the probability measure $\mathbb{P}^\alpha$.

NB. In general, $\mathbf{X}^0 = (X_t^0)_{0 \leq t \leq T}$ *is not adapted* to the filtration generated by $\mathbf{W}^\alpha = (W_t^\alpha)_{0 \leq t \leq T}$, so it cannot be a strong solution of the SDE (3.2); it is only a *weak solution* of the SDE (3.2).

Cost Functional

We assume that we are given a function $g : \Omega \times \mathbb{R}^d \ni (\omega,x) \hookrightarrow g(\omega,x) \in \mathbb{R}$ such that $g(x)$ is an $\mathcal{F}_T$-measurable random variable for each $x \in \mathbb{R}^d$, and a function $f : [0,T] \times \Omega \times \mathbb{R}^d \times A \hookrightarrow \mathbb{R}$ satisfying the same assumption **(A.3)** as the drift b. The cost functional J is defined by

$$J(\alpha) = \mathbb{E}^{\mathbb{P}^\alpha}\left[\int_0^T f(s,X_s^0,\alpha_s)ds + g(X_T^0)\right], \qquad \alpha \in \mathbb{A}. \tag{3.23}$$

For the sake of the present discussion, we assume that f is bounded, but various forms of polynomial boundedness would do as well. As explained earlier, the goal of a stochastic control problem is to find an admissible control $\alpha \in \mathbb{A}$ which minimizes the cost functional

$J(\alpha)$. Also, recall that as we explained earlier, we use the reduced Hamiltonian $\tilde{H}$ instead of the full Hamiltonian H when the volatility is not controlled.

Remark 3.6. *What we call weak formulation in this section does not fit quite cleanly in the strong versus weak dichotomy of the optimal control problem. Indeed, while we keep the measure space $(\Omega, \mathcal{F})$ and the set $\mathbb{A}$ of admissible controls fixed, we vary the probability measure, giving the likelihoods of the possible scenarios of the controlled state. In fact, the probability measure $\mathbb{P}^\alpha$ could be viewed as the control of the system. The above expression for the value of the cost functional can be replaced by an expectation over the single probability measure $\mathbb{P}$ if one uses the definition (3.19)–(3.20) of the density of the measure $\mathbb{P}^\alpha$. However, the resulting expression is not of the standard form (3.9)!*

3.2 ▪ First Financial Applications

For the sake of illustration, this chapter is sprinkled with applications of stochastic optimization to financial problems, which we describe at various levels of detail depending upon the technicalities involved. Rather than presenting models in their greatest generality, we emphasize motivation and solution implementation strategies. Some of these applications may not even need special stochastic optimization techniques when the optimum can be computed scenario by scenario. This is indeed the case for the two examples presented in this section. They are borrowed from [36].

Other applications, requiring more sophisticated tools from stochastic analysis (such as the classical Merton problem), will be presented later in the chapter.

3.2.1 ▪ Greenhouse Gas Emission Control

We describe the optimization problem of a power producer facing carbon regulation in the form of a liquid market for pollution permits. For the sake of simplicity, we assume that $[0, T]$ is a single phase of the regulation, and that no banking or borrowing of the certificates is possible at the end of the phase. We consider the single firm problem, leaving the multifirm optimization and the equilibrium issues for later. For the sake of illustration, we consider two simple models which do not require the sophisticated techniques of optimal stochastic control presented in this chapter.

Modeling First the Emissions Dynamics

We assume that the source of randomness in the model is given by the time evolution $\mathbf{W} = (W_t)_{0 \leq t \leq T}$ of a (possibly infinite) sequence of independent one-dimensional Wiener processes $\mathbf{W}^j = (W_t^j)_{0 \leq t \leq T}$. In other words, $W_t = (W_t^0, W_t^1, \cdots, W_t^i, \cdots)$ for each fixed-$t \in [0, T]$. All these Wiener processes are assumed to be defined on a complete probability space $(\Omega, \mathcal{F}, \mathbb{P})$, and we denote by $\mathbb{F} = \{\mathcal{F}_t, t \geq 0\}$ the Brownian filtration they generate. Here, $T > 0$ is a fixed-time horizon representing the end of the regulation period.

We will eventually extend the model to include N firms, but for the time being, we consider only the problem of one single firm whose electricity production generates emissions of carbon dioxide, and we denote by E_t the cumulative emissions up to time t of the firm. We also denote by $\tilde{E}_t$ the perception at time t (say, the conditional expectation, for example) of what the total cumulative emission E_T will be at the end of the time horizon. Clearly, $\mathbf{E}$ and $\tilde{\mathbf{E}}$ can be different stochastic processes, but they have the same terminal values at time T, i.e., $E_T = \tilde{E}_T$. We will assume that the dynamics of the proxy $\tilde{\mathbf{E}}$ for the cumulative emissions of the firm are given by an Itô process of the form

$$\tilde{E}_t = \int_0^t (b_s - \xi_s)ds + \sigma_t W_t \qquad (3.24)$$

(so $\tilde{E}_0 = 0$), where b represents the (conditional) expectation of what the rate of emission change would be in a world without carbon regulation, in other words, in what is called *business as usual*, while ξ is the rate of abatement chosen by the firm. In mathematical terms, ξ represents the control on emission reduction implemented by the firm. Clearly, the firm only acts on the drift of its *perceived* emissions. For the sake of simplicity, we assume that the processes b and σ are adapted and bounded. Because of the vector nature of the Wiener process W, the volatility process σ is in fact a sequence of scalar volatility processes $(\sigma^j)_{j \geq 0}$. We identify two extreme examples for the sake of illustration. For the purpose of this section, we could use one single scalar Wiener process and one single scalar volatility process, as long as we allow the filtration $\mathbb{F}$ to be larger than the filtration generated by the single Wiener process. This fact will be needed when we study a model with different firms.

Continuing with the description of the model, we assume that the abatement decision is based on a cost function $c : \mathbb{R} \to \mathbb{R}$, which is assumed to be continuously differentiable (C^1 in notation), nondecreasing, strictly convex, and satisfying the Inada-like conditions

$$c'(-\infty) = -\infty \quad \text{and} \quad c'(+\infty) = +\infty. \qquad (3.25)$$

Note that $(c')^{-1}$ exists because of the assumption of strict convexity. Since $c(x)$ can be interpreted as the cost to the firm for an abatement rate of level x, without any loss of generality we will also assume $c(0) = \min c = 0$. Notice that (3.25) implies that $\lim_{x \to \pm\infty} c(x) = +\infty$.

Remark 3.7. A typical example of abatement cost function is given by the quadratic cost function $c(x) = \alpha x^2$ for some $\alpha > 0$, or more generally, the power cost function $c(x) = \alpha |x|^{1+\beta}$ for $\alpha > 0$ and $\beta > 0$.

The firm controls its destiny by choosing its own abatement schedule ξ as well as the quantity θ of pollution permits it holds as a result of trading in the allowance market. For these controls to be admissible, ξ and θ need only be adapted processes satisfying the integrability condition (3.3). Given its initial wealth x, the terminal wealth X_T of the firm is given by

$$X_T = X_T^{\xi,\theta} = x + \int_0^T \theta_t dY_t - \int_0^T c(\xi_t)dt - E_T Y_T, \qquad (3.26)$$

where we use the notation $(Y_t)_{0 \leq t \leq T}$ for the evolution of the price of an emission allowance. We are not interested in the existence or the formation of this price. We assume that there exists a liquid frictionless market for emission allowances, and that Y_t is the price at time t of one allowance. The interpretation of the four terms in the right-hand side is clear: (1) x is the initial wealth of the firm; (2) the first integral give the proceeds of trading in the allowance market; (3) the second integral represents the costs associated to the production; and (4) the last term is the terminal cost for the emissions throughout the period. The risk preferences of the firm are given by a utility function $U : \mathbb{R} \to \mathbb{R}$, which is assumed to be C^1, increasing, strictly concave, and satisfying the Inada conditions

$$(U)'(-\infty) = +\infty \quad \text{and} \quad (U)'(+\infty) = 0. \qquad (3.27)$$

3.2. First Financial Applications

The optimization problem of the firm can be written as the computation of

$$V(x) := \sup_{(\xi,\theta)\in \mathbb{A}} \mathbb{E} U(X_T^{\xi,\theta}), \tag{3.28}$$

where $\mathbb{E}$ denotes the expectation under the *historical* measure $\mathbb{P}$, and $\mathbb{A}$ is the set of abatement and trading strategies (ξ,θ) admissible to the firm. The following simple result holds.

Proposition 3.8. *The optimal abatement strategy of the firm is given by*

$$\xi_t^* = [c']^{-1}(Y_t).$$

Remark 3.9. *The beauty of this simple result is its powerful intuitive meaning: given a price Y_t for an emission allowance, the firm implements all the abatement measures which make economic sense, namely, all those costing less than the current market price of emissions.*

Proof. If we rewrite the last term in expression (3.26) of the terminal wealth by replacing E_T by $\tilde{E}_T$, a simple integration by parts gives

$$E_T Y_T = Y_T \left(\int_0^T b_t dt + \int_0^T \sigma_t dW_t \right) - Y_T \int_0^T \xi_t dt$$

$$= Y_T \left(\int_0^T b_t dt + \int_0^T \sigma_t dW_t \right) - \int_0^T Y_t \xi_t dt - \int_0^T \left(\int_0^t \xi_s ds \right) dY_t$$

so that $X_T = A_T^{\tilde{\theta}} + B_T^{\xi}$ with

$$A_T^{\tilde{\theta}} = \int_0^T \tilde{\theta}_t dY_t - Y_T \left(\int_0^T b_t dt + \int_0^T \sigma_t dW_t \right),$$

where the modified control $\tilde{\theta}$ is defined by $\tilde{\theta}_t = \theta_t + \int_0^t \xi_s ds$, and

$$B_T^{\xi} = x - \int_0^T [c(\xi_t) + Y_t(b_t - \xi_t)] dt.$$

Notice that B_T^{ξ} depends only upon ξ without depending upon $\tilde{\theta}$, while $A_T^{\tilde{\theta}}$ depends only upon $\tilde{\theta}$ without depending upon ξ. Since the set $\mathbb{A}$ of admissible controls is equivalently described by varying the couples (θ,ξ) or $(\tilde{\theta},\xi)$, when computing the maximum

$$\sup_{(\theta,\xi)\in\mathbb{A}} \mathbb{E}\{U(X_T)\} = \sup_{(\tilde{\theta},\xi)\in\mathbb{A}} \mathbb{E}\{U(A_T^{\tilde{\theta}} + B_T^{\xi})\}$$

one can perform the optimizations over $\tilde{\theta}$ and ξ separately, for example by fixing $\tilde{\theta}$ and optimizing with respect to ξ before maximizing the result with respect to $\tilde{\theta}$. The proof is complete once we notice that U is increasing and that for each $t \in [0,T]$ and each $\omega \in \Omega$, the quantity B_T^{ξ} is maximized by the choice $\xi_t^* = (c')^{-1}(Y_t)$. □

Modeling Power Price First

We consider a second model for which stochastic optimization reduces to a mere path by path optimization. As before, the model is simplistic, especially in the case of a single firm in a regulatory environment with a liquid frictionless market for emission allowances. However, this model will become interesting later on when we consider N firms interacting on the same market. As before, the model concerns an economy with one production good (say electricity) whose production is the source of a negative externality (say greenhouse gas emissions). Its price $(P_t)_{0 \leq t \leq T}$ evolves according to the following Itô stochastic differential equation:

$$\frac{dP_t}{P_t} = \mu(P_t)dt + \sigma(P_t)dW_t, \qquad (3.29)$$

where the deterministic functions μ and σ are assumed to be C^1 with bounded derivatives. At each time $t \in [0,T]$, the firm chooses its instantaneous rate of production q_t and its production costs are $c(q_t)$, where c is a function $c : \mathbb{R}_+ \hookrightarrow \mathbb{R}$ which is assumed to be C^1 and strictly convex. With these notations, the profits and losses from the production at the end of the period $[0,T]$ are given by the integral

$$\int_0^T [P_t q_t - c(q_t)]dt.$$

The emission regulation mandates that at the end of the period $[0,T]$, the cumulative emissions of each firm be measured, and that one emission permit be redeemed per unit of emission. As before, we denote by $(Y_t)_{0 \leq t \leq T}$ the process giving the price of one emission permit at time t. For the sake of simplicity, we assume that the cumulative emissions E_t up to time t are proportional to the production in the sense that $E_t = \epsilon Q_t$, where the positive number ϵ represents the rate of emission of the production technology used by the firm, and Q_t denotes the cumulative production up to and including time t:

$$Q_t = \int_0^t q_s ds.$$

At the end of the time horizon, the cost incurred by the firm because of the regulation is given by $E_T Y_T = \epsilon Q_T Y_T$. The firm may purchase allowances: we denote by θ_t the amount of allowances held by the firm at time t. Under these conditions, the terminal wealth of the firm is given by

$$X_T = X_T^{q,\theta} = x + \int_0^T \theta_t dY_t + \int_0^T [P_t q_t - c(q_t)]dt - \epsilon Q_T Y_T, \qquad (3.30)$$

where we used the notation x for the initial wealth of the firm. The first integral in the right-hand side of the above equation gives the proceeds from trading in the allowance market, the next term gives the profits from the production and sale of the good, and the last term gives the costs of the emission regulation. We assume that the risk preferences of the firm are given by a utility function $U : \mathbb{R} \to \mathbb{R}$, which is assumed to be C^1, increasing, strictly concave, and satisfying the Inada conditions (3.27) stated earlier. The optimization problem of the firm can be written as

$$V(x) := \sup_{(\mathbf{q},\boldsymbol{\theta}) \in \mathbb{A}} \mathbb{E}U(X_T^{q,\theta}), \qquad (3.31)$$

where $\mathbb{E}$ denotes the expectation under the *historical* measure $\mathbb{P}$, and $\mathbb{A}$ is the set of admissible production and trading strategies $(\mathbf{q}, \boldsymbol{\theta})$. For these controls to be admissible, $\mathbf{q}$ and $\boldsymbol{\theta}$ need only be adapted processes satisfying the integrability condition (3.3).

Proposition 3.10. *The optimal production strategy of the firm is given by*

$$q_t^* = (c')^{-1}(P_t - \epsilon Y_t).$$

Remark 3.11. *As before, the optimal production strategy* $\mathbf{q}^*$ *is independent of the risk aversion (i.e., the utility function) of the firm. The intuitive interpretation of this result is clear: once a firm observes both prices P_t and Y_t, it computes the price for which it can sell the good minus the price it will have to pay because of the emission regulation, and the firm uses this corrected price to choose its optimal rate of production in the usual way.*

Proof. A simple integration by parts (notice that Q_t is of bounded variations) gives

$$Q_T Y_T = \int_0^T Y_t dQ_t + \int_0^T Q_t dY_t = \int_0^T Y_t q_t dt + \int_0^T Q_t dY_t \qquad (3.32)$$

so that $X_T = A_T^{\tilde{\theta}} + B_T^q$ with

$$A_T^{\tilde{\theta}} = \int_0^T \tilde{\theta}_t dY_t \qquad \text{with} \qquad \tilde{\theta}_t = \theta_t - \epsilon^i \int_0^t q_s ds,$$

which depends only upon $\tilde{\boldsymbol{\theta}}$ and

$$B_T^q = x - \int_0^T [(P_t - \epsilon Y_t)q_t - c(q_t)]dt,$$

which depends only upon $\mathbf{q}$ without depending upon $\tilde{\boldsymbol{\theta}}$. Since the set $\mathbb{A}$ of admissible controls is equivalently described by varying the couples $(\mathbf{q}, \boldsymbol{\theta})$ or $(\mathbf{q}, \tilde{\boldsymbol{\theta}})$, when computing the maximum

$$\sup_{(\mathbf{q},\boldsymbol{\theta})\in\mathbb{A}} \mathbb{E}U(X_T) = \sup_{(\mathbf{q},\tilde{\boldsymbol{\theta}})\in\mathbb{A}} \mathbb{E}U(A_T^{\tilde{\theta}} + B_T^q),$$

one can perform the optimizations over $\mathbf{q}$ and $\tilde{\boldsymbol{\theta}}$ separately, for example by fixing $\tilde{\boldsymbol{\theta}}$ and optimizing with respect to $\mathbf{q}$ before maximizing the result with respect to $\tilde{\boldsymbol{\theta}}$. The proof is complete once we notice that U is increasing and that for each $t \in [0, T]$ and each $\omega \in \Omega$, the quantity B_T^q is maximized by the choice $q_t^* = (c')^{-1}(P_t - \epsilon Y_t)$. $\square$

3.3 • Dynamic Programming and the HJB Equation

In this section, we restrict ourselves to the Markovian case. For the sake of convenience, we use the following standard notations from the theory of Markov diffusions. We denote by $X_s^{t,x,\alpha}$ the value at time $s \in [t, T]$, of the strong solution of (3.2) over the time interval $[t, T]$ when the initial condition is $X_t^{t,x,\alpha} = x$. Similarly, we denote by $\mathbb{A}_t$ the set of admissible controls over the interval $[t, T]$. Unless otherwise stated, $\mathbb{A}_t$ will be the set of A-valued progressively measurable processes $\boldsymbol{\alpha} = (\alpha_s)_{t \leq s \leq T}$ satisfying

$$\mathbb{E} \int_t^T (|b(s, x, \alpha_s)|^2 + |\sigma(s, x, \alpha_s)|^2)ds < \infty \qquad (3.33)$$

for all $x \in \mathbb{R}^d$. We then define the cost after time t as

$$J(t, x, \boldsymbol{\alpha}) = \mathbb{E}\left[\int_t^T f(s, X_s^{t,x,\alpha}, \alpha_s)ds + g(X_T^{t,x,\alpha})\right] \qquad (3.34)$$

for $(t, x) \in [0, T] \times \mathbb{R}^d$ and $\boldsymbol{\alpha} \in \mathbb{A}_t$, and the HJB value function v by

$$v(t, x) = \inf_{\boldsymbol{\alpha} \in \mathbb{A}_t} J(t, x, \boldsymbol{\alpha}) \qquad (3.35)$$

for $(t, x) \in [0, T] \times \mathbb{R}^d$.

Remark 3.12. *The function v will often be called the HJB value function in order to avoid confusion with the decoupling field of an FBSDE, which we sometimes call the FBSDE value function. As defined, it is the infimum over a continuum of functions, and as such it may not be smooth, or even measurable for that matter. The two examples below show that undesirable situations can indeed happen. In the following, we will explicitly make various forms of regularity assumptions on the HJB value function v. At the least, we will implicitly assume that it is jointly measurable in the variables s and x. This innocent-looking assumption is not always easy to check and may require the use of measurable selection theorems in some cases.*

Example 1. Let us consider the very simple (one-dimensional deterministic) example where $d = 1$, $A = [-1, +1]$ (so that $k = 1$), $\sigma = 0$, $f = 0$, $b(t, x, \alpha) = \alpha$, and $g(x) = -x^2$. For (s, x) fixed, $X_s^{t,x,\alpha}$ is given by

$$X_s^{t,x,\alpha} = x + \int_t^s \alpha_r dr$$

and since

$$v(t,x) = \inf_{\alpha \in \mathbb{A}_t} -X_T^2 = -\sup_{\alpha \in \mathbb{A}_t} \left(x + \int_t^T \alpha_r dr\right)^2,$$

the supremum is attained for $\alpha_s = 1$ or $\alpha_s = -1$ for all $s \in [t, T]$ and

$$v(t,x) = \begin{cases} -(x+T-t)^2 & \text{if } x \geq 0, \\ -(x-T+t)^2 & \text{otherwise.} \end{cases}$$

In the present situation, the value function is continuous but not differentiable at $x = 0$.

Example 2. We now consider a stochastic case. We assume $d = k = 1$, $b = 0$, and $\sigma(t, x, \alpha) = \alpha$ so that

$$X_T^{t,x,\alpha} = x + \int_t^T \alpha_r dW_r.$$

We also assume that $f = 0$ and that g is measurable and bounded from above. We claim that if it is $C^{1,2}$, the HJB value function is independent of t and equal to the convex envelope, g^{conv} of g—in other words, the largest convex function h satisfying $h \leq g$. First we notice that

$$v(t,x) = \inf_{\alpha \in \mathbb{A}_t} \mathbb{E}g(X_T^{t,x,\alpha})$$
$$\geq \inf_{\alpha \in \mathbb{A}_t} \mathbb{E}g^{conv}(X_T^{t,x,\alpha})$$
$$\geq \inf_{\alpha \in \mathbb{A}_t} g^{conv}(\mathbb{E}X_T^{t,x,\alpha})$$
$$\geq g^{conv}(x),$$

where we successively used the fact that $g \geq g^{conv}$, Jensen's inequality, and the fact that $(X_s^{t,x,\alpha})_{t \leq s \leq T}$ is a martingale. In order to prove the reverse inequality, we assume that $v \in C^{1,2}$. In this case, the HJB inequality proven in Theorem 3.15 below implies that

$$\partial_t v + \frac{1}{2}\alpha^2 \partial_{xx}^2 v \geq 0 \tag{3.36}$$

for all $(t, x, \alpha) \in [0, T] \times \mathbb{R} \times \mathbb{R}$. Choosing $\alpha = 0$ shows that $v(t, x)$ is a nondecreasing

3.3. Dynamic Programming and the HJB Equation

function of t whenever x is fixed, and

$$v(t,x) \leq \lim_{t \nearrow T} v(t,x) = v(T-,x).$$

But since g is bounded from above, we can use Fatou's lemma and conclude that

$$\mathbb{E} \limsup_{t \nearrow T} g(X_T^{t,x,\alpha}) \geq \limsup_{t \nearrow T} \mathbb{E} g(X_T^{t,x,\alpha}),$$

and since this is true for all $\alpha \in \mathbb{R}$, we conclude that

$$g(x) \geq \limsup_{t \nearrow T} \mathbb{E} g(X_T^{t,x,\alpha})$$

$$\geq \limsup_{t \nearrow T} \inf_{\alpha \in \mathbb{A}} \mathbb{E} g(X_T^{t,x,\alpha})$$

so that $v(t,x) \leq g(x)$. Now, letting α go to ∞ in (3.36) gives the fact that $\partial_{xx}^2 v \geq 0$, showing that $v(t,x)$ is a convex function of x when t is fixed, and by definition of the convex envelope we get

$$v(t,x) \leq g^{conv}(x),$$

and consequently $v(t,x) = g^{conv}(x)$. Now, choosing a function g so that g^{conv} is not C^2 gives an example of an HJB value function v which is not $C^{1,2}$.

Not all results concerning the HJB value function are negative. For the sake of illustration, we state the following result, whose proof can be found in [114].

Proposition 3.13. *If $f = 0$ and g is Lipschitz, then*
  (i) v is Lipschitz in x uniformly in $t \in [0,T]$;
  (ii) if $\mathbb{A}$ is bounded, then there exists a constant $c > 0$ such that for all $x \in \mathbb{R}^d$ and $t, t' \in [0,T]$ we have

$$|v(t,x) - v(t',x)| \leq c(1+|x|)\sqrt{|t-t'|}.$$

3.3.1 ▪ The Dynamic Programming Principle

The dynamic programming principle (DPP) is an induction formula providing a form of time consistency for the HJB value function of a stochastic control problem. It states that, given any fixed time $t \in [0,T]$, and any stopping time τ with values in the interval $[t,T]$, in order to solve the optimal control problem over the interval $[t,T]$, we can look for an optimal control over the interval $[t,\tau]$, using for terminal cost at time τ the value function of the optimal control problem over the interval $[\tau,T]$ and initial condition $X_\tau^{t,x,\alpha}$.

Theorem 3.14 (dynamic programming principle). *If the HJB value function v is continuous, then for any initial condition $(t,x) \in [0,T] \times \mathbb{R}^d$, and any stopping time τ with values in $[t,T]$, we have*

$$v(t,x) = \inf_{\alpha \in \mathbb{A}_t} \mathbb{E}\left[\int_t^\tau f(s, X_s^{t,x,\alpha}, \alpha_s)ds + v(\tau, X_\tau^{t,x,\alpha})\right]. \tag{3.37}$$

For further discussion and references on the proof, see the Notes & Complements section at the end of this chapter.

3.3.2 • The Hamilton–Jacobi–Bellman Equation

The Hamilton–Jacobi–Bellman equation is the infinitesimal form of the dynamic programming principle when the stopping time is chosen to be $\tau = s + \Delta s$ with $\Delta s \searrow 0$. The following notation makes the statement of this result more convenient. We define the operator symbol L by

$$L(t, x, y, z, \alpha) = b(t, x, \alpha) \cdot y + \frac{1}{2} \text{trace}[\sigma(t, x, \alpha) \sigma(t, x, \alpha)^\dagger z] + f(t, x, \alpha) \qquad (3.38)$$

and by $\hat{L}$ the minimized symbol

$$\hat{L}(t, x, y, z) = \inf_{\alpha \in A} L(t, x, y, z, \alpha). \qquad (3.39)$$

Theorem 3.15 (HJB equation). *Let us assume that $v \in C^{1,2}([0,T] \times \mathbb{R}^d)$ and that the running cost $f(\cdot, \cdot, \alpha)$ is continuous on $[0,T] \times \mathbb{R}^d$ for each fixed $\alpha \in A$. Then for each $(t, x) \in [0, T] \times \mathbb{R}^d$ we have*

$$\partial_t v(t, x) + \hat{L}(t, x, \partial_x v(t, x), \partial^2_{xx} v(t, x)) \geq 0. \qquad (3.40)$$

Proof. We assume that (t, x) is fixed, and we choose $\tau = t + h$ for some small constant $h > 0$, and a deterministic and constant control equal to a specific element $\alpha \in A$. Then the dynamic programming principle implies that

$$v(t, x) \leq \mathbb{E}\left[\int_t^{t+h} f(s, X_s^{t,x,\alpha}, \alpha) ds + v(t + h, X_{t+h}^{t,x,\alpha})\right]. \qquad (3.41)$$

The function v being smooth, we apply Itô's formula on the time interval $[t, t + h]$ and obtain

$$v(t+h, X_{t+h}^{t,x,\alpha}) = v(t, x) + \int_t^{t+h} (\partial_t v + \mathscr{L}_t^\alpha v)(s, X_s^{t,x,\alpha}) ds + \text{martingale}, \qquad (3.42)$$

where $\mathscr{L}_t^\alpha$ is the generator of the diffusion semigroup associated to the fixed control $\alpha \in A$:

$$\mathscr{L}_t^\alpha = b(t, x, \alpha) \cdot \partial_x v + \frac{1}{2} \text{trace}[\sigma(t, x, \alpha) \sigma(t, x, \alpha)^t \partial^2_{xx} v]. \qquad (3.43)$$

Plugging (3.42) into (3.41), we get

$$0 \leq \mathbb{E}\left[\int_t^{t+h} [(\partial_t v + \mathscr{L}_t^\alpha v)(s, X_s^{t,x,\alpha}) + f(s, X_s^{t,x,\alpha}, \alpha)] dt\right],$$

and dividing both sides by h and letting $h \searrow 0$, we get

$$0 \leq \partial_t v(t, x) + \mathscr{L}_t^\alpha v(t, x) + f(t, x, \alpha),$$

and since this holds for any $\alpha \in A$, we have $0 \leq \partial_t v(t, x) + \inf_{\alpha \in A}[\mathscr{L}_t^\alpha v(t, x) + f(t, x, \alpha)]$. □

It is possible to prove that the reverse inequality holds, and hence that

$$\partial_t v(t, x) + \hat{L}(t, x, \partial_x v(t, x), \partial^2_{xx} v(t, x)) = 0 \qquad (3.44)$$

3.3. Dynamic Programming and the HJB Equation

whenever the HJB value function v is jointly continuous. The proof of such a result can be found in [114]. An (incomplete) intuitive argument goes as follows. If we assume that α^* is an optimal control, then the dynamic programming principle gives

$$v(t,x) = \mathbb{E}\left[\int_t^{t+h} f(t, X_s^{t,x,\alpha^*}, \alpha_s^*)ds + v(t+h, X_{t+h}^{t,x,\alpha^*})\right]$$

from which we get

$$0 \geq \partial_t v(s,x) + \mathscr{L}_t^{a_t^*} v(s,x) + f(s, x, a_t^*)$$

by a similar argument, showing that the above inequality is in fact an equality.

Solution Strategy

The results of this section suggest the following strategy when it comes to solving a stochastic control problem.

- Construct a solution v of the HJB partial differential equation (3.44) and check that it is $C^{1,2}$.
- For each (t,x), determine a minimizer $\hat{\alpha}(t,x)$ for the function $\mathscr{L}^\alpha v(t,x) + f(t,x,\alpha)$ of α.
- Check that the control $\hat{\alpha}_t = \hat{\alpha}(t, X_t^{t,x,\hat{\alpha}})$ is admissible.

If the above three steps can be taken successfully, the function v is the value function of the stochastic control problem and the control $(\hat{\alpha})_{t\geq 0}$ is optimal. The first bullet point typically requires uniform ellipticity, stating that the lowest eigenvalue of the nonnegative definite matrix $\sigma(t,x,\alpha)\sigma(t,x,\alpha)^\dagger$ is uniformly bounded away from 0. In some cases, hypoellipticity in the sense of Hörmander suffices. The second bullet point usually relies on convex analysis arguments and implicit function theorems, while the last bullet point requires stochastic analysis arguments. See the Notes & Complements section at the end of the chapter for references.

3.3.3 ▪ Pontryagin Maximum Principle and the HJB Equation

This subsection is informal in nature. Its purpose is to argue that, in some very loose sense, the HJB equation can also be viewed as a consequence of the Pontryagin maximum principle for the optimal control of deterministic systems. Even though we go over the Pontryagin maximum principle in detail only later in Chapter 4, we use this principle in the classical case of deterministic systems, and we refer the reader unfamiliar with this principle to the references given in the Notes & Complements section at the end of the chapter.

Control of the Kolmogorov Equation

Consistent with the Markovian setup considered in this section, we limit ourselves to Markovian controls in closed loop feedback form $\alpha_t = \alpha(t, X_t)$ for some deterministic function $(t,x) \hookrightarrow \alpha(t,x)$. For the purpose of this subsection, for each $t \in [0,T]$, we denote by $\alpha_t(\cdot)$ the function $x \hookrightarrow \alpha(t,x)$, and by $\mathbf{A}$ the set of those functions of x appearing in this way. In other words, we view the function $(t,x) \hookrightarrow \alpha(t,x)$ as $[0,T] \ni t \hookrightarrow \alpha_t(\cdot) \in \mathbf{A}$. Since our discussion is informal, we shall not specify assumptions on the running cost function $f : [0,T] \times \mathbb{R}^d \times A \ni (t,x,\alpha) \hookrightarrow f(t,x,\alpha) \in \mathbb{R}$ or the terminal cost function

$g : \mathbb{R}^d \ni x \hookrightarrow g(x) \in \mathbb{R}$. Denoting as usual by $\mu_t = \mathcal{L}(X_t)$ the law of the state at time t, exchanging the expectation and the integral sign, we can rewrite the objective functional which the controller tries to minimize as

$$J(\boldsymbol{\alpha}) = \mathbb{E}\left[\int_0^T f(t, X_t, \alpha_t(X_t))dt + g(X_T)\right]$$
$$= \int_0^T F(t, \mu_t, \alpha_t(\cdot))dt + G(\mu_T) \tag{3.45}$$

with $F(t, \mu, \alpha(\cdot)) = \langle f(t, \cdot, \alpha(\cdot)), \mu\rangle$ and $G(\mu) = \langle g, \mu\rangle$ whenever $t \in [0, T]$, $\alpha(\cdot) \in \mathbf{A}$, and $\mu \in \mathcal{P}_2(\mathbb{R}^d)$. We used the duality between function and measure spaces to denote by $\langle \varphi, \mu\rangle$ the integral of the function φ with respect to the measure μ. Both functions F and G are deterministic and linear in the variable μ. We want to look at them as the running and terminal cost functions of a deterministic control problem on the space of measures, the admissible controls being deterministic functions $[0, T] \ni t \hookrightarrow \alpha_t(\cdot) \in \mathbf{A}$. This is quite natural because once the control $\boldsymbol{\alpha}$ is chosen, the measure $\mu_t = \mathcal{L}(X_t)$ evolves over time as the solution of the forward Kolmogorov equation. So we view the distribution μ_t as the state at time t of a system whose dynamics are given by an ordinary differential equation (ODE) in the infinite-dimensional space $\mathcal{M}(\mathbb{R}^d)$ of measures on $\mathbb{R}^d$. Recalling the discussion of Subsection 1.2.4 in Chapter 1, and assuming for the sake of simplicity that the volatility σ is not controlled and constant, i.e., $\sigma(t, x, \alpha) \equiv \sigma$, we have

$$\frac{d}{dt}\mu_t = \mathscr{L}_t^{\alpha, *}\mu_t, \tag{3.46}$$

where the operator $\mathscr{L}_t^{\alpha, *}$ is the adjoint of the operator $\mathscr{L}_t^{\alpha}$ defined as

$$[\mathscr{L}_t^{\alpha}\varphi](x) = b(t, x, \alpha(t, x)) \cdot \partial_x \varphi(x) + \frac{\sigma^2}{2}\Delta\varphi(x)$$

on smooth functions φ. Plain integration by parts gives (at least formally)

$$\mathscr{L}_t^{\alpha, *}\mu_t = \frac{\sigma^2}{2}\Delta\mu_t - \text{div}[\mu_t b(t, \cdot, \alpha(t, \cdot))],$$

at least in the sense of Schwartz distributions, or when μ_t has a smooth density.

Anticipating the contents of Chapter 4, or piggybacking on the classical theory of deterministic systems, we propose approaching the minimization of the objective functional (3.45) under the dynamic constraint (3.46) via the Pontryagin maximum principle. Notice that the space of controls is now a space of functions of x with values in the original control space A, so strictly speaking, we are dealing with an infinite-dimensional dynamical system with an infinite-dimensional control space. Still, the classical approach can be followed. We introduce the Hamiltonian of this deterministic control problem:

$$H(t, \mu, p, \alpha) = \langle \mathscr{L}_t^{\alpha, *}\mu, p\rangle + F(t, \mu, \alpha(\cdot))$$
$$= \langle \mathscr{L}_t^{\alpha} p + f(t, \cdot, \alpha(\cdot)), \mu\rangle,$$

where the adjoint variable p is a function of x. The necessary part of the deterministic version of the Pontryagin maximum principle says that the optimal control $\hat{\boldsymbol{\alpha}}$ (if any) should satisfy

$$H(t, \mu_t, p_t, \hat{\alpha}_t(\cdot)) = \inf_{\alpha(\cdot) \in \mathbf{A}} H(t, \mu_t, p_t, \alpha(\cdot)) \tag{3.47}$$

along the trajectory $t \hookrightarrow (\mu_t, p_t)$, where μ_t is the solution of the forward dynamics (3.46) for the optimal control $\hat{\alpha}$ and p_t is the solution of the adjoint equation

$$\dot{p}_t = -\partial_\mu H(t, \mu_t, p_t, \hat{\alpha}_t(\cdot)) \tag{3.48}$$

with terminal condition $p_T = dG(\mu_T)/d\mu$. Here and throughout the book, we use the dot notation $\dot{p}$ for the derivative of a function of time. Notice that the adjoint equation is very simple. Indeed, both the Hamiltonian H and the terminal cost function G are linear in μ, so their "functional derivatives" are easy to compute. The adjoint equation reads

$$-\dot{p}_t = \mathscr{L}_t^{\hat{\alpha}} p + f(t, \cdot, \hat{\alpha}(\cdot))$$
$$= b(t, \cdot, \hat{\alpha}_t(\cdot))\partial_x p_t + \frac{\sigma^2}{2}\Delta p_t + f(t, \cdot, \hat{\alpha}(\cdot)).$$

Let us now assume that there exists a function $[0, T] \times \mathbb{R}^d \times \mathbb{R}^d \ni (t, x, y) \hookrightarrow \hat{\alpha}(t, x, y) \in A$ such that

$$b(t, x, \hat{\alpha}(t, x, y)) \cdot y + f(t, x, \hat{\alpha}(t, x, y)) = \inf_{\alpha \in A} b(t, x, \alpha) \cdot y + f(t, x, \alpha)$$

for all (t, x, y). In this case, the HJB equation of the original stochastic control problem with state variable $x \in \mathbb{R}^d$ reads

$$0 = \partial_t v(t, x) + b(t, x, \hat{\alpha}(t, x, \partial_x v(t, x))) \cdot \partial_x v(t, x)$$
$$+ \frac{\sigma^2}{2}\Delta v(t, x) + f(t, x, \hat{\alpha}(t, x, \partial_x v(t, x)))$$

with terminal condition $v(T, x) = g(x)$. This is exactly the adjoint equation for the control of the (deterministic) Kolmogorov equation with state variable $\mu \in \mathcal{M}(\mathbb{R}^d)$, if we set $\hat{\alpha}_t(\cdot) = \hat{\alpha}(t, \cdot, \partial_x v(t, \cdot))$. Notice that for this particular control $\hat{\alpha}$ we have

$$H(t, \mu, p, \hat{\alpha}_t(\cdot)) = \inf_{\alpha(\cdot) \in \mathbf{A}} H(t, \mu, p, \alpha(\cdot)),$$

not only along an optimal trajectory $t \hookrightarrow (\mu_t, p_t)$, but for every μ and p.

3.4 • Infinite Horizon Case

We now discuss the case of infinite horizon ($T = \infty$) problems which often arise in economic applications. A typical example of such a problem occurs when the objective is to minimize or maximize the expected present (discounted) value of future cash flows. So, using the same notation as before, we assume that the terminal cost g vanishes and the object of the optimization becomes

$$J(\boldsymbol{\alpha}) = \mathbb{E}\int_0^\infty e^{-\beta t} f(X_t, \alpha_t) dt$$

for some discount factor $\beta > 0$. The presence of the discount factor is of crucial importance from a mathematical standpoint. Indeed, it will guarantee (e.g., when f is bounded or satisfies mild growth conditions) that $J(\boldsymbol{\alpha})$ is finite. This form of objective functional appears naturally in many applications to economic and financial models. Note also that most parts of the discussion in this section still hold when the discounting factor β is a function $t \hookrightarrow \beta(t, X_t, \alpha_t)$ of time, state, and control. We shall refrain from working at this level of generality to avoid heavier notation and longer lists of assumptions.

Restricting ourselves to control processes $\boldsymbol{\alpha}$ satisfying (3.3) for all $0 \leq t \leq T < \infty$, the controlled state process $(X_s^{t,x,\alpha})_{t \leq s}$ starting from x at time t exists for all times $s \geq t$, and we can define, for each $x \in \mathbb{R}^d$, the set $\mathbb{A}_t(x)$ of admissible controls after t as the set of those controls for which

$$\mathbb{E} \int_t^\infty e^{-\beta s} |f(X_s^{t,x,\alpha}, \alpha_s)| ds < \infty.$$

The goal is to compute

$$v_t(x) = \inf_{\boldsymbol{\alpha} \in \mathbb{A}_t(x)} J_t(x, \boldsymbol{\alpha}), \qquad (3.49)$$

where the expected cost to be minimized is defined as

$$J_t(x, \boldsymbol{\alpha}) = \mathbb{E} \int_t^\infty e^{-\beta(s-t)} f(X_s^{t,x,\alpha}, \alpha_s) ds. \qquad (3.50)$$

The arguments given earlier in the case of finite horizon problems provide versions of the dynamic programming principle and the Hamilton–Jacobi–Bellman equation for the infinite horizon setting. However, the latter is particularly convenient when the coefficients (drift, volatility, and running cost functions) do not depend upon the time variable t, providing some form of stationarity to the problem. Indeed, in this case, the set of admissible controls $\mathbb{A}_t(x) = \mathbb{A}(x)$ and the value function $v_t(x) = v(x)$ do not depend upon the time variable t. Accordingly, the DPP reads: for every $\boldsymbol{\alpha} \in \mathbb{A}(x)$ and stopping time τ

$$v(x) \leq \mathbb{E}\left[\int_0^\tau e^{-\beta s} f(X_s^{0,x,\alpha}, \alpha_s) ds + e^{-\beta \tau} v(X_\tau^{0,x})\right], \qquad (3.51)$$

and for every $\epsilon > 0$, there exists $\boldsymbol{\alpha} \in \mathbb{A}(x)$ such that for any stopping time τ

$$v(x) + \epsilon \geq \mathbb{E}\left[\int_0^\tau e^{-\beta s} f(X_s^{0,x,\alpha}, \alpha_s) ds + e^{-\beta \tau} v(X_\tau^{0,x})\right]. \qquad (3.52)$$

In this setting, the operator symbol has the form

$$L(x, y, z, \alpha) = b(x, \alpha) \cdot y + \frac{1}{2} \mathrm{trace}[\sigma(x, \alpha) \sigma(x, \alpha)^\dagger z] + f(x, \alpha), \qquad (3.53)$$

and its minimization leads to

$$\hat{L}(x, y, z) = \inf_{\alpha \in A} L(x, y, z, \alpha). \qquad (3.54)$$

Finally, the HJB equation reads

$$-\beta v(x) + \hat{L}(x, \nabla v(x), \nabla^2 v(x)) = 0. \qquad (3.55)$$

We used the gradient notation ∇ instead of ∂_x because x is the only variable of the function v. The infinite horizon verification theorem takes the following form: if v is C^2 with bounded derivatives and satisfies (3.55), if $\boldsymbol{\alpha}$ is any admissible control process and $\mathbf{X} = \mathbf{X}^\alpha$ is the associated controlled state with $X_0 = x$, then

$$v(x) \leq \mathbb{E} \int_0^\infty e^{-rt} f(X_t, \alpha_t) dt.$$

The procedure to determine the optimal control follows the same lines as in the finite horizon case. Indeed, if we can identify a function $\hat{\alpha}$ such that, for each $x \in \mathbb{R}^d$,

$$\hat{\alpha}(x) \in \arg\inf_{\alpha \in A} L(x, \nabla v(x), \nabla^2 v(x), \alpha),$$

where v is a solution of (3.55), and if we can solve the corresponding state equation

$$dX_t = b(X_t, \hat{\alpha}(X_t))dt + \sigma(X_t, \hat{\alpha}(X_t))dW_t,$$

then a simple application of Itô's lemma shows that

$$v(x) = \mathbb{E} \int_0^\infty e^{-rt} f(X_t, \hat{\alpha}_t) dt$$

with $\hat{\alpha}_t = \hat{\alpha}(X_t)$, proving that v is the value function of the problem and $\hat{\alpha}$ is an optimal control.

3.5 ▪ Constraints and Singular Control Problems

In the first part of this section, we assume that the drift and volatility functions, as well as the running and terminal cost functions, are deterministic, and we consider a class of optimal control problems of Markov diffusion processes for which the operator symbol cannot be minimized over the entire space. Later, we discuss models which do not quite fit this setup exactly, but which we still call singular because the space of admissible control contains controls which can be singular measures in the time domain. We try to explain the commonality in structure of the two classes of models, in so doing justifying the rationale for the use of the same terminology of *singular control* for these two different classes of models.

3.5.1 ▪ A First Class of Singular Control Problems

For the purpose of this first attempt to delineate a family of singular control problems, we say that a model is singular when the minimized operator symbol $\hat{L}$ defined by

$$\hat{L}(t, x, y, z) = \inf_{\alpha \in A} L(t, x, y, z, \alpha)$$

is equal to $-\infty$ on part of the space. In order to be able to analyze such cases, we assume the existence of a continuous function G on $[0, T] \times \mathbb{R}^d \times \mathbb{R}^d \times \mathbb{R}^{d^2}$ such that

$$\hat{L}(t, x, y, z) > -\infty \iff G(t, x, y, z) \geq 0.$$

As we are about to show, in such a case, the parabolic HJB partial differential equation representing the infinitesimal form of the dynamic programming principle should be replaced by a quasi-variational inequality (QVI). The derivation of the HJB equation (3.40) shows that

$$G(t, x, \partial_x v(t, x), \partial_{xx}^2 v(t, x)) \geq 0$$

and

$$\partial_t v(t, x) + \hat{L}(t, x, \partial_x v(t, x), \partial_{xx}^2 v(t, x)) \geq 0. \tag{3.56}$$

Moreover, if $G(t, x, \partial_x v(t, x), \partial_{xx}^2 v(t, x)) > 0$ for some $(t, x) \in [0, T] \times \mathbb{R}^d$, then there exists a neighborhood of (t, x) in $[0, T] \times \mathbb{R}^d$ on which $\hat{L}(t, x, \partial_x v(t, x), \partial_{xx}^2 v(t, x))$ is finite. Recall that we assume that $v \in C^{1,2}$. Then, the optimal control is still attained in a neighborhood of (t, x) in $[0, T] \times \mathbb{R}^d$, and the argument of the proof of Theorem 3.15 shows that the equality should hold in (3.56). In other words, the HJB partial differential equation (3.44) becomes the QVI

$$0 = \min\left[\partial_t v(t, x) + \hat{L}(t, x, \partial_x v(t, x), \partial_{xx}^2 v(t, x)), \ G(t, x, \partial_x v(t, x), \partial_{xx}^2 v(t, x))\right]. \tag{3.57}$$

While the terminal condition $v(T,x) = g(x)$ holds in most applications, it may not be the right terminal condition in the case of singular control problems. We shall discuss this delicate issue on a case by case basis.

3.5.2 ▪ The Classical Merton Portfolio Optimization Problem

We use the market model introduced in Subsection 2.1.4 of Chapter 2, and for the sake of simplicity, we consider only the case of one risky asset. Moreover, we assume that the coefficients are deterministic and constant over time. So the model reduces to

$$\begin{cases} dP_t^0 = rP_t^0 dt, & P_0^0 = 1, \\ dP_t^1 = P_t^1[\mu^1 dt + \sigma^1 dW_t], & P_0^1 = p. \end{cases} \quad (3.58)$$

Terminal Utility of Wealth Maximization

We first consider the problem of the optimization of the expected utility of terminal wealth. We examine the optimization of the utility of consumption later on in this section. Let us denote by α_t the proportion of wealth invested in the stock at time t. In the notation of Chapter 2 we would have $\alpha_t = \pi_t^1/V_t$. From now on, since we are only considering the case of one risky asset, we drop the superscript ^{1} and use the notation μ and σ for μ^1 and σ^1. Also, we assume that $\mu > r$. So, if we now denote by X_t the wealth (i.e., the value of the portfolio which we denoted by V_t in Chapter 2), then we have

$$dX_t = X_t[\alpha_t\mu + (1-\alpha_t)r]dt + \alpha_t X_t \sigma dW_t. \quad (3.59)$$

Since in the present context the proportion of wealth invested in the stock can be any real number, an admissible control $\boldsymbol{\alpha} = (\alpha_t)_t$ is a progressively measurable process such that $\int_0^T |\alpha_t|^2 dt < \infty$. This form of admissibility guarantees the existence and uniqueness of a solution of (3.64). Indeed, it is plain to see that the equation satisfied by the logarithm of the wealth has a unique solution if the control is admissible in this sense. If we denote by $(X_s^{t,x})_{t \leq s \leq T}$ the solution of (3.64) starting from x at time $s = t$, then the Merton portfolio choice problem is to compute

$$v(t,x) = \sup_{\alpha \in \mathbb{A}} \mathbb{E}[U(X_T^{t,x})], \quad (t,x) \in [0,T] \times [0,\infty),$$

where the utility function U is a concave increasing function on $(0,\infty)$.

Lemma 3.16. *For each $t \in [0,T]$ the function $x \hookrightarrow v(t,x)$ is nondecreasing and concave (and strictly concave if there exists an optimal control).*

Proof. Let us assume that $t \in [0,T]$ is fixed as well as x and y such that $0 < x < y < \infty$. If $\alpha \in \mathbb{A}$ is a generic admissible control process, and if we denote by $X_s^{t,x}$ and $X_s^{t,y}$ the solutions at time s of (3.64) starting from x and y at time t, then

$$d(X_s^{t,y} - X_s^{t,x}) = (X_s^{t,y} - X_s^{t,x})[(\alpha_s(\mu-r)+r)ds + \alpha_s \sigma dW_s],$$

and since $X_t^{t,y} - X_t^{t,x} = y - x > 0$, we conclude that $X_s^{t,y} - X_s^{t,x} > 0$ for all $s \in [t,T]$. Now, U being nondecreasing, we have

$$\mathbb{E}[U(X_T^{t,x})] \leq \mathbb{E}[U(X_T^{t,y})] \leq v(t,y),$$

and since this is true for all $\alpha \in \mathbb{A}$, we conclude that $v(t, x) \leq v(t, y)$, proving that $v(t, \cdot)$ is nondecreasing. We now assume that $t \in [0, T]$ is fixed, $0 < x_1 < x_2 < \infty$, α^1 and α^2 are admissible controls, and $\lambda \in [0, 1]$. We write x_λ for $\lambda x_1 + (1 - \lambda)x_2$ and define the process α^λ by

$$\alpha_s^\lambda = \frac{\lambda X_s^{t,x_1}\alpha_s^1 + (1-\lambda)X_s^{t,x_2}\alpha_s^2}{\lambda X_s^{t,x_1} + (1-\lambda)X_s^{t,x_2}}, \qquad t \leq s \leq T,$$

where X_s^{t,x_1} and X_s^{t,x_2} are the solutions at time s of (3.64) starting from x_1 and x_2 at time t, and controlled by the admissible control processes α^1 and α^2 respectively. Clearly, α^λ is an admissible control and the associated controlled process is the process $\mathbf{X}^\lambda$ defined by $X_s^\lambda = \lambda X_s^{t,x_1} + (1-\lambda)X_s^{t,x_2}$ for $t \leq s \leq T$. The utility function being concave, we have

$$U(X_T^\lambda) \geq \lambda U(X_T^{t,x_1}) + (1-\lambda)U(X_T^{t,x_2}),$$

and by definition of the value function,

$$v(t, \lambda x_1 + (1-\lambda)x_2) \geq \lambda \mathbb{E}[U(X_T^{t,x_1})] + (1-\lambda)\mathbb{E}[U(X_T^{t,x_2})].$$

Since this is true for all admissible control processes α^1 and α^2, we can conclude that

$$v(t, \lambda x_1 + (1-\lambda)x_2) \geq \lambda v(t, x_1) + (1-\lambda)v(t, x_2),$$

proving concavity of the value function. Finally, notice that this argument also proves strict concavity if the supremum defining the value function is always attained and the utility function U is strictly concave. □

We now write the HJB equation of the problem. The operator symbol L reads

$$L(x, y, z, \alpha) = x[\alpha(\mu - r) + r]y + \frac{1}{2}\alpha^2 x^2 \sigma^2 z.$$

The corresponding maximized operator symbol $\hat{L}(x, y, z)$ is thus equal to $+\infty$ whenever $z > 0$ or $xy(\mu - r) \neq 0$ if $z = 0$. Otherwise, it is given by the formula

$$\hat{L}(x, y, z) = \frac{1}{2}\frac{(\mu - r)^2}{\sigma^2}\frac{y^2}{z} + rxy,$$

the minimum being attained for the value

$$\hat{\alpha} = -\frac{\mu - r}{\sigma^2}\frac{y}{xz}$$

when $z > 0$. According to the discussion of Subsection 3.5.1, the HJB equation should be written as a QVI, but instead, we look for a strictly concave classical solution of the nonlinear PDE

$$\partial_t v(t, x) - \frac{1}{2}\frac{(\mu - r)^2}{\sigma^2}\frac{\partial_x v(t, x)^2}{\partial_{xx}^2 v(t, x)} + rx\partial_x v(t, x) = 0 \qquad (3.60)$$

with terminal condition $v(T, x) = U(x)$. This nonlinear PDE cannot be solved in general. However, it can be solved in the particular case of the constant relative risk aversion (CRRA) utility function

$$U(x) = \frac{1}{1-\gamma}x^{1-\gamma}, \qquad x > 0, \qquad (3.61)$$

considered by Merton. The coefficient $\gamma > 0$ has the interpretation of a coefficient of risk aversion. In this case, we look for a function v of the form $v(t,x) = \varphi(t)U(x)$. With this *ansatz*, the HJB equation (3.60) is equivalent to

$$\dot{\varphi}(t) = \rho\varphi(t), \qquad \varphi(T) = 1$$

with

$$\rho = \frac{1}{2}\frac{(\mu-r)^2}{\sigma^2}\frac{1-\gamma}{\gamma} + r(1-\gamma). \tag{3.62}$$

This simple ODE can be solved explicitly. We get $\varphi(t) = e^{-\rho(T-t)}$, and the solution of the HJB equation (3.60) reads

$$v(t,x) = \frac{1}{1-\gamma}e^{-\rho(T-t)}x^{1-\gamma}.$$

This solution is strictly convex and smooth enough to apply the verification theorem, from which we learn that an investor behaving optimally to maximize his/her expected terminal utility of wealth should hold the constant proportion

$$\alpha^* = \frac{\mu - r}{\gamma\sigma^2} \tag{3.63}$$

of his/her wealth in stock.

Remark 3.17. *Since the market model used in this section is complete, there is a one-to-one correspondence between the $\mathcal{F}_T$-measurable random variables Y of order 2 and the terminal wealth W_T^π associated to an investment portfolio. So an alternative approach to the solution of the Merton problem can be based on the following successive steps:* (1) *solve the optimization problem*

$$Y = \arg\sup_{Z \in \mathcal{F}_T} \mathbb{E}\{U(Z)\},$$

typically by a duality argument, and (2) *find the replicating portfolio π for which $Y = W_T^\pi$. This approach was used, for example, in [43], among others. We only mention this paper because we shall review some of its results in our discussion of hedge fund manager incentives.*

Utility of Consumption Maximization

As before, we denote by α_t the proportion of wealth invested in the risky asset, and now by c_t the consumption rate at time t. In this subsection, we consider the infinite horizon problem for the sake of simplicity. An admissible control is now a pair of progressively measurable processes $(\alpha_t, c_t)_{t\geq 0}$ with $c_t \geq 0$, so that $\mathbb{A} = \mathbb{R} \times [0, \infty)$, such that $\int_0^\infty (|\alpha_t|^2 + c_t)dt < \infty$ almost surely. Given such an admissible control, the corresponding wealth process $\mathbf{X} = (X_t)_t$ is the unique strong solution of the equation

$$dX_t = X_t[\alpha_t(\mu - r) + r]dt + \alpha_t \sigma dW_t] - c_t dt, \tag{3.64}$$

and our goal is to compute the value function

$$v(x) = \sup_{(\boldsymbol{\alpha},\mathbf{c})\in\mathbb{A}} \mathbb{E}\left[\int_0^\infty e^{-\beta t}U(c_t)dt\right]$$

3.5. Constraints and Singular Control Problems

for some discounting factor $\beta > 0$. Here, $U(c)$ represents the utility of consumption at rate c. The (time-independent) operator symbol L reads

$$L(x, y, z, \alpha, c) = \left(x[\alpha(\mu - r) + r] - c\right)y + \frac{1}{2}\alpha^2 x^2 \sigma^2 z - U(c),$$

the running cost having a minus sign because the value function is the result of a maximization instead of a minimization. The corresponding minimized operator symbol is thus

$$\hat{L}(x, y, z) = \inf_{\alpha \in \mathbb{R}} \left(xy[\alpha(\mu - r) + r] + \frac{1}{2}\alpha^2 x^2 \sigma^2 z\right) + \inf_{c \geq 0} -yc - U(c)$$

$$= \begin{cases} -\infty & \text{whenever } z \leq 0 \text{ or } y \geq 0, \\ -\frac{1}{2}\frac{(\mu-r)^2}{\sigma^2}\frac{y^2}{z} + rxy + \tilde{U}(-y) & \text{otherwise,} \end{cases}$$

where $\tilde{U}$ denotes the Fenchel–Legendre transform of the convex function $-U$. Recall that the transform of a function f is the function $\tilde{f}$ defined by

$$\tilde{f}(y) = \inf_{c \geq 0} f(c) - cy. \tag{3.65}$$

Notice that the infimums are attained for the values

$$\hat{\alpha} = -\frac{\mu - r}{\sigma^2}\frac{y}{xz} \quad \text{and} \quad \hat{c} = (U')^{-1}(-y).$$

As a result, the HJB equation can be written as the QVI:

$$\min\left[\beta v(x) + \frac{1}{2}\frac{(\mu-r)^2}{\sigma^2}\frac{v'(x)^2}{v''(x)} - rxv'(x) - \tilde{U}(v'(x)), v'(x), -v''(x)\right] = 0, \tag{3.66}$$

which we will solve for the value function by finding a strictly concave and strictly increasing function v solving

$$\beta v(x) + \frac{1}{2}\frac{(\mu-r)^2}{\sigma^2}\frac{v'(x)^2}{v''(x)} - rxv'(x) - \tilde{U}(v'(x)) = 0. \tag{3.67}$$

Even though we are now dealing with an ODE instead of a PDE, a significant factor in the popularity of the infinite horizon models, this ODE is still nonlinear and cannot be solved in general. As before, we solve it in the particular case of the CRRA power utility function. In this case, the Legendre transform writes as

$$\tilde{U}(y) = -\frac{\gamma}{1-\gamma}y^{(\gamma-1)/\gamma}, \qquad y > 0.$$

As before, we look for a solution by making an *ansatz* on its form. So we search for v among the functions $v(x) = \kappa U(x)$. With this ansatz, the HJB equation (3.67) is equivalent to

$$\beta - \rho + \frac{\gamma}{1-\gamma}\kappa^{-1/\gamma} = 0,$$

with ρ given as before by (3.62). This solution is smooth enough to apply the verification theorem as long as the *boundary condition* at ∞ is satisfied. From the result of the verification argument, we learn that an investor behaving optimally to maximize his/her expected utility of consumption should hold the constant proportion α^* of his wealth in stock, with α^* given by

$$\alpha^* = \hat{\alpha} = -\frac{\mu - r}{\sigma^2}\frac{v'(x)}{xv''(x)} = \frac{\kappa(\mu - r)}{\gamma\sigma^2},$$

and consume according to the optimal schedule given by the optimal rate of consumption

$$c_t^* = \kappa^{-1/\gamma} X_t.$$

In other words, he/she should consume the constant proportion $\kappa^{-1/\gamma}$ of his/her wealth.

3.5.3 ▪ Markowitz Mean-Variance Portfolio Selection

Again, we use the market model introduced in Subsection 2.1.4 of Chapter 2 to present a continuous time version of the mean-variance criterion for the selection of an optimal portfolio originally proposed by Harry Markowitz in one-period Gaussian models. Recall that transaction costs are ignored in the market model of Subsection 2.1.4. As in the previous section, we denote by X_t the wealth of the investor at time t (which was denoted by V_t in Subsection 2.1.4 of Chapter 2), but we denote by $\alpha_t^i = \pi_t^i$ the amount of wealth invested in the ith risky asset at time t. In this way, the dynamics of the wealth of the investor are given by

$$dX_t = \left[X_t r_t + \sum_{i=1}^m \alpha_t^i [(\mu_t^i - r_t)]\right] dt + \left(\sum_{i=1}^m \alpha_t^i \sigma_t^i\right) dW_t. \qquad (3.68)$$

Roughly speaking, the objective of the investor is to maximize the expected terminal wealth $\mathbb{E}[X_T]$ while minimizing the variance $\text{var}[X_T] = \mathbb{E}[X_T^2] - \mathbb{E}[X_T]^2$. A portfolio is said to be efficient, or to lie on the efficient frontier if there is no other admissible portfolio with both a greater expected terminal wealth and a smaller variance of terminal wealth. Formulated in this way, the objective function is multivalued, and standard multiobjective optimization theory suggests searching for efficient portfolios by solving single objective optimization problems given by *weighted averages* of the separate objectives. In other words, one may want to minimize the objective

$$J^\gamma(\boldsymbol{\alpha}) = -\mathbb{E}[X_T] + \gamma \text{var}[X_T] \qquad (3.69)$$

over all the admissible portfolios $\boldsymbol{\alpha}$ for each fixed $\gamma \in \mathbb{R}$. Note that, even for γ fixed, this is not a standard optimal control problem because of the term $\mathbb{E}[X_T]^2$ hidden in the variance term. As such, the problem belongs to the class of optimal control problems for the dynamics of McKean–Vlasov type. We shall solve it as such in Subsection 4.4.6. However, we shall show in Subsection 4.3.4 that, as suggested by the form (3.69), it is possible to reduce the solution of the mean-variance portfolio optimization problem to LQ (linear-quadratic) optimal control problems.

3.5.4 ▪ Commodity Inventory Valuation

In this example, we consider the valuation of the inventory of a storable (though possibly perishable) commodity. For the purpose of the present discussion, the reader can think of the commodity as wheat in order to justify the terminology we use throughout. We assume that the commodity is harvested in continuous time, and we denote by x_t^1 the harvest at time t. We shall assume that $\mathbf{x}^1 = (x_t^1)_{t \geq 0}$ is a Markov process in $[0, \infty)$. For the sake of the present discussion we assume that

$$dx_t^1 = \mu(x_t^1) dt + \sigma(x_t^1) dW_t \qquad (3.70)$$

for some coefficients μ and σ, to be specified. One could choose to work with a geometric Brownian motion for tractability reasons, but it would be more realistic, though not easy to handle, to assume that $\mathbf{x}^1$ is recurrent and that it could be equal to 0. At each time $t \geq 0$, we

3.5. Constraints and Singular Control Problems

denote by c_t the rate of consumption and by x_t^2 the inventory level. We assume that the rate of consumption is given by a nonnegative progressively measurable process $\mathbf{c} = (c_t)_{t\geq 0}$. Naturally

$$dx_t^2 = (x_t^1 - c_t)dt. \tag{3.71}$$

Since x_t^2 represents the actual level of the physical storage of the commodity at time t, we should have $x_t^2 \geq 0$ at all times.

According to standard partial equilibrium arguments, optimal storage decisions should maximize the expected present value of the economy net surplus (defined as the difference between consumer and producer surplus), subject to remaining inventories being nonnegative. Formally, consumption is used as the control a central planner (or representative agent) applies to the state $X_t = (x_t^1, x_t^2)$ of the system at time t, in order to maximize the overall expected present value

$$J(\mathbf{c}) = \mathbb{E}\left[\int_0^\tau e^{-\delta t} S(c_t) dt\right],$$

where $\delta > 0$ is a discount factor, τ is the first time X_t exits the domain $[0, \infty) \times [0, \infty)$, and the excess surplus function S is defined as

$$S(c) = \int_0^c D(c') dc',$$

where D is a nonnegative decreasing function on $(0, \infty)$ representing the inverse demand function for the commodity.

Notice that we could have used x_t^2 as the state of the system, x_t^1 merely representing a source of exogenous randomness. Nevertheless, we will work with the higher dimensional state X_t to take advantage of the Markov property. Except for the constraint forcing the controlled state X_t to stay in the domain $\mathcal{X} = [0, \infty) \times [0, \infty) \subset \mathbb{R}^2$ (and hence the presence of the stopping time τ in the definition of the objective function to maximize), this problem seems to belong to the class of infinite horizon problems considered in this chapter. Notice that the stopping time τ (which only depends upon the inventory since $x_t^1 \geq 0$ by assumption) is identically infinite when the nonnegative adapted process $\mathbf{c} = (c_t)_{t\geq 0}$ is such that $x_t^2 \geq 0$ for all $t \geq 0$. Instead of including this property in the definition of the set $\mathbb{A}$ of admissible controls, we included the stopping time in the definition of the objective function.

The function

$$D(x) = \beta x^{-\gamma}, \qquad x > 0, \tag{3.72}$$

for some $\beta > 0$ and $\gamma \in (0, 1)$, is often used as a model for the inverse demand function. With this choice

$$D^{-1}(c) = \beta^{1/\gamma} c^{-1/\gamma} \qquad \text{and} \qquad S(c) = \frac{\beta}{1-\gamma} c^{1-\gamma}.$$

Singular Control Approach

Another possible approach is to treat the problem as a singular control problem with the QVI strategy outlined earlier. Indeed, the operator symbol of the problem is given by

$$L(t, x_1, x_2, y_1, y_2, z_1, z_2, c) = \frac{1}{2}\sigma(x_1)^2 z_1 + \mu(x_1) y_1 + (x_1 - c) y_2 - S(c)$$

(notice that we changed the sign of the *running cost* $S(c)$ because we are maximizing instead of minimizing, as is the case most of the time in these notes). From now on, we drop the variables t and z_2 which do not appear in the right-hand side. The minimized operator symbol $\hat{L}$ reads

$$\hat{L}(x_1, x_2, y_1, y_2, z_1) = \frac{1}{2}\sigma(x_1)^2 z_1 + \mu(x_1)y^1 + x_1 y_2 + \tilde{S}(y_2),$$

where $\tilde{S}$ is the Legendre transform whose definition for the convex function $f = -S$ was given in (3.65). Notice that $\tilde{S}(y) = -\infty$ (and hence $\hat{L}(t, x_1, x_2, y_1, y_2, z_1) = -\infty$) whenever $y > 0$, and $\tilde{S}(y) < \infty$ whenever $y \geq 0$. Consequently, the value function of our control problem should solve the QVI

$$\min\left[\frac{1}{2}\sigma(x_1)^2 \partial_{x_1 x_1}^2 v + \mu(x_1)\partial_{x_1} v + x_1 \partial_{x_2} v - \tilde{S}(\partial_{x_2} v),\ \partial_{x_2} v\right] = 0.$$

Despite its apparent simplicity, and even in the case of simple and explicit coefficient functions μ and σ, this QVI does not seem to have explicit solutions, even in the case of the power inverse demand function. As a result, it will have to be solved numerically.

3.5.5 ▪ Allowing for Singular Controls

We now describe a second class of stochastic optimal control problems which are naturally called singular. We assume that the dynamics of the controlled state are given by a diffusion-like equation of the form

$$dX_t = b(t, X_t, \alpha_t)dt + \sigma(t, X_t, \alpha_t)dW_t + \kappa(t, X_{t-})d\xi_t, \qquad 0 \leq t \leq T, \quad (3.73)$$

with initial condition $X_0 = x$, drift and diffusion coefficients given by deterministic functions $(b, \sigma) : [0, T] \times \mathbb{R}^d \times A \ni (t, x, \alpha) \hookrightarrow (b, \sigma)(t, x, \alpha) \in \mathbb{R}^d \times \mathbb{R}^{d \times m}$ (involving a progressively measurable process $\alpha = (\alpha_t)_{0 \leq t \leq T}$ with values in $A \subset \mathbb{R}^k$, which we want to think of as the absolutely continuous control), the third term in the right-hand side given by a continuous function $\kappa : [0, T] \times \mathbb{R}^d \ni (t, x) \hookrightarrow \kappa(t, x) \in \mathbb{R}^{d \times k'}$ (assumed to be Lipschitz in the x-variable), and a progressively measurable process $\boldsymbol{\xi} = (\xi_t)_{0 \leq t \leq T}$ in $\mathbb{R}^{k'}$ with right continuous nondecreasing components with left limits. Since the components of ξ_t are of bounded variations, the components of $d\xi_t$ can be viewed as positive measures on $[0, T]$, and integrals of $\kappa(t, X_{t-})d\xi_t$ interpreted as standard Stieltjes integrals computed path by path. We set $\xi_{0-} = 0$ by convention, and we think of $\boldsymbol{\xi} = (\xi_t)_{0 \leq t \leq T}$ as of a singular control. Under these conditions, for each $(t, x) \in [0, T] \times \mathbb{R}^d$ and $(\alpha, \boldsymbol{\xi})$ as above, there exists a unique strong solution of (3.73) on the interval $[t, T]$, which satisfies $X_t = x$. Even though this result looks more general than the existence and uniqueness result given in Chapter 1 for standard SDEs, its proof follows the same Picard iteration argument. We shall denote the solution process by $(X_s^{t,x,\alpha,\boldsymbol{\xi}})_{t \leq s \leq T}$.

We assume as before that the running and terminal cost functions f and g are deterministic, $f : [0, T] \times \mathbb{R}^d \times A \ni (t, x, \alpha) \hookrightarrow f(t, x, \alpha) \in \mathbb{R}$ and $g : \mathbb{R}^d \ni x \hookrightarrow g(x) \in \mathbb{R}$, and we consider further (running) costs given by a deterministic function $\theta : [0, T] \times \mathbb{R}^d \ni (t, x) \hookrightarrow \theta(t, x) \in \mathbb{R}^{k'}$. Now, for each $(t, x) \in [0, T] \times \mathbb{R}^d$, we denote by $\mathbb{A}_t(x)$ the set of couples of processes $(\alpha, \boldsymbol{\xi})$ for which

$$\mathbb{E}\left[\int_t^T |f(s, X_s^{t,x,\alpha,\boldsymbol{\xi}}, \alpha_s)|ds + |g(X_T^{t,x,\alpha,\boldsymbol{\xi}})| + \int_t^T \sum_{j=1}^{k'} |\theta^j(s, X_{s-}^{t,x,\alpha,\boldsymbol{\xi}})|d\xi_s^j\right] < \infty.$$

For each $(t, x) \in [0, T] \times \mathbb{R}^d$ and each couple $(\alpha, \boldsymbol{\xi}) \in \mathbb{A}_t(x)$ of admissible controls, we

3.5. Constraints and Singular Control Problems

define the cost to the controller by

$$J(t,x,\boldsymbol{\alpha},\boldsymbol{\xi}) = \mathbb{E}\left[\int_t^T f(s,X_s^{t,x,\alpha,\xi},\alpha_s)ds + g(X_T^{t,x,\alpha,\xi}) + \int_t^T \sum_{j=1}^{k'} \theta^j(s,X_{s-}^{t,x,\alpha,\xi})d\xi_s^j\right]$$

and the value function of the (singular) control problem by

$$v(t,x) = \inf_{(\boldsymbol{\alpha},\boldsymbol{\xi})\in\mathbb{A}_t(x)} J(t,x,\boldsymbol{\alpha},\boldsymbol{\xi}).$$

We now prove a version of the verification theorem for the present situation. We work with the Markovian (absolutely continuous) control $\alpha_t = \alpha(t,x)$ given by a deterministic function α on $[0,T]\times\mathbb{R}^d$. The following notation will be useful. For each such function $(t,x) \hookrightarrow \alpha(t,x)$, we denote by $\mathscr{L}_t^\alpha$ the infinitesimal generator of the controlled diffusion (for the singular control $\boldsymbol{\xi}\equiv 0$)

$$dX_s = b(s,X_s,\alpha(s,X_s))ds + \sigma(s,X_s,\alpha(s,X_s))dW_s, \quad t\le s\le T,$$

starting from $X_t = x$, namely, the operator defined by

$$\mathscr{L}_s^\alpha\phi = b(s,x,\alpha(s,x))\partial_x\phi(t,x) + \frac{1}{2}\text{trace}[\sigma(s,x,\alpha(s,x))\sigma(s,x,\alpha(s,x))^\dagger \partial_{xx}^2\phi(t,x)].$$

This notation was introduced earlier in the case of a constant function $\alpha(t,x)\equiv\alpha\in A$.

Theorem 3.18. *Let us assume that the function $\phi\in C^{1,2}([0,T]\times\mathbb{R}^d)$ satisfies $\phi(T,x) = g(x)$ for all $x\in\mathbb{R}^d$ and*
  (i) $\partial_t\phi(t,x) + [\mathscr{L}_t^\alpha\phi](t,x) + f(t,x,\alpha) \ge 0$ *for all $\alpha\in A$, and $(t,x)\in[0,T]\times\mathbb{R}^d$;*
  (ii) $\sum_{i=1}^d \kappa_{i,j}(t,x)\partial_{x_i}\phi(t,x) + \theta_j(t,x) \ge 0$ *for all $j\in\{1,\cdots,k'\}$, and $(t,x)\in[0,T]\times\mathbb{R}^d$; then*

$$\phi(t,x) \le v(t,x) \qquad \text{for all} \quad (t,x)\in[0,T]\times\mathbb{R}^d. \tag{3.74}$$

Moreover, if we define the nonintervention region D by

$$D = \{(t,x)\in[0,T]\times\mathbb{R}^d; \min_{1\le j\le k'}\sum_{i=1}^d \kappa_{i,j}(t,x)\partial_{x_i}\phi(t,x) + \theta_j(t,x) > 0\} \tag{3.75}$$

and if we assume in addition that:
  (iii) *there exists a deterministic function $\hat\alpha:[0,T]\times\mathbb{R}^d\hookrightarrow\hat\alpha(t,x)\in A$ such that*

$$\partial_t\phi(t,x) + [\mathscr{L}_t^{\hat\alpha}\phi](t,x) + f(t,x,\hat\alpha(t,x)) = 0, \qquad (t,x)\in[0,T]\times\mathbb{R}^d;$$

  (iv) *there exists a process $\hat{\boldsymbol{\xi}}$ such that $(\hat{\boldsymbol{\alpha}},\hat{\boldsymbol{\xi}})\in\mathbb{A}_0(x_0)$ and $(t,X_t^{0,x_0,\hat\alpha,\hat\xi})\in D$ for all $t\in[0,T]$ and*

$$\sum_{j=1}^{k'}\left[\theta_j(t,X_{t-}) + \sum_{i=1}^d \partial_{x_i}\phi(t,X_{t-})\kappa_{i,j}(t,X_{t-})\right]d\hat\xi_t^{c,j} = 0,$$

and $\phi(t_n,X_{t_n}) - \phi(t_n,X_{t_n-}) + \theta(t_n,X_{t_n-})\Delta\hat\xi_{t_n} = 0$, where we denote by $t_1 < t_2 < \cdots$ the jump times of $\hat{\boldsymbol{\xi}}$.
  Then $\phi(t,x) = v(t,x)$ for all $(t,x)\in[0,T]\times\mathbb{R}^d$ and the control $(\hat{\boldsymbol{\alpha}},\hat{\boldsymbol{\xi}})$ is optimal.

Proof. If $(\alpha, \xi) \in \mathbb{A}_t(x)$, since the process $(X_s^{t,x,\alpha,\xi})_{t \leq s \leq T}$ is a semimartingale which we denote simply by X_t for the sake of notation, the generalized Itô formula gives

$$\phi(T, X_T) = \phi(t, X_t) + \int_t^T \partial_t \phi(s, X_s) ds + \int_t^T \partial_x \phi(s, X_s) \sigma(s, X_s, \alpha_s) dW_s$$
$$+ \frac{1}{2} \int_t^T \text{trace}[\sigma(s, X_s, \alpha_s) \sigma(s, X_s, \alpha_s)^\dagger \partial_{xx}^2 \phi(s, X_s)] ds$$
$$+ \int_t^T \partial_x \phi(s, X_s) b(s, X_s, \alpha_s) ds$$
$$+ \int_t^T \partial_x \phi(s, X_s) \kappa(s, X_s) d\xi_s^c + \sum_{t < s \leq T} \Delta \phi(s, X_s).$$

If we denote by $t_1 < t_2 < \cdots$ the jump times of ξ, then we can write

$$\Delta \phi(t_n, X_{t_n}) = \phi(t_n, X_{t_n}) - \phi(t_n, X_{t_n-})$$
$$= \partial_x \phi(t_n, \tilde{X}_n) \cdot \Delta X_{t_n}$$
$$= \sum_{i=1}^d \partial_{x_i} \phi(t_n, \tilde{X}_n) \sum_{j=1}^{k'} \kappa_{i,j}(t_n, X_{t_n}) \Delta \xi_{t_n}^i,$$

where $\tilde{X}_n$ is a point in the interval $[X_{t_n-}, X_{t_n}]$ whose existence is guaranteed by the mean-value theorem. Using the terminal condition $\phi(T, X_T) = g(X_T)$, we can rewrite the first set of equalities as

$$\phi(t, x) = g(X_T) - \int_t^T \left(\partial_t \phi(s, X_s) + b(s, X_s, \alpha_s) \partial_x \right) ds$$
$$- \int_t^T \partial_x \phi(s, X_s) \sigma(s, X_s, \alpha_s) dW_s$$
$$- \int_t^T \partial_x \phi(s, X_s) \kappa(s, X_s) d\xi_s^c - \int_t^T \theta(s, X_s) d\xi_s^c$$
$$- \sum_{t < s \leq T} \Delta \phi(s, X_s) - \sum_{t < s \leq T} \theta(s, X_s) \Delta \xi_s$$
$$+ \int_t^T f(s, X_s, \alpha_s) ds + \int_t^T \theta(s, X_s) d\xi_s^c,$$

and taking expectations on both sides we get

$$\phi(t, x) = \mathbb{E}\left[g(X_T) + \int_t^T f(s, X_s, \alpha_s) ds + \int_t^T \theta(s, X_{s-}) d\xi_s \right]$$
$$+ \mathbb{E}[\text{martingale}]$$
$$\geq 0,$$

which proves the desired inequality. The rest of the proof follows by inspection, checking that all the above inequalities are saturated when $\alpha = \hat{\alpha}$ and $\xi = \hat{\xi}$. □

Remark 3.19. $\hat{\xi}$ is the local time of (t, X_t) on the boundary ∂D of the nonintervention region, reflected back into $\overline{D}$ in the direction given by the rows of $\kappa(t, x)$.

3.5.6 ▪ Merton Problem with Transaction Costs

We revisit the problem of the maximization of the expected utility of consumption, discussed earlier in Subsection 3.5.2. However, we add a new twist—the inclusion of transaction costs. Even though we restrict ourselves to the particular case of proportional transaction costs, this added feature renders the mathematical problem much more challenging. The optimization problem cannot be solved if we restrict the class of control processes to adapted functions of time giving the rates of trading and consumption. A solution can only be found if we extend the class of admissible controls to include the singular controls introduced earlier in Subsection 3.5.5. While we regard this application as a great example of the importance of the theory of singular control problems, the level of technicality and the lengths of the existing proofs are the main reasons for our decision not to provide a complete solution here. So with our apology, the reader interested in complete and detailed proofs is referred to the Notes & Complements section at the end of the chapter for references in which these proofs can be found.

As before, we denote by $\alpha_t \geq 0$ the rate of consumption from the bank account at time t. We also denote by ξ_t^1 (resp., ξ_t^2) the cumulative purchase (resp., sale) of stock up to and including time t. So both $\boldsymbol{\xi}^1 = (\xi_t^1)_{t \geq 0}$ and $\boldsymbol{\xi}^2 = (\xi_t^2)_{t \geq 0}$ are nonnegative adapted processes with nondecreasing sample paths. For the sake of consistency with the notations used in this chapter, we denote by X_t^0 the amount of money in the bank account at time t (denoted by π_t^0 when we discussed the valuation of the European contingent claim in Chapter 2), and by X_t^1 the amount of money invested in the stock at time t (formerly denoted by π_t^1). We assume that the portfolio manager needs to pay transaction costs proportional to the amount of stock he/she buys or sells, and to the amount of money he/she deposits into or withdraws from the bank account. If we denote by $r_1 > 0$ and $r_2 > 0$ these transaction cost rates, the dynamics of the portfolio manager holdings can be written as

$$\begin{cases} dX_t^0 &= (rX_t^0 - \alpha_t)dt - (1+r_1)d\xi_t^1 + (1-r_2)d\xi_t^2, \quad X_{0-}^0 = x_0^0, \\ dX_t^1 &= X_t^1[\mu_t^1 dt + \sigma_t^1 dW_t] + d\xi_t^1 - d\xi_t^2, \quad X_{0-}^1 = x_0^1. \end{cases} \quad (3.76)$$

Finally, we define the solvency region as the set of couples (x^0, x^1) of positions which can be liquidated instantly without running out of funds:

$$S = \{x = (x^0, x^1) \in \mathbb{R}^2; \; x^0 + (1+r_1)x^1 \geq 0, \; x^0 + (1-r_2)x^1 \geq 0\}.$$

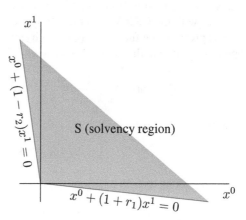

Given starting values x^0 and x^1 for the amounts in the bank account and invested in the stock, the set $\mathbb{A}(x_0^0, x_0^1)$ of admissible controls is then chosen to be the set of couples $(\boldsymbol{\alpha}, \boldsymbol{\xi})$

of adapted processes such that $\alpha_t \geq 0$ and $\xi_t = (\xi_t^1, \xi_t^2)$, where both $\boldsymbol{\xi}^1$ and $\boldsymbol{\xi}^2$ are adapted processes with nondecreasing sample paths starting from 0 at time $t = 0$, and such that the process solving (3.76) satisfies $(X_t^0, X_t^1) \in S$ at all times. Note that this set is not empty if $(x_0^0, x_0^1) \in S$. Indeed, if $(x^0, x^1) \in S$ and $x^0 + (1 + r_1)x^1 = 0$, then necessarily $x^1 \geq 0$, and by choosing $\alpha_t \equiv 0$, $\xi_t^2 \equiv 0$, and $\xi_t^1 \equiv x^1$, we have $X_t = (0,0)$ for all $t > 0$. Similarly, if $(x^0, x^1) \in S$ and $x^0 + (1 - r_2)x^1 = 0$, then by setting $\alpha_t \equiv 0$, $\xi_t^1 \equiv 0$, and $\xi_t^2 \equiv x^0$, we have $X_t = (0,0)$ for all $t > 0$ as before. Now if $(x_0^0, x_0^1) \in \mathring{S}$ the interior of the solvency region, then we choose α_t arbitrary, and, for example, ξ_t^1 and ξ_t^2 zero, and solve (3.76) until the solution (X_t^0, X_t^1) reaches the boundary of S, time at which we start using the constant controls used previously to take the controlled state to $(0,0)$.

For a given discount factor $\beta > 0$, the goal of the infinite horizon problem is to maximize

$$J((x^0, x^1), \boldsymbol{\alpha}, \boldsymbol{\xi}) = J(\boldsymbol{\alpha}, \boldsymbol{\xi}) = \mathbb{E}\left[\int_0^\infty e^{-\beta t} U(\alpha_t) dt\right]$$

for $(\boldsymbol{\alpha}, \boldsymbol{\xi}) \in \mathbb{A}(x_0^0, x_0^1)$. We denote by $v(x^0, x^1)$ the value function giving the value of this maximum. Notice that $\boldsymbol{\alpha} \equiv 0$ is the only admissible consumption plan if $(x_0^0, x_0^1) \in \partial S$. This implies the Dirichlet boundary condition

$$v(x) = 0, \quad x \in \partial S.$$

We recover the setup of Subsection 3.5.5 above if we set

$$X_t = \begin{bmatrix} X_t^0 \\ X_t^1 \end{bmatrix}, \quad b(t, X, \alpha) = \begin{bmatrix} rX^0 - \alpha \\ \mu_t^1 X^1 \end{bmatrix},$$

$$\sigma(t, X, \alpha) = \begin{bmatrix} 0 \\ \sigma_t^1 X^1 \end{bmatrix}, \quad \kappa(t, X) = \begin{bmatrix} -(1 + r_1) & 1 - r_2 \\ 1 & -1 \end{bmatrix}.$$

It was proven by Davis and Norman in [48] (see also [112]) that the value function is concave, smooth, and solves the QVI

$$\min\left[\beta v - \frac{1}{2}\sigma^2(x^1)^2 \partial^2_{x^1 x^1} v + rx^0 \partial_{x^0} v + \mu x^1 - \tilde{U}(\partial_{x^0} v), \partial_{x^1} v + (1 - r_2)\partial_{x^0} v, \partial_{x^1} v - (1 + r_1)\partial_{x^0} v\right]$$
$$= 0$$

in the classical sense. As before, $\tilde{U}$ denotes the Fenchel–Legendre transform of U. Again, we choose to work with the CRRA power utility function. The candidate for the optimal Markov control is given by the function $\hat{\alpha}(y)$ identifying the argument of the minimization of the first operator symbol in the QVI. In other words, for $y^0 > 0$,

$$\hat{\alpha}(y^0) = \arg\sup_{\alpha > 0} \frac{1}{1 - \gamma}\alpha^{1-\gamma} - \alpha y^0 = (y^0)^{-1/\gamma},$$

which gives the Markov control

$$\alpha^*(x) = [\partial_{x^0} v(x)]^{-1/\gamma}.$$

We now identify the nonintervention region

$$D = \{x = (x^0, x^1) \in S; \partial_{x^1} v(x) + (1 - r_2)\partial_{x^0} v(x) > 0$$

and

$$\partial_{x^1} v(x) - (1 + r_1)\partial_{x^0} v(x) > 0\}.$$

3.5. Constraints and Singular Control Problems

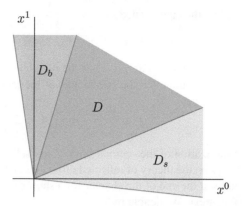

Figure 3.1. *Split of the solvency region into the regions D, D_b, and D_s identified in the text.*

First we observe that for any $\rho > 0$ we have

$$J((\rho x^0, \rho x^1), \rho\boldsymbol{\alpha}, \rho\boldsymbol{\xi}^1, \rho\boldsymbol{\xi}^2) = \rho^{1-\gamma} J((x^0, x^1), \boldsymbol{\alpha}, \boldsymbol{\xi}^1, \boldsymbol{\xi}^2),$$

and this scale invariance of the objective function translates into the scale invariance of the value function

$$v(\rho x^0, \rho x^1) = \rho^{1-\gamma} v(x^0, x^1),$$

which in turn implies that the value function v is of the form

$$v(x) = (x^0 + x^1)^{1-\gamma} w\left(\frac{x^0}{x^0 + x^1}\right),$$

where the function w is defined by $w(z) = v(z, 1-z)$ for $z \in \left(1 - 1/r_2, r_1/(1+r_1)\right)$. The concavity of v and the above scale invariance imply the existence of numbers z_1 and z_2 in the interval $\left[1 - 1/r_2, r_1/(1+r_1)\right]$ such that

$$D = \left\{ x = (x^0, x^1) \in S; z_1 < \frac{x^0}{x^0 + x^1} < z_2 \right\}.$$

In fact, this cone naturally splits the solvency region into three disjoint cones (see Figure 3.1). Indeed, the complement in S of D is the union of two cones D_b and D_s, which can be identified as follows.

$$D_b = \left\{ x = (x^0, x^1) \in S; \frac{x^0}{x^0 + x^1} > z_2 \right\}$$

is characterized by the equivalence

$$\partial_{x^1} v(x) - (1+r_1)\partial_{x^0} v(x) = 0 \iff x \in D_b,$$

which shows that D_b can be interpreted as the region in which the rational investor should borrow money from the bank account. On the other hand

$$D_s = \left\{ x = (x^0, x^1) \in S; z_1 > \frac{x^0}{x^0 + x^1} \right\}$$

is characterized by the equivalence

$$\partial_{x^1} v(x) + (1 - r_2)\partial_{x^0} v(x) = 0 \iff x \in D_s,$$

which shows that D_s can be interpreted as the region in which the rational investor should sell the stock. We assume that the parameters of the model are such that these cones are not empty, namely that

$$1 - \frac{1}{r_2} < z_1 < z_2 < \frac{r_1}{1 + r_1}.$$

The characterization of the optimal buying and selling strategies $\boldsymbol{\xi}^{1*}$ and $\boldsymbol{\xi}^{2*}$ is more involved. It relies on the theory of reflected diffusions and their local times at the boundary. See the appendix at the end of the chapter for a short review of the essential results needed in the present context. We denote by

$$\partial^1 D = \{x \in \partial D; x^0 = z_1(x^0 + x^1)\} \quad \text{and} \quad \partial^2 D = \{x \in \partial D; x^0 = z_2(x^0 + x^1)\}$$

the straight pieces of the boundary ∂D which are disjoint except for the origin.

Let us first assume that the starting point $x = (x^0, x^1)$ is in $D \setminus \{0\}$. A smoothing argument, followed by the control of the limit when the smoothing is removed, allows us to ignore the fact that the boundary ∂D is not smooth everywhere due to the presence of the wedge at the origin. We apply the generalization of the Skorohod lemma, recalled in the appendix at the end of the chapter, to the diffusion process x_t obtained by choosing the optimal consumption $\alpha^*(x)$ identified above and $\xi_t^1 \equiv \xi_t^2 \equiv 0$, and reflecting this process on ∂D in the direction given by the inward-pointing vector

$$\vec{n}(x) = (1 - r_2, -1) \quad \text{when } x \in \partial^1 D \quad \text{and} \quad \vec{n}(x) = (-1, 1 - r_1) \quad \text{when } x \in \partial^2 D,$$

properly normalized to be of length 1. We get the reflected process $X_t^* = y(t)$ with local time on the boundary $|k|(t)$, which we split into two parts in order to get the optimal buying and selling strategies $\boldsymbol{\xi}^{1*}$ and $\boldsymbol{\xi}^{2*}$:

$$\xi_t^{1*} = \int_0^t \mathbf{1}_{\{X_s \in \partial^1 D\}} d|k|(s) \quad \text{and} \quad \xi_t^{2*} = \int_0^t \mathbf{1}_{\{X_s \in \partial^2 D\}} d|k|(s).$$

Using the verification argument to prove that these controls are optimal is straightforward when ∂D is included in the open first quadrant of the (x^0, x^1)-plane because the value function is actually C^2 in this quadrant. A modicum of care is needed to go through the verification argument when the x^0 axis and/or the x^1 axis are contained in D. The reader interested in the details is referred to the exhaustive analysis of Shreve and Soner [112].

If on the other hand, the starting point $x = (x^0, x^1)$ is in $S \setminus D$, then the optimal strategy is to push it *instantaneously* to ∂D and to use the optimal controls constructed above from time 0+ onward. More precisely, if $(x^0, x^1) \in D_b$, then

$$z_2 < \frac{x^0}{x^0 + x^1} < \frac{r_1}{1 + r_1},$$

and we can find $\delta > 0$ such that if we choose $\xi_{0+}^1 = \delta$ and $\xi_{0+}^2 = 0$, then computing X_{0+}^0 and X_{0+}^1 from the dynamics of X_t^0 and X_t^1, we have $X_{0+}^0 = z_2(X_{0+}^0 + X_{0+}^1)$, namely, $X_{0+} = (X_{0+}^0, X_{0+}^1) \in \partial D$. After this initial jump, we can use the optimal control $(\alpha^*, \boldsymbol{\xi}^{1*}, \boldsymbol{\xi}^{2*})$ constructed above when the starting point (x^0, x^1) is in ∂D. Similarly, if $(x^0, x^1) \in D_s$, then

$$1 - \frac{1}{r_2} < \frac{x^0}{x^0 + x^1} < z_1,$$

and we can find $\delta > 0$ such that if we choose $\xi_{0+}^1 = 0$ and $\xi_{0+}^2 = \delta$, then computing X_{0+}^0 and X_{0+}^1 from the dynamics of X_t^0 and X_t^1, we have $X_{0+}^0 = z_1(X_{0+}^0 + X_{0+}^0)$, and consequently $X_{0+} = (X_{0+}^0, X_{0+}^1) \in \partial D$. As before, after this initial jump, we can use the optimal control $(\alpha^*, \xi^{1*}, \xi^{2*})$ constructed above when the starting point (x^0, x^1) is in ∂D.

3.6 • Viscosity Solutions of HJB Equations and QVIs

Motivated by the discussion of singular control problems in Section 3.5, we investigate the solution of parabolic QVIs of the type

$$\min\left[\partial_t v(t,x) + \hat{L}_1(t, x, \partial_x v(t,x), \partial_{xx}^2 v(t,x)),\ G_1(t, x, \partial_x v(t,x), \partial_{xx}^2 v(t,x))\right] = 0$$

for $(t, x) \in [0, T] \times \mathbb{R}^d$ for finite horizon problems, and of elliptic type

$$\min\left[\beta v(x) + \hat{L}_2(x, \partial_x v(x), \partial_{xx}^2 v(x)),\ G_1(x, \partial_x v(x), \partial_{xx}^2 v(x))\right] = 0$$

for $x \in \mathbb{R}^d$ for infinite horizon problems. Here, the functions $\hat{L}_1$ and $\hat{L}_2$ are the minimized operator symbols

$$\hat{L}_1(t, x, y, z) = \inf_{\alpha \in A}\left[b(t, x, \alpha) \cdot y + \frac{1}{2}\text{trace}[\sigma(t, x, \alpha)\sigma(t, x, \alpha)^\dagger z] + f(t, x, \alpha)\right] \quad (3.77)$$

and

$$\hat{L}_2(x, y, z) = \inf_{\alpha \in A}\left[b(t, x, \alpha) \cdot y + \frac{1}{2}\text{trace}[\sigma(t, x, \alpha)\sigma(t, x, \alpha)^\dagger z] + f(x, \alpha)\right] \quad (3.78)$$

whenever these infimums are finite, while G_1 and G_2 are continuous functions such that

$$\hat{L}_1(t, x, y, z) > -\infty \iff G_1(t, x, y, z) \geq 0$$

and

$$\hat{L}_2(x, y, z) > -\infty \iff G_2(x, y, z) \geq 0.$$

In what follows, we include the two cases under the same umbrella of an equation of the form

$$F(x, \partial_x v(x), \partial_{xx}^2 v(x)) = 0, \qquad x \in \mathcal{O}, \quad (3.79)$$

where $\mathcal{O}$ is an open set in $\mathbb{R}^N$. We denote by $\mathcal{S}_N$ the space of symmetric real $N \times N$ matrices. Our first assumption will be to require that the function F is elliptic in the sense that

⋄ **(E)** for every $x \in \mathcal{O}$, $v \in \mathbb{R}$, $y \in \mathbb{R}^N$, and $z, z' \in \mathcal{S}_N$, we have

$$z \leq z' \implies F(x, v, y, z) \geq F(x, v, y, z').$$

The function $\hat{L}_1$ defined in (3.77) satisfies the ellipticity assumption **(E)**. In the case of parabolic equations ($N = 1 + d$ and $\mathcal{O}$ of the form $\mathcal{O} = [0, T) \times \mathcal{O}_d$ for some open set $\mathcal{O}_d \subset \mathbb{R}^d$) we shall assume that the function F is parabolic in the sense that

⋄ **(P)** for every $t \in [0, T)$, $x \in \mathcal{O}_d$, $v \in \mathbb{R}$, $y_t, y_t' \in \mathbb{R}$, $y \in \mathbb{R}^d$, and $z \in \mathcal{S}_d$, we have

$$y_t \leq y_t' \implies F(t, x, v, y_t, y, z) \geq F(t, x, v, y_t', y, z).$$

The function $\hat{L}_2$ defined in (3.78) satisfies the parabolicity assumption **(P)**.

3.6.1 ▪ Generalities on Viscosity Solutions

Definition 3.20. *For any real-valued function v on $\mathcal{O}$ we define its upper semicontinuous envelope v^* by*
$$v^*(x) = \limsup_{x' \to x} v(x')$$
and its lower semicontinuous envelope v_* by
$$v_*(x) = \liminf_{x' \to x} v(x').$$

v^* (resp., v_*) is the smallest upper (resp., largest lower) semicontinuous function greater (resp., smaller) than or equal to v on $\mathcal{O}$. Also, v is upper (resp., lower) semicontinuous on $\mathcal{O}$ if and only if $v = v^*$ (resp., $v = v_*$) on $\mathcal{O}$.

Definition 3.21. *A locally bounded function v on $\mathcal{O}$ is said to be a viscosity subsolution of (3.79) if for every $x \in \mathcal{O}$ and every $\varphi \in C^2(\mathcal{O})$ such that $0 = v^*(x) - \varphi(x) = \max_{\mathcal{O}}(v^* - \varphi)$, it holds that*
$$F(x, \partial_x \varphi(x), \partial^2_{xx}\varphi(x)) \leq 0.$$

One says that v is a viscosity supersolution of (3.79) if for every $x \in \mathcal{O}$ and every $\varphi \in C^2(\mathcal{O})$ such that $0 = v^*(x) - \varphi(x) = \min_{\mathcal{O}}(v^* - \varphi)$, it holds that
$$F(x, \partial_x \varphi(x), \partial^2_{xx}\varphi(x)) \geq 0.$$

Finally, one says that v is a viscosity solution of (3.79) if it is a sub- and a supersolution of (3.79).

Remark 3.22. *Requiring that the minima and maxima appearing in the above definition be strict or even local would not change the notions of sub- and supersolutions in the viscosity sense.*

Remark 3.23. *Notice also that, if v is a sub- (resp., super-) solution of (3.79), then v^* (resp., v_*) is a upper (resp., lower) semicontinuous sub- (resp., super-) solution of (3.79).*

We now check that the notion of the viscosity solution is a generalization of the notion of the classical solution.

Proposition 3.24. *If $v \in C^2(\mathcal{O})$, v is a viscosity sub- (resp., super-) solution of (3.79) if and only if v is a classical sub- (resp., super-) solution of (3.79).*

Proof. We give the proof in the parabolic case only. The elliptic case is similar. Let us assume that $v \in C^2(\mathcal{O})$ is a viscosity supersolution of (3.79). Since v is smooth, we can choose the test function $\varphi = v$, and any $(t, x) \in [0, T) \times \mathcal{O}_d$ is a global minimum of $v - \varphi \equiv 0$. By definition of a viscosity supersolution, we have
$$F(t, x, \varphi(t,x), \partial_t \varphi(t,x), \partial_x \varphi(t,x), \partial^2_{xx}\varphi(t,x)) \geq 0,$$
showing that $v = \varphi$ is a classical supersolution. If we now assume that v is a classical supersolution, if $\varphi \in C^2(\mathcal{O})$ and $(t, x) \in [0, T) \times \mathcal{O}_d$ is a global minimum of $v - \varphi$, then we have
$$\partial_t(v - \varphi)(t, x) \geq 0$$

3.6. Viscosity Solutions of HJB Equations and QVIs

(with equality if $t > 0$), and

$$\partial_x v(t,x) = \partial_x \varphi(t,x), \qquad \partial^2_{xx} v(t,x)) = \partial^2_{xx} \varphi(t,x),$$

from which we conclude that

$$F(t, x, \varphi(t,x), \partial_t \varphi(t,x), \partial_x \varphi(t,x), \partial^2_{xx}\varphi(t,x))$$
$$\geq F(t, x, v(t,x), \partial_t v(t,x), \partial_x v(t,x), \partial^2_{xx} v(t,x))$$
$$\geq 0$$

using the parabolicity assumption, which proves that v is a viscosity supersolution. The equivalence of the notions of classical and viscosity subsolutions for $v \in C^2(\mathcal{O})$ is proven in exactly the same way. □

3.6.2 ▪ Comparison Principles

Definition 3.25. *We say that the comparison principle holds if whenever v is a viscosity subsolution and w is a viscosity supersolution of (3.79) such that $v^* \leq w_*$ on the boundary $\partial \mathcal{O}$, then the same inequality holds on the entire domain $\overline{\mathcal{O}}$.*

In the case of unbounded domains (e.g., $\mathcal{O} = \mathbb{R}^N$ in the elliptic case so that $\partial \mathcal{O} = \emptyset$, and $[0,T) \times \mathbb{R}^d$ in the parabolic case for which $\partial \mathcal{O} = \{T\} \times \mathbb{R}^d$), bounds on the rate of increase at infinity replace the values of the functions at the points of the boundary.

Remark 3.26. *Equivalently, the comparison principle holds if whenever v is a upper semicontinuous viscosity subsolution and w is a lower semicontinuous viscosity supersolution of (3.79) such that $v \leq w$ on the boundary $\partial \mathcal{O}$, then the same inequality holds on the entire domain $\overline{\mathcal{O}}$.*

Here are some examples of functions F for which the comparison principle holds.

- $F(x, v, y, z) = \beta v + H(x,y) + (1/2)\text{trace}[\sigma(x)\sigma(x)^\dagger z]$ for some $\beta > 0$, $\sigma : \mathcal{O} \to \mathbb{R}^{N \times m}$ Lipschitz, and $H : \mathcal{O} \times \mathbb{R}^N \to \mathbb{R}$ satisfying

    (A1) $|H(x,y) - H(x',y)| \leq m(|x - x'|)(1 + |y|)$ for some function m such that $\lim_{x \searrow 0} m(x) = 0$,

    (A2) $\lim_{|y| \to \infty} H(x,y) = +\infty$ uniformly in $x \in \mathcal{O}$.

- $F(x, v, y, z) = H(x,y)$ with $H : \mathcal{O} \times \mathbb{R}^N \to \mathbb{R}$ satisfying (A1) and (A2) above together with

    (A3) $y \hookrightarrow H(x,y)$ is convex for each fixed $x \in \mathcal{O}$;

    (A4) there exist $\varphi \in C^1(\mathcal{O})$ continuous on $\overline{\mathcal{O}}$ and $\delta > 0$ such that $H(x, \partial_x \varphi(x)) \leq -\delta$ for all $x \in \mathcal{O}$.

- If A is compact and $F(x, v, y, z) = \beta v + H_2(x, v, y, z)$ for some $\beta > 0$ and H_2 defined in (3.78) with b, σ, and f as they are there.

- If A is compact and $F(t, x, v, y_t, y, z) = y_t + H_1(t, x, y, z)$ with H_1 defined in (3.77) with b, σ, and f as they are there.

In the last two cases, the comparison principle holds in the class of functions with at most quadratic growth, proving uniqueness of the viscosity solution of the HJB equation in this class of functions. In the general case of finite horizon problems, the terminal condition $v_*(T, \cdot) = v^*(T, \cdot) = g$ needs to have at most quadratic growth.

3.6.3 ▪ Viscosity Solutions of QVIs

As usual, we denote by L the operator symbol

$$L(t,x,y,z,\alpha) = b(t,x,\alpha) \cdot y + \frac{1}{2}\text{trace}[\sigma(t,x,\alpha)\sigma(t,x,\alpha)^\dagger z] + f(t,x,\alpha),$$

and by $\hat{L}$ the corresponding minimized operator symbol

$$\hat{L}(t,x,y,z) = \inf_{\alpha \in A} L(t,x,y,z,\alpha),$$

and we assume that $\hat{L}$ is continuous on the interior of the set of (t,x,y,z) for which $\hat{L}(t,x,y,z) > -\infty$, and that there exists a continuous function G on $[0,T) \times \mathbb{R}^d \times \mathbb{R}^d \times \mathbb{R}^{d \times d}$ satisfying

$$\hat{L}(t,x,y,z) > -\infty \iff G(t,x,y,z) \geq 0.$$

Theorem 3.27. *Under the above assumptions, the value function is a viscosity solution of the HJB QVI*

$$\min\left[\partial_t v(t,x) + \hat{L}(t,x,\partial_x v(t,x), \partial_{xx}^2 v(t,x)), G(t,x,\partial_x v(t,x), \partial_{xx}^2 v(t,x))\right] = 0$$
(3.80)

for $(t,x) \in [0,T) \times \mathbb{R}^d$ in the finite horizon case, and

$$\min\left[\beta v(t,x) + \hat{L}(x,\partial_x v(x), \partial_{xx}^2 v(x)), G(x,\partial_x v(x), \partial_{xx}^2 v(x))\right] = 0 \quad (3.81)$$

for $x \in \mathbb{R}^d$ in the infinite horizon case.

In the finite horizon case, in order to be solved, the parabolic equation (3.80) should be complemented with a terminal condition at time T. The obvious candidate should be given by the fact that by its very definition, the value function satisfies $v(T,x) = g(x)$ for $x \in \mathbb{R}^d$. However, the singularity of the operator symbol can be the source of discontinuities at $t \nearrow T$, and the left limit of $v(t,x)$ as $t \nearrow T$ should be the appropriate terminal condition. This left limit is characterized in the next theorem.

Theorem 3.28. *If we assume that the cost functions f and g have at most linear growth, the function*

$$v_*(T,x) = \liminf_{t \nearrow T} v(t,x)$$

is a viscosity supersolution of the equation

$$\min\left[v_*(T,x) - g(x), G(T,x,\partial_x v_*(T,x), \partial_{xx}^2 v_*(T,x))\right] = 0, \quad x \in \mathbb{R}^d,$$

and furthermore, if g is upper semicontinuous, then the function

$$v^*(T,x) = \limsup_{t \nearrow T} v(t,x)$$

is a viscosity subsolution of the equation

$$\min\left[v^*(T,x) - g(x), G(T,x,\partial_x v^*(T,x), \partial_{xx}^2 v^*(T,x))\right] = 0, \quad x \in \mathbb{R}^d.$$

Moreover, when a comparison principle holds for this type of PDE, any upper semicontinuous subsolution is always less than or equal to any lower semicontinuous supersolution,

3.6. Viscosity Solutions of HJB Equations and QVIs

and in the present situation we would have $v(T-,x) = v^*(T,x) = v_*(T,x)$, which is a viscosity solution of

$$\min\left[v(T-,x) - g(x), G(T,x,\partial_x v(T-,x), \partial^2_{xx} v(T-,x))\right] = 0, \qquad x \in \mathbb{R}^d.$$

In fact, $v(T-,x)$ can be characterized as the smallest viscosity supersolution of this equation which is greater than or equal to g. We shall see several examples where this terminal condition is different from g.

3.6.4 ▪ Uncertain Volatility

We consider the controlled state dynamics

$$dX_t = \alpha_t X_t dW_t \qquad (3.82)$$

in $A = (0, \infty)$, where the set $\mathbb{A}$ of admissible controls is chosen to be the set of processes adapted to the filtration generated by $\boldsymbol{W} = (W_t)_{0 \le t \le T}$, with values in the interval $[\underline{a}, \overline{a}]$ with $0 \le \underline{a} \le \overline{a} \le \infty$. We assume that the running cost function is 0 and we choose a continuous function g with linear growth for the terminal cost, which we interpret as the payoff of a European option. The value function

$$v(t,x) = \sup_{\boldsymbol{\alpha} \in \mathbb{A}} \mathbb{E}[g(X^{t,x}_T)], \qquad (t,x) \in [0,T] \times \mathbb{R}^d,$$

has at most linear growth since $(X^{t,x}_s)_{t \le s \le T}$ is a positive supermartingale. The operator symbol reads

$$L(x, z, \alpha) = \frac{1}{2}\alpha^2 x^2 z.$$

⋄ When $\overline{a} < \infty$, its maximized form is

$$\hat{L}(x,z) = \frac{1}{2}\tilde{a}(z)x^2 z \quad \text{for} \quad \tilde{a}(z) = \begin{cases} \overline{a} & \text{if } z \ge 0, \\ \underline{a} & \text{if } z < 0. \end{cases}$$

It then follows that v is continuous on $[0,T] \times \mathbb{R}^d$, and the unique viscosity solution of

$$\partial_t v(t,x) + \frac{1}{2}\tilde{a}(\partial^2_{xx} v(t,x))x^2 \partial^2_{xx} v(t,x) = 0 \quad \text{for} \quad (t,x) \in [0,T) \times \mathbb{R}^d$$

with linear growth satisfying the terminal condition $v(T,x) = g(x)$ for $x \in \mathbb{R}^d$. When $\underline{a} > 0$, v is in fact a smooth solution in the classical sense because of uniform ellipticity. It is interesting to notice that, when g is convex (resp., concave), $v(t,x) = \mathbb{E}[g(\tilde{X}^{t,x}_T)]$ where the process $(\tilde{X}^{t,x}_s)_{t \le s \le T}$ is the geometric Brownian motion obtained by replacing the uncertain volatility α_t in (3.82) by the constant deterministic volatility $\overline{a}$ (resp., $\underline{a}$).

⋄ When $\overline{a} = \infty$, the maximized operator symbol $\hat{L}$ is given by

$$\hat{L}(x,z) = \begin{cases} \frac{1}{2}\underline{a}^2 x^2 z & \text{if } z \le 0, \\ \infty & \text{if } z > 0, \end{cases}$$

so we can use the regularizing symbol $G(t,x,y,z) = z$, and the value function v is now a viscosity solution of the QVI

$$\min\left[\partial_t v(t,x) + \frac{1}{2}\underline{a}^2 x^2 \partial^2_{xx} v,\ \partial^2_{xx} v\right] = 0, \qquad (t,x) \in [0,T) \times \mathbb{R}^d,$$

with terminal condition $v(T-,x) = \hat{g}(x)$ given by the concave envelope of g, namely, the smallest concave function greater than or equal to g. As before, but this time with significantly more work, one can check that $v(t,x) = \mathbb{E}[\hat{g}(\tilde{X}_T^{t,x})]$, where the process $(\tilde{X}_s^{t,x})_{t \leq s \leq T}$ is the geometric Brownian motion obtained by replacing the uncertain volatility α_t in (3.82) by the constant deterministic volatility $\underline{a}$.

3.6.5 • High Water Marks and Hedge Fund Management

We denote by X_t the value at time $t \geq 0$ of a fund invested in a riskless asset with constant rate of return $r \geq 0$ (for example the bank account P_t^0 used so far), and a risky proprietary technology S_t returning

$$dS_t = S_t[\mu dt + \sigma dW_t], \tag{3.83}$$

where the rate of return $\mu > 0$ and the volatility $\sigma > 0$ will be assumed to be constant in the numerical experiments reported below (for which we shall also assume $\mu > r$). So, from a mathematical point of view, S_t will play a role very similar to the risky stock P_t^1 used earlier. At each time $t \geq 0$, we denote by α_t the proportion of the fund value invested in the risky technology and by π_t the amount of money actually invested, so that $\pi_t = \alpha_t X_t$ as before.

We now describe a stylized, though rather realistic, form of manager compensation structure over a fixed period $[0, T]$. We assume that the fund manager owns the fraction a of the fund, and that on the remaining part $(1-a)$ of the fund, he/she earns a management fee paid at an annual rate b, and an incentive fee paid at the annual rate c on the amount by which the terminal value X_T of the fund exceeds a high water mark H_T. The typical values of b and c used in the industry are $b = 2\%$ and $c = 20\%$. The high water mark is usually computed from the value of a benchmark index defined in the contract. For the sake of definiteness, as a realistic choice for the high water mark, we pick $H_T = e^{rT} H_0$, and we will set the initial value of the high water mark to the value of the fund at the beginning of the management period. Assuming that the entire wealth of the manager is in the fund, his/her wealth at the end of the management period would be

$$\tilde{X}_T = aX_T + (1-a)[bTX_T + c(X_T - H_T)^+]. \tag{3.84}$$

The case of a proprietary trader with no fund ownership corresponds to $a = 0$, while the case of a hedge fund manager with a 10% stake in the fund is captured by setting $a = 0.1$. The major difference from the standard Merton problem discussed earlier is the fact that the terminal wealth $\tilde{X}_T$ is not a concave function of the terminal value of the fund. Figure 3.2 gives the plot of the function

$$\psi(x) = ax + (1-a)[bTx + c(x-h)^+],$$

giving the terminal wealth of the fund manager for typical values of the parameters (left pane) and of its utility together with its concave envelope (right pane). As we can guess from the discussion of the classical Merton portfolio problem, this lack of concavity will raise delicate issues (both at the level of the theory and of the numerical computations), which we address below.

Furthermore, in order to make our hedge fund management model more realistic, we consider the possibility of liquidation of the fund in the case of poor performance. Again, as in the case of the computation of the high water mark, in practice, the liquidation of the fund is decided after comparison with a performance benchmark. For the sake of definiteness, we shall consider a function $[0,T] \ni t \hookrightarrow \zeta(t)$ and monitor the time

$$\tau = \inf\{t > 0;\ X_t \leq \zeta(t)\} \wedge T, \tag{3.85}$$

3.7. Impulse Control Problems

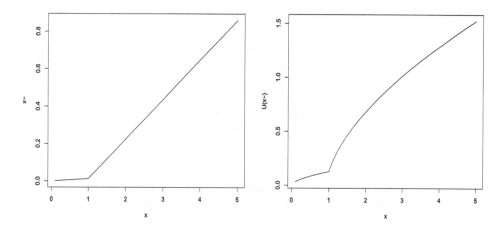

Figure 3.2. *Plot of the fund manager terminal wealth as a function of the terminal value of the fund (left pane), and its utility and concave envelope (right pane).*

giving the first time the value of the fund goes below the liquidation barrier ψ if this occurs during the management period $[0, T]$, or T otherwise. In the numerical experiments reported below we use

$$\zeta(t) = \alpha X_0 e^{rT} \tag{3.86}$$

for some positive number $\alpha < 1$, typically $\alpha = 0.5$. In the case of premature liquidation, the manager keeps his share aX_τ of the fund, and gets the management fees accumulated up until the time of liquidation. Because of our definition of the stopping time τ, in all cases, the terminal wealth of the manager is given by the formula

$$\tilde{X}_T = aX_{\tau \wedge T} + (1-a)[b(\tau \wedge T)X_{\tau \wedge T} + c(X_{\tau \wedge T} - H_T)^+]. \tag{3.87}$$

As in the classical Merton problem, we assume that the manager seeks to maximize his/her expected utility of terminal wealth, and as before, we will limit our discussion to the case of the CRRA power utility function (3.61).

Remark 3.29. *The case $a = 1$ corresponds to a* standard *investor without the management and incentive fees, and in the absence of a liquidation barrier, the current problem reduces to the classical Merton problem discussed earlier in Subsection 3.5.2.*

3.7 ▪ Impulse Control Problems

In this section, the actions of the controller will create jumps in the state trajectories. The elements α of the set $\mathbb{A}$ of admissible controls will be random point measures on $[0, \infty) \times E$. To be more specific, an admissible control α will be given by a sequence $\alpha = (\tau_n, \zeta_n)_{n \geq 0}$, where for each $n \geq 0$, τ_n is an $\mathbb{F}$-stopping time and ζ_n is an $\mathcal{F}_{\tau_n}$-random variable with values in a measurable space $(E, \mathcal{E})$. For the sake of definiteness, we shall restrict ourselves to the case $E = \mathbb{R}^k$. The τ_n's, which form a nondecreasing sequence of stopping times, give the times at which actions—which we shall call impulses—are taken by the controller, and the ζ_n's represent the nature/size of the actions taken. It will be convenient to view the control α as the random point measure $\nu = \sum_{n \geq 0} \delta_{(\tau_n, \zeta_n)}$ on $[0, \infty) \times E$. These actions impact the dynamics of the state of the system through a continuous function

$$\Gamma : \mathbb{R}^d \times E \ni (x, \zeta) \hookrightarrow \Gamma(x, \zeta) \in \mathbb{R}^d,$$

which gives the size of the jump of the state when the controller acts on the system. To be more specific, at each time τ_n, the state of the system jumps from X_{τ_n-} to $X_{\tau_n} = \Gamma(X_{\tau_n-}, \zeta_n)$.

We shall assume that, in addition to the running costs given as usual by a function f, which for the sake of simplicity we assume to be independent of the control α, the costs of the impulses are given by a continuous function

$$c : \mathbb{R}^d \times E \ni (x, \zeta) \hookrightarrow \Gamma(x, \zeta) \in \mathbb{R}.$$

We give the definition of the objective function in the infinite horizon case as well as in the finite horizon case, for which we introduce a terminal cost function g.

Finite Horizon T

The goal is to minimize the objective function

$$J(\alpha) = \mathbb{E}\left[\int_0^T f(t, X_t) ds + g(X_T) - \int\int_{(0,T] \times E} c(X_{t-}, \zeta) \nu(dt, d\zeta)\right]$$

so for each $(t, x) \in [0, T] \times \mathbb{R}^d$ we define the value function

$$v(t, x) = \inf_{\alpha \in \mathbb{A}} \mathbb{E}\left[\int_t^T f(s, X_s^{t,x}) ds + g(X_T^{t,x}) - \int\int_{(t,T] \times E} c(X_{s-}^{t,x}, \zeta) \nu(ds, d\zeta)\right],$$

where we used the notation $\mathbf{X}^{t,x} = (X_s^{t,x})_{t \leq s \leq T}$ for the controlled state process starting from x at time t, namely, the process satisfying

$$X_s^{t,x} = x + \int_t^s b(u, X_u^{t,x}) du + \int_t^s \sigma(u, X_u^{t,x}) dW_u + \sum_{t < \tau_i \leq s} c(X_{\tau_i-}^{t,x}, \zeta_i), \quad (3.88)$$

where we used the definition of the point measure ν to rewrite the double integral with respect to ν as a sum:

$$\int\int_{(t,T] \times E} c(X_{s-}^{t,x}, \zeta) \nu(ds, d\zeta) = \sum_{t < \tau_n \leq T} c(X_{\tau_n-}^{t,x}, \zeta_n).$$

Remark 3.30. *Under generic Lipschitz assumptions on the coefficients, the classical Picard fixed-point argument can be used to prove existence and uniqueness of the solution $\mathbf{X}^{t,x} = (X_s^{t,x})_{t \leq s \leq T}$ of the integral equation (3.88).*

Infinite Horizon

In this case, we assume that the terminal cost g is zero, and the running cost function f as well as the coefficients b and σ of the state dynamics do not depend upon t. The goal is to minimize the objective function

$$J(\alpha) = \mathbb{E}\left[\int_0^\infty e^{-\beta t} f(X_t) ds - \int\int_{(0,\infty) \times E} e^{-\beta t} c(X_{t-}, \zeta) \nu(dt, d\zeta)\right],$$

3.7. Impulse Control Problems

where $\beta > 0$ is a discount factor. For each $x \in \mathbb{R}^d$, the value $v(x)$ of the problem is defined by

$$v(x) = \inf_{\alpha \in \mathcal{A}} \mathbb{E}\left[\int_0^\infty e^{-\beta t} f(X_t^x) ds - \int\int_{(0,\infty) \times E} e^{-\beta t} c(X_{t-}^x, \zeta) \nu(dt, d\zeta)\right],$$

where we used the notation $\mathbf{X}^x = (X_t^x)_{t \geq 0}$ for the controlled state process starting from x at time 0. As before, the integral with respect to the random point measure can be written as a sum:

$$\int\int_{(0,\infty) \times E} e^{-\beta t} c(X_{t-}, \zeta) \nu(dt, d\zeta) = \sum_{n \geq 0} e^{-\beta \tau_n} c(X_{\tau_n-}^x, \zeta_n).$$

3.7.1 • Dynamic Programming Principle

In the case of finite horizon T, the dynamic programming principle (DPP) states that for any stopping time $\tau \in \mathcal{T}_{t,T}$, we have

$$v(t,x) = \inf_{\alpha \in \mathcal{A}} \mathbb{E}\left[\int_t^\tau f(s, X_s^{t,x}) ds) - \int\int_{(t,\tau] \times E} c(X_{s-}^{t,x}, \zeta) \nu(ds, d\zeta) + v(\tau, X_\tau^{t,x})\right]. \tag{3.89}$$

A similar statement holds in the infinite horizon case.

3.7.2 • Quasi-Variational Inequality (QVI)

Recalling our discussion of the impact of the admissibility of singular controls and the appearance of QVIs as a result, the fact that the HJB equation for impulse control problems takes the form of a QVI should not come as a surprise. To understand the form of the relevant QVIs, we appeal to our earlier discussion of BSDEs with barriers. When $x \hookrightarrow \phi(x)$ is a smooth function playing the role of an obstacle, the solution of the QVI

$$\min\left[\partial_t v + \mathcal{L}v + f, v - \phi\right] = 0, \qquad v(T, \cdot) = g,$$

is related to the solution of an optimal stopping problem (and the pricing of an American option) as its solution is given by

$$v(t, X_t) = \operatorname*{ess\,sup}_{\tau \in \mathcal{T}_{t,T}} \mathbb{E}\left[\int_t^T f(X_s) ds + \phi(X_\tau) \mathbf{1}_{\{\tau < T\}} + g(X_T) \mathbf{1}_{\{\tau = T\}} \Big| \mathcal{F}_t\right],$$

and as we proved in Section 2.3, the process $Y_t = v(t, X_t)$ (which can be identified as a Snell envelope) is given by the solution $(\mathbf{Y}, \mathbf{Z}, \mathbf{K})$ of the reflected BSDE

$$Y_t = g(X_T) + \int_t^T f(X_s) ds - \int_t^T Z_s dW_s + K_T - K_t, \qquad Y_t \geq \phi(X_t),$$

and $\mathbf{Y}$ is minimal in the sense that $\int_0^T (Y_t - \phi(X_t)) dK_t = 0$.

Impulse control problems offer a natural generalization of these QVIs, where the unknown function v appears in the definition of the obstacle ϕ via a nonlocal operator. We discuss the finite and infinite horizon cases in parallel.

- In the finite horizon case, the dynamic programming equation (3.89) takes the infinitesimal form

$$\min\left[\partial_t v + \mathcal{L}v + f, \mathcal{G}v - v\right] = 0 \quad \text{on } [0,T) \times \mathbb{R}^d,$$

where $\mathcal{G}$ is the nonlocal operator defined for any function φ by

$$[\mathcal{G}\varphi](x) = \inf_{e \in E}[\varphi(\Gamma(x,e)) - c(x,e)].$$

- As we saw in the case of singular control problems, identifying the right terminal condition can be a tricky business. In the present situation it reads

$$\min\left[v - g, v - \mathcal{G}v\right] = 0 \quad \text{on } \{T\} \times \mathbb{R}^d.$$

This terminal condition says that the terminal value function $v(T, \cdot)$ is the smallest function above g satisfying the impulse condition $v(T,x) \geq [\mathcal{G}v(T,\cdot)](x)$. This last quantity is often written as $\mathcal{G}v(T,x)$ for notational convenience. In other words, when $t \nearrow T$, the controller may decide to do nothing, or to make an intervention by choosing the best decision possible.

- Continuation (or nonintervention) region:

$$\mathcal{C} = \{(t,x) \in [0,T) \times \mathbb{R}^d; \ v(t,x) < \mathcal{G}v(t,x)\}.$$

- Impulse region:

$$\mathcal{D} = \{(t,x) \in [0,T) \times \mathbb{R}^d; \ v(t,x) = \mathcal{G}v(t,x)\}.$$

3.7.3 ▪ Standard Approach to QVIs

When the definition of the obstacle contains the unknown function as above, a standard approach is to decouple this dependence through an iterative procedure which reduces the problem (at least when the approximation procedure converges), to a sequence of simpler problems of the *optimal stopping* type. To be more specific, assuming that the approximations $v_0, v_1, \cdots, v_n$ have been constructed, one can construct v_{n+1} by solving the simpler QVI (of the optimal stopping type) where the obstacle function is independent of the unknown function

$$\min\left[\partial_t v_{n+1} + \mathcal{L}v_{n+1} + f, \mathcal{G}v_n - v_{n+1}\right] = 0 \quad \text{on } [0,T) \times \mathbb{R}^d, \quad v_{n+1}(T, \cdot) = g.$$

While theoretically appealing, this method has its shortcomings, not the least of which is the fact that the function v_n needs to be known/computed on the entire domain before one can solve for v_{n+1}.

3.7.4 ▪ Optimal Impulse Control

If we assume the existence of a function $\hat{e}$ which satisfies

$$\hat{e}(t,x) \in \arg\inf_{e \in E}[v(t, \Gamma(x,e)) - c(x,e)],$$

then given a fixed initial condition $(t, x) \in [0, T] \times \mathbb{R}^d$, an optimal control is given by the impulse control $\hat{\alpha} = (\hat{\tau}_n, \hat{\zeta}_n)_{n \geq 0}$ defined by

$$\hat{\tau}_1 = \inf\{s \geq t;\ (s, \hat{X}_s) \in \mathcal{D}\}, \qquad \hat{\zeta}_1 = \hat{e}(\hat{\tau}_1, \hat{X}_{\tau_1 -}),$$
$$\cdots \qquad\qquad\qquad\qquad \cdots$$
$$\hat{\tau}_{n+1} = \inf\{s \geq \hat{\tau}_n;\ (s, \hat{X}_s) \in \mathcal{D}\}, \qquad \hat{\zeta}_{n+1} = \hat{e}(\hat{\tau}_{n+1}, \hat{X}_{\tau_{n+1}-}),$$
$$\cdots \qquad\qquad\qquad\qquad \cdots$$

where $\hat{\mathbf{X}}$ is the state process starting from x at time t, controlled by $\hat{\alpha}$.

3.7.5 • Optimal Switching

We say that we are dealing with an optimal switching control problem when the set E is finite, say $E = \{1, \cdots, m\}$, whose elements are called the regimes of the system, and the state X_t has two components; one is a pure jump component I_t satisfying

$$I_t = \zeta_0 \mathbf{1}_{\{0 \leq t < \tau_0\}} + \sum_{n \geq 0} \zeta_n \mathbf{1}_{\{\tau_n \leq t < \tau_{n+1}\}},$$

and the other component is a diffusion process $\tilde{\mathbf{X}}$ with m regimes, with dynamics given by drift and volatility coefficients $(b_i)_{1 \leq i \leq m}$ and $(\sigma_i)_{1 \leq i \leq m}$ in such a way that

$$d\tilde{X}_t = b_{I_t}(\tilde{X}_t)dt + \sigma_{I_t}(\tilde{X}_t)dW_t.$$

In other words, $\tilde{X}_t$ has drift b_i and volatility σ_i on any time interval $[\tau_n, \tau_{n+1})$ on which $\zeta_n = i$.

We discuss the case of infinite horizon, the case of finite horizon being treated similarly. The value function is a function of the state $\tilde{x}$ and the regime i, so we write it as a set of m value functions v_i. Namely,

$$v_i(\tilde{x}) = \inf_{\alpha \in \mathbb{A}} \mathbb{E}\left[\int_0^\infty e^{-\beta t} f(\tilde{X}_t^{\tilde{x},i}, I_t^i) dt - \sum_{n \geq 0} e^{-\beta \tau_n} c(\tilde{X}_{\tau_n -}^{x,i}, \zeta_{n-1}, \zeta_n)\right],$$

where $\tilde{X}_t^{\tilde{x},i}$ denotes the state at time t if it started from x in regime i at time $t = 0$. We also assume nondegenerescence of the switching costs

$$\sup_{x \in \mathbb{R}^d} \left[c(x, i, j) + c(x, j, k) - c(x, i, k)\right] < 0, \qquad j \neq i, k.$$

We also implicitly assume that $c(x, i, i) = 0$. The dynamic programming QVI becomes a system of QVIs

$$\min\left[-\beta v_i + \mathcal{L}_i v_i + f_i,\ \min_{j \neq i}[v_j - c(\cdot, i, j)] - v_i\right] = 0, \qquad x \in \mathbb{R}^d,\ 1 \leq i \leq m, \quad (3.90)$$

where $\mathcal{L}_i$ is the infinitesimal generator of the diffusion in regime i, namely,

$$[\mathcal{L}_i v](x) = b_i(x) \cdot \partial_x v(x) + \frac{1}{2} \text{trace}[\sigma_i(x) \sigma_i(x)^\dagger \partial_{xx}^2 v(x)].$$

If the coefficients b_i, σ_i, f_i, and $c(\cdot, i, j)$ satisfy Lipschitz conditions, then the value functions $(v_i)_{1 \leq i \leq m}$ are the unique continuous viscosity solutions of the system (3.90) with linear growth. Analogously, we define the switching regions and the continuation regions for each regime i

$$\mathcal{S}_i = \{x \in \mathbb{R}^d;\ v_i(x) = \min_{j \neq i}[v_j(x) - c(x,i,j)]\}$$

and its complement

$$\mathcal{C}_i = \{x \in \mathbb{R}^d;\ v_i(x) > \min_{j \neq i}[v_j(x) - c(x,i,j)]\}.$$

Remark 3.31. *In the one-dimensional case $d = 1$, if $\sigma_i > 0$ and the switching costs $C(\cdot, i, j)$ are C^1, then the value functions v_i satisfy the so-called smooth fit condition in the sense that they are continuously differentiable along the boundary of the switching region.*

3.8 ▪ Ergodic Control

In this section, we restrict ourselves to the control of autonomous diffusions (in the sense that the coefficients are time-independent) and we consider the large time behavior of the controlled system. More precisely, we consider the controlled diffusion process **X** given by

$$dX_t = b(X_t, \alpha_t)dt + \sigma(X_t, \alpha_t)dW_t$$

and for each admissible control $\boldsymbol{\alpha} \in \mathbb{A}$ and finite horizon $T > 0$, we define the cost

$$J_T(\boldsymbol{\alpha}) = \mathbb{E}\left[\int_0^T f(X_t, \alpha_t)dt\right]$$

(obviously we assume zero terminal cost, i.e., $g = 0$) and we investigate its large time behavior. Clearly, several questions arise. If we suspect that the controlled process is ergodic, $J_T(\boldsymbol{\alpha})$ is expected to be asymptotically proportional to T, the constant of proportionality being equal to the integral of the running cost function with respect to the invariant measure of the process. This suggests controlling and optimizing over $\boldsymbol{\alpha}$ the quantity

$$\limsup_{T \to \infty} \frac{1}{T}\mathbb{E}\left[\int_0^T f(X_t, \alpha_t)dt\right].$$

Of a similar nature is the analysis of the large time behavior of the quantity J_T defined by

$$J_T = \inf_{\boldsymbol{\alpha} \in \mathbb{A}} J_T(\boldsymbol{\alpha}).$$

These problems are clearly related to the infinite horizon problem described in Section 3.4 and for which we recall some notation. For each discount factor $\beta > 0$ and each admissible control $\boldsymbol{\alpha}$, we defined the cost

$$J^{(\beta)}(\boldsymbol{\alpha}) = \mathbb{E}\left[\int_0^\infty e^{-\beta t}f(X_t, \alpha_t)dt\right],$$

and with this notation at hand, the goal would be to relate the properties of the large T behavior of J_T to the small β behavior of the quantity

$$J^{(\beta)} = \inf_{\boldsymbol{\alpha} \in \mathbb{A}} J^{(\beta)}(\boldsymbol{\alpha}).$$

3.8.1 • Examples

The following examples should shed some light on the possible forms of relationship between the two asymptotic behaviors of value functions.

Example 1

We first consider the simple LQ model

$$dX_t = \alpha_t dt + dW_t, \qquad J_T(\boldsymbol{\alpha}) = \mathbb{E}\left[\int_0^T \frac{1}{2}(X_t^2 + \alpha_t^2)dt\right]. \qquad (3.91)$$

This optimal control problem can be solved explicitly using the stochastic maximum approach presented in Chapter 4. It is easy to see that the (reduced) Hamiltonian is minimized for $\hat{\alpha} = -y$. Consequently, the dynamics of the optimally controlled state are given by the solution of the forward SDE of the FBSDE system

$$\begin{cases} dX_t = -Y_t dt + dW_t, \\ dY_t = -X_t dt + Z_t dW_t, \qquad Y_T = 0, \end{cases} \qquad (3.92)$$

where the second equation is the so-called adjoint equation introduced and studied in Chapter 4. This FBSDE is affine and can be solved with the approach presented in Section 2.6. One finds that the adjoint process giving the negative of the optimal control is equal to

$$Y_t = \frac{\sinh(T-t)}{\cosh(T-t)},$$

and the dynamics of the optimally controlled state read

$$dX_t = -\frac{\sinh(T-t)}{\cosh(T-t)} X_t dt + dW_t.$$

Notice that in the limit $T \to \infty$, these dynamics look like $dX_t = -X_t dt + dW_t$ representing an ergodic Ornstein–Uhlenbeck process. In any case,

$$\begin{aligned} J_T &= \inf_{\boldsymbol{\alpha} \in \mathbb{A}} J_T(\boldsymbol{\alpha}) \\ &= \mathbb{E}\left[\int_0^T \left(1 + \frac{\sinh^2(T-t)}{\cosh^2(T-t)}\right) X_t^2 dt\right] \\ &= \int_0^T \left(1 + \frac{\sinh^2(T-t)}{\cosh^2(T-t)}\right) \mathbb{E}[X_t^2] dt \end{aligned}$$

can be computed explicitly since X_t is Gaussian. This leads to the existence of a positive constant $\overline{\lambda} > 0$ such that

$$\lim_{T \to \infty} \frac{1}{T} J_T = \overline{\lambda}$$

independently of the initial condition X_0. This is an instance of ergodic control.

Example 2

We consider the same dynamical equation (3.21) for the controlled state, but we now consider the minimization of the objective function

$$J_T(\boldsymbol{\alpha}) = \mathbb{E}\left[\int_0^T \frac{1}{2}(X_t^2 + \alpha_t^2)dt\right]. \qquad (3.93)$$

As above, $\hat{\alpha} = -y$ minimizes the reduced Hamiltonian. The equation for the optimally controlled state dynamics of (3.92) remains the same, while the backward component becomes
$$dY_t = -dt + Z_t dW_t, \qquad Y_T = 0,$$
which can be easily solved and gives $Y_t = T - t$. Consequently, the dynamics of the optimally controlled state are given by $dX_t = -(T-t)dt + dW_t$. This state process is definitely not ergodic. The optimal cost
$$J_T = \mathbb{E}\left[\int_0^T \left(X_t + \frac{1}{2}(T-t)^2\right) dt\right]$$
does not scale with T, even asymptotically, and remembers the value of the initial condition X_0. So a minor change in the running cost (replacing X_t^2 by X_t) can dramatically change the asymptotic behavior of the solution of the problem. The loss of convexity is the major reason why the problem suddenly *left* the class of ergodic control problems.

Example 3

For this third example, we keep the objective function of (3.93) and slightly change the forward dynamics of (3.92) by assuming
$$dX_t = -X_t dt + \alpha_t dt + dW_t.$$
Again, $\hat{\alpha} = -y$ minimizes the reduced Hamiltonian and the FBSDE solving the problem is now
$$\begin{cases} dX_t = -(X_t + Y_t)dt + dW_t, \\ dY_t = (Y_t - 1)dt + Z_t dW_t, \qquad Y_T = 0. \end{cases} \tag{3.94}$$
The solution of the backward component is $Y_t = 1 - e^{t-T}$, and the dynamics of the optimally controlled state are given by the equation
$$dX_t = [-X_t + e^{t-T} - 1]dt + \alpha_t dt + dW_t,$$
which is mean-reverting and ergodic.

Concluding Remarks

This introductory subsection was devoted to the discussion of three simple examples, showing that the ergodicity of an optimally controlled process can come from the convexity of the Hamiltonian (and in particular of the running cost function) with respect to the state variable x, but also that this ergodicity may come from a strong ergodicity already present in the dynamics of the controlled states even before the optimization is performed.

3.8.2 ▪ Further Discussion of Ergodic Control Problems

Trying to cover the full spectrum of asymptotic behaviors of optimally controlled systems would take us far beyond the scope of this book, so for the sake of simplicity, we limit ourselves to an informal discussion of some of the issues raised by the above examples. For the purpose of this discussion, we restrict ourselves to dynamics of the form
$$dX_t = [b(X_t) + \alpha_t]dt + dW_t \tag{3.95}$$

3.8. Ergodic Control

for a function b satisfying

$$[b(x) - b(x')] \cdot (x - x') \leq -c|x - x'|^2$$

for a positive constant c and for all x and x'. This hypothesis guarantees that the uncontrolled dynamics given by

$$dX_t = b(X_t)dt + dW_t \tag{3.96}$$

are ergodic. Indeed, if X_t^x and $X_t^{x'}$ denote the solutions starting from x and x' at time $t = 0$, then

$$X_t^x - X_t^{x'} = x - x' + \int_0^t [b(X_s^x) - b(X_s^{x'})]ds$$

so that

$$d(X_t^x - X_t^{x'})^2 = 2[b(X_t^x) - b(X_t^{x'})] \cdot (X_t^x - X_t^{x'})dt \leq -2c(X_t^x - X_t^{x'})^2 dt,$$

which implies that

$$(X_t^x - X_t^{x'})^2 \leq (x - x')^2 e^{-2ct},$$

which shows that the trajectories rapidly forget their initial conditions. If we denote by v_T the HJB value function for the control problem with finite horizon T, the HJB equation reads

$$\partial_t v_T(t, x) + \frac{1}{2}\Delta v_T(t, x) + b(x)\partial_x v_T(t, x) + \inf_\alpha [\alpha \partial_x v_T(t, x) + f(x, \alpha)] = 0$$

with terminal condition $v_T(T, x) = 0$. We analyze the large T behavior of v_T. Based on the intuition developed through the discussion of the three particular cases considered in the previous section, we expect that the dependence upon time will disappear in this asymptotic regime, and the solution of the above HJB equation will become asymptotic to a function of x only, say $v(x)$, which should satisfy an equation of the form

$$\overline{\lambda} + \frac{1}{2}\Delta v(x) + b(x)\partial_x v(x) + \inf_\alpha [\alpha \partial_x v(x) + f(x, \alpha)] = 0 \tag{3.97}$$

for a constant $\overline{\lambda}$ to be determined. To determine $\overline{\lambda}$ and to understand its role, we assume existence and uniqueness of a minimizer, say $\hat{\alpha}(x, y)$, of the function $\alpha \hookrightarrow \alpha y + f(x, \alpha)$. Then, the solution $\tilde{\mathbf{X}} = (\tilde{X}_t)_{t \geq 0}$ of

$$d\tilde{X}_t = [b(\tilde{X}_t) + \hat{\alpha}(\tilde{X}_t, \partial_x v(\tilde{X}_t))]dt + dW_t$$

is still ergodic, and by the law of large numbers for ergodic processes we get

$$\overline{\lambda} = \lim_{T \to \infty} \frac{1}{T}\mathbb{E}\left[\int_0^T f(\tilde{X}_t, \hat{\alpha}(\tilde{X}_t, \partial_x v(\tilde{X}_t)))dt\right].$$

We can check this claim by a formal verification argument. Indeed, assuming that v is a C^2 with derivatives growing at most linearly, then we can apply Itô's formula to $v(X_t)$, where X_t is the solution of (3.95) for a generic control α, and get

$$dv(X_t) = \left[\frac{1}{2}\Delta v(X_t) + b(X_t)\partial_x v(X_t) + \alpha_t \partial_x v(X_t)\right]dt + \partial_x v(X_t)dW_t$$

$$= \left[\overline{\lambda} - \inf_\alpha [\alpha \partial_x v(x) + f(x, \alpha)]|_{x=X_t} + \alpha_t \partial_x v(X_t)\right]dt + \partial_x v(X_t)dW_t$$

so that

$$\mathbb{E}\left[v(X_T) + \int_0^T f(X_t, \alpha_t)dt\right] \geq \mathbb{E}[v(X_0) + T\overline{\lambda}] = \mathbb{E}[v(X_0)] + T\overline{\lambda},$$

and consequently,

$$\frac{1}{T}\mathbb{E}\left[\int_0^T f(X_t, \alpha_t)dt\right] \geq \frac{1}{T}\mathbb{E}[v(X_0) - v(X_T)] + \overline{\lambda},$$

from which we conclude by proving first that v has linear growth, and then that $\sup_T \mathbb{E}[|X_T|] < \infty$. The latter is easily proved when, for example, the space of controls is bounded. In fact, the same verification argument shows that for any $\beta > 0$, any classical solution of the HJB equation

$$-\beta v^\beta(x) + \frac{1}{2}\Delta v^\beta(x) + b(x)\nabla v^\beta(x) + \inf_\alpha[\alpha \nabla v^\beta(x) + f(x,\alpha)] = 0$$

is the value function of the infinite horizon problem.

Appendix: Skorohod Lemma and Reflected Diffusions

Because of their importance in the interpretation of the solution of singular control problems, we state without proof several real analysis results which had (and still have) a significant impact on the theory of reflected diffusion processes. The first result is the original of the series.

Given a continuous function $x : [0,\infty) \to \mathbb{R}$ such as $x(0) = x_0 > 0$, there exists a unique couple of continuous functions y and k from $[0,\infty)$ into $[0,\infty)$ satisfying
(1) $y(t) = x(t) + k(t)$ and $y(t) \geq 0$ for all > 0;
(2) $k(0) = 0$ and $t \hookrightarrow k(t)$ is nondecreasing;
(3) $k(t)$ increases only when $y(t) = 0$ and $\int y(t)\, dk(t) = 0$.
The function k is given by

$$k(t) = \sup_{0 \leq s \leq t} x(s)^-.$$

This result is known as Skorohod's lemma. See for example [87]. What is truly remarkable about this result is the fact that it provides an explicit formula for k, and that if one applies the above construction to each individual sample path of a linear Brownian motion starting from $x > 0$, the corresponding set of continuous functions would provide realizations of the process of Brownian motion reflected at the origin, with $k(t)$ being its local time. This result led to many generalizations, the most notable being presumably the result of Lions and Sznitman [87], constructing the normal reflection of a diffusion process at the boundary of a smooth domain. The deterministic extension of Skorohod's lemma they used is the following: given an open bounded set in $\mathbb{R}^d$ with smooth boundary ∂D, and given a continuous function $[0,\infty) \ni t \hookrightarrow x(t) \in \mathbb{R}^d$ such that $x(0) \in D$, there exists a unique couple of continuous functions y and k from $[0,\infty)$ into $\mathbb{R}^d$ satisfying
(1) $y(t) = x(t) + k(t)$, $y(t) \in \overline{D}$ for all $t > 0$;
(2) $k(0) = 0$ and $t \hookrightarrow k(t)$ has bounded variations on each time interval $[0,T]$;
(3) the total variation $|k|$ increases only when $y(t) \in \partial D$, namely,

$$|k|(t) = \int_0^t \mathbf{1}_{y(s) \in \partial D}\, d|k|(s) \quad \text{and} \quad k(t) = \int_0^t \vec{n}(y(s))\, d|k|(s),$$

where $\vec{n}(y)$ is an inward-pointing unit vector at $y \in \partial D$ varying smoothly with y, and $|k|$ denotes the total variation of k.

When applied to the sample paths of a diffusion process in $\mathbb{R}^d$ starting from a point inside the domain D, this result provides a path by path construction of the diffusion process with normal reflection at the boundary.

For a more recent account of the versatility of the Skorohod lemma, the interested reader is referred to the work [25] of Burdzy, Kang, and Ramanan.

Notes & Complements

Typical proofs of the dynamic programming principle (DPP) involve highly technical measure-theoretical arguments and measurable selection theorems. Theorem 3.14 given in the text can be viewed as a simple corollary of a result of Touzi [114], who gives a weak form of the dynamic programming which can be proven directly without having to appeal to measurable selection theorems. He defines the upper and lower envelopes

$$v_*(t,x) = \liminf_{(t',x') \to (t,x)} v(t',x') \quad \text{and} \quad v^*(t,x) = \limsup_{(t',x') \to (t,x)} v(t',x'),$$

which we introduced to identify the terminal condition of singular HJB equations solved in the viscosity sense, and to show that if v is locally bounded, then for any $(t,x) \in [0,T] \times \mathbb{R}^d$ and any family $\{\tau^\alpha; \alpha \in \mathbb{A}_t\}$ of finite stopping times independent of $\mathcal{F}_t$ for which $(X_s^{t,x,\alpha} \mathbf{1}_{[t,\tau^\alpha]}(s))_{t \leq s \leq T}$ is L^∞ bounded for all $\alpha \in \mathbb{A}_t$

$$v(t,x) \leq \inf_{\alpha \in \mathbb{A}_t} \mathbb{E}\left[\int_t^{\tau^\alpha} f(s, X_s^{t,x,\alpha}, \alpha_s)ds + v^*(\tau^\alpha, X_{\tau^\alpha}^{t,x,\alpha})\right],$$

$$v(t,x) \geq \inf_{\alpha \in \mathbb{A}_t} \mathbb{E}\left[\int_t^{\tau^\alpha} f(s, X_s^{t,x,\alpha}, \alpha_s)ds + v_*(\tau^\alpha, X_{\tau^\alpha}^{t,x,\alpha})\right].$$

Indeed, when v is continuous, $v(t,x) = v_*(t,x) = v^*(t,x)$, and the strong form of the dynamic programming principle given in Theorem 3.14 follows from the above result.

Our terse discussion of risk sensitive control is merely intended as an introduction to the terminology. The reader interested in finding out more about the subject is referred to the paper [52] by El Karoui and Hamadène for a presentation in the spirit of this chapter. Bensoussan, Frehse, and Nagai's paper [18] and the book by Bertsekas [20] can also be consulted for further references on the subject.

For the control problems considered in this chapter, the HJB equation is a highly nonlinear partial differential equation (PDE), and conditions for the existence of classical solutions are restrictive. Classical references include [57], [61], and [75]. Because of the importance of applications of optimal control, approximation schemes were developed, leading to the introduction of the notion of the viscosity solution by P. L. Lions in [84].The analysis of the stability of these schemes led to some of the most important developments in the theory of viscosity solutions of PDEs. For a thorough discussion of this theory in relation to optimal control problems (including the deterministic case) the reader is referred to the book by Fleming and Soner [58], and Touzi's lecture notes [114] for the stochastic case.

Pontryagin's maximum principle will be presented in Chapter 4 as a cornerstone of the probabilistic approach to stochastic control problems. Without hesitation, we chose to use a deterministic version of this maximum principle to shed some light on how the HJB equation relates to the maximum principle. This was done for the purpose of illustration, as we believe this exercise to be informative as it highlights the similarities and differences between the control of the dynamics of a state in $\mathbb{R}^d$ and its distribution in $\mathcal{P}_2(\mathbb{R}^d)$.

Most of the material from Section 3.3 is borrowed from Touzi's Fields Institute lectures [114], where the interested reader will find complete proofs and a thorough introduction to viscosity solutions as they pertain to problems in stochastic control.

The mean-variance approach to portfolio selection was introduced by Markowitz in his groundbreaking work [92]; see also [93] for an expanded version in book form. The main originality of this contribution was to allow investors to seek the highest returns while controlling the risk via the variance of the portfolio. This idea inspired hundreds of extensions and applications and can be regarded as one of foundational building blocks of modern finance theory. Many multiperiod generalizations of Markowitz's one-period model were proposed and analyzed. The setup described in the text is borrowed from [119], in which the authors Zhou and Li propose a dynamic generalization in continuous time. The model will be solved in Chapter 4.

Uncertain volatility models were first treated by T. J. Lyons [89] and Avellaneda and Paras [10]. The discussion in the text is modeled after the account given by Cvitanic, Pham, and Touzi in [47] which can also be found in Pham's book [106]. Our discussion of the Merton problem with transaction costs is modeled after the corresponding section of Fleming and Soner's book [58]. While the results are originally due to Davis and Norman [48], the standard reference for detailed and complete proofs is Shreve and Soner's paper [112]. The reader interested in a discussion of more general mathematical models for transaction costs is referred to Kabanov and Safarian's book [73].

The discussion of averaged cost/reward objective functions of the form

$$\lim_{t \to \infty} \frac{1}{t} \mathbb{E} \int_0^t f(s, X_s, \alpha_s) ds$$

is central to what is known as ergodic control problems. The interested reader is referred to [110] for a general presentation of these problems. The optimal control of ergodic Markov diffusions was already considered by Bensoussan and Frehse [17] as early as 1987. The solution of the HJB equation was done in the early 1990s by Arisawa [8, 9] and P. L. Lions [85, 86]. This was used by Lasry and Lions in their first *full length paper* on mean-field games [81], in which they treat a stationary model on the torus. More recently, they considered the full ergodic mean-field game problem in a paper with Cardaliaguet and Porretta [28].

Ergodic properties of McKean–Vlasov diffusions were first studied by Benachour, Roynette, Talay, and Vallois in [12, 13]. This difficult problem received a renewal of interest with the works of Carrillo, McCann, and Villani [44]. However, the theory of the ergodic control of McKean–Vlasov dynamics in the setting of Carmona and Delarue [34] is still an open problem.

Chapter 4
Probabilistic Approaches to Stochastic Control

This chapter explains when and how the solution of a stochastic control problem can be recast in a probabilistic framework, reducing the analysis to the solutions of stochastic equations. These equations provide probabilistic representations of the value function and its derivative. In the first case, the value function itself appears as the solution of a BSDE. This is known as the weak formulation of the control problem. In the second approach, the derivative of the value function appears as the decoupling field of an FBSDE derived from the Pontryagin stochastic maximum principle. We introduce the adjoint equation as a BSDE and prove a duality relationship for the variation process. This leads to a necessary condition for optimality, and under slightly stronger conditions, a sufficient condition as well. We also explain how, when the Isaacs condition is satisfied, the adjoint BSDE becomes an FBSDE. We implement the approach in the case of linear-quadratic (LQ) models. The last section is devoted to the generalization of these classical results to the control of McKean–Vlasov dynamics introduced in Chapter 1.

4.1 ▪ BSDEs and Stochastic Control

In order to demonstrate a first connection between the theory of BSDEs and the analysis of optimal control problems, we revisit the setup of the weak formulation introduced in Subsection 3.1.7 and whose notation we use freely. Recall that the major assumption is that the volatility of the state dynamics is given by an invertible square matrix, the norm of its inverse being uniformly bounded. Under this assumption, for each admissible control α, we used Girsanov's theorem to introduce a new Wiener process W^α and a new probability measure $\mathbb{P}^\alpha$ used to write the cost functional. We refer to Subsection 3.1.7 for details.

4.1.1 ▪ Value Function as a Solution to a BSDE

Most often, and especially when dealing with Markovian models, the search for optimal controls in feedback (closed loop) form has been through the solution of the HJB equation. Our earlier discussion of the connection between PDEs and BSDEs suggests that stochastic control problems could possibly be approached by searching for the value function of the problem as the solution of a BSDE. This section builds on this idea.

We assume that there exists a function

$$\hat{\alpha} : [0,T] \times \Omega \times \mathbb{R}^d \times \mathbb{R}^d \ni (t, \omega, x, y) \hookrightarrow \hat{\alpha}(t, \omega, x, y) \in A$$

such that for every $(t, \omega, x, y, \alpha) \in [0, T] \times \Omega \times \mathbb{R}^d \times \mathbb{R}^d \times A$, we have

$$\tilde{H}(t, \omega, x, y, \hat{\alpha}(t, \omega, x, y)) \leq \tilde{H}(t, \omega, x, y, \alpha),$$

In other words, we assume that for each $(t, \omega, x, y) \in [0, T] \times \Omega \times \mathbb{R}^d \times \mathbb{R}^d$, we have

$$\hat{\alpha}(t, \omega, x, y) \in \arg\inf_{\alpha \in \mathbb{A}_t(\omega, x)} \tilde{H}(t, \omega, x, y, \alpha),$$

where we can think of $\mathbb{A}_t(\omega, x)$ as the set of actions which could be taken at time t in the scenario $\omega \in \Omega$ when the state variable is x, i.e., $X_t(\omega) = x$. Remember that in most applications considered in this text, we have $\mathbb{A}_t(\omega, x) = A$ for a fixed subset of $\mathbb{R}^k$.

Proposition 4.1. *Let $\mathbf{X}^0 = (X_t^0)_{t \in [0,T]}$ be the strong solution of the driftless dynamics (3.17), and let us assume that $(\hat{\mathbf{Y}}, \hat{\mathbf{Z}})$ is a solution of the BSDE*

$$d\hat{Y}_t = -\tilde{H}(t, X_t^0, -\sigma(t, X_t^0)^{\dagger -1} \hat{Z}_t, \hat{\alpha}(t, X_t^0, -\sigma(t, X_t^0)^{\dagger -1} \hat{Z}_t))dt - \hat{Z}_t \cdot dW_t, \quad 0 \leq t \leq T,$$

with terminal condition $\hat{Y}_T = g(X_T^0)$. Then the control $\hat{\boldsymbol{\alpha}}$ defined by $\hat{\alpha}_t = \hat{\alpha}(t, X_t^0, \hat{Z}_t)$ is an optimal control and the value of the problem is given by

$$J(\hat{\boldsymbol{\alpha}}) = \hat{Y}_0. \tag{4.1}$$

Proof. Let us set $\xi = g(X_T^0)$ for the sake of definiteness. Notice that this terminal condition is independent of the choice of the admissible control $\boldsymbol{\alpha}$. For each admissible control $\boldsymbol{\alpha} \in \mathbb{A}$, we denote by $(\mathbf{Y}^\alpha, \mathbf{Z}^\alpha)$ the solution of the BSDE

$$dY_t^\alpha = -\tilde{H}(t, X_t^0, -\sigma(t, X_t^0)^{\dagger -1} Z_t^\alpha, \alpha_t)dt - Z_t^\alpha \cdot dW_t, \quad 0 \leq t \leq T, \quad Y_T^\alpha = \xi.$$

Existence and uniqueness hold for this BSDE. Indeed, the driver is affine in (y, z), the coefficient of y is zero, and the coefficient of z is uniformly bounded, as is the constant term which is given by f. Notice that assuming that f had linear growth in the control variable $\alpha \in A$ would be enough. In any case, we have

$$Y_t^\alpha = \xi + \int_t^T \tilde{H}(s, X_s^0, -\sigma(s, X_s^0)^{\dagger -1} Z_s^\alpha, \alpha_s)ds + \int_t^T Z_s^\alpha \cdot dW_s$$

$$= \xi + \int_t^T \left[f(s, X_s^0, \alpha_s) - [\sigma(s, X_s^0)^{\dagger -1} Z_s^\alpha] \cdot b(s, X^0, \alpha_s) \right] ds + \int_t^T Z_s^\alpha \cdot dW_s$$

$$= \xi + \int_t^T f(s, X_s^0, \alpha_s)ds + \int_t^T Z_s^\alpha \cdot [-\sigma(s, X^0)^{-1} b(s, X^0, \alpha_s)ds + dW_s]$$

$$= \xi + \int_t^T f(s, X^0, \alpha_s)ds + \int_t^T Z_s^\alpha dW_s^\alpha.$$

So, by taking $\mathbb{P}^\alpha$-conditional expectations with respect to $\mathcal{F}_t$, we get

$$Y_t^\alpha = \mathbb{E}^{\mathbb{P}^\alpha}\left[\xi + \int_t^T f(s, X^0, \alpha_s)ds \bigg| \mathcal{F}_t \right],$$

and since $\mathcal{F}_0$ is trivial, we get

$$Y_0^\alpha = \mathbb{E}^{\mathbb{P}^\alpha}\left[\xi + \int_0^T f(s, X^0, \alpha_s)ds \right] = J(\boldsymbol{\alpha}).$$

In order to conclude the proof, we notice that $(\hat{Y}, \hat{Z})$ is the solution of the BSDE with

4.1. BSDEs and Stochastic Control

terminal condition ξ and driver Ψ^* defined by

$$\Psi^*(t,\omega,y,z) = \tilde{H}(t, X_t^0, -\sigma(t, X_t^0)^{\dagger-1}z, \hat{\alpha}(t, X_t^0, -\sigma(t, X_t^0)^{\dagger-1}z)),$$

while $(\mathbf{Y}^\alpha, \mathbf{Z}^\alpha)$ is the solution of the BSDE with the same terminal condition ξ and driver Ψ defined by

$$\Psi(t,\omega,y,z) = \tilde{H}(t, X_t^0, -\sigma(t, X_t^0)^{\dagger-1}z, \alpha_t),$$

and by assumption on the function $\hat{\alpha}$, we have

$$\Psi^*(t,\omega,y,z) \leq \Psi(t,\omega,y,z) \quad \text{a.s.}$$

for every t, y, and z. From this, we conclude that $\hat{Y}_0 \leq Y_0^\alpha$ by the comparison theorem (Theorem 2.4) for scalar BSDEs. $\square$

Remark 4.2. *Notice that we can get rid of all the adjoints (as given by the exponent †) if we assume that the volatility matrices σ are symmetric.*

4.1.2 • Optimization of a Family of BSDEs

In the previous subsection, we considered the case of uncontrolled volatility (σ independent of the control), and we proceeded in the following way: we first represented the value function of the problem for a generic control as the solution of a BSDE whose coefficients depend upon the control, and then we performed the optimization from there. We now disentangle these two steps, and formalize this optimization procedure in a far more general setup than in the above proof. But before stating and proving the main result of this subsection, we recall the notion of the essential infimum of a (possibly uncountable) family of random variables. The interested reader may consult [96] or [50] for details.

Theorem 4.3. *Let $\mathbb{A}$ be a set and let $(X^\alpha)_{\alpha \in \mathbb{A}}$ be a (not necessarily countable) family of random variables defined on a common probability space. Then there exists a unique (up to almost sure equality) random variable X such that*

(i) *for each $\alpha \in \mathbb{A}$ we have $X \leq X^\alpha$, $\mathbb{P}$-a.s.;*
(ii) *if $\tilde{X}$ is another random variable such that for each $\alpha \in \mathbb{A}$ we have $\tilde{X} \leq X^\alpha$, $\mathbb{P}$-a.s., then $\tilde{X} \leq X$, $\mathbb{P}$-a.s.*

Notice that the random variable X cannot be defined as the pointwise infimum

$$X(\omega) = \inf_{\alpha \in \mathbb{A}} X^\alpha(\omega)$$

since X defined in this way would most likely NOT be measurable!

Definition 4.4. *The random variable whose existence is given in the above theorem is called the essential infimum and is denoted by $X = \text{ess inf}_{\alpha \in \mathbb{A}} X^\alpha$.*

The above result is due to Neveu [96]. It was extended to *càdlàg* (i.e., right continuous with left limits) adapted processes by Dellacherie [50]. The following special case will be useful in the construction of solutions of BSDEs.

Theorem 4.5. *Let $\mathbb{A}$ be a set of progressively measurable processes with values in $(A, \mathcal{A})$, and for each $\alpha \in \mathbb{A}$ let us assume that $(\mathbf{Y}^\alpha, \mathbf{Z}^\alpha)$ solves $BSDE(\Psi^\alpha, \xi^\alpha)$, where the driver*

Ψ^α and the terminal condition ξ^α satisfy the assumption of the existence and uniqueness result of Theorem 2.2. Let us also assume that $(\mathbf{Y}, \mathbf{Z})$ solves $BSDE(\Psi, \xi)$ and that there exists $\hat{\alpha} \in \mathbb{A}$ such that

$$\Psi^{\hat{\alpha}}(t, Y_t, Z_t) = \Psi(t, Y_t, Z_t), \qquad dt \otimes d\mathbb{P}\text{-}a.s.,$$
$$\xi^{\hat{\alpha}} = \xi, \qquad d\mathbb{P}\text{-}a.s.,$$

and that

$$\Psi^{\hat{\alpha}}(t, Y_t, Z_t) = \operatorname{ess\,inf}_{\alpha \in \mathbb{A}} \Psi^\alpha(t, Y_t, Z_t), \qquad dt \otimes d\mathbb{P}\text{-}a.s.,$$
$$\xi^{\hat{\alpha}} = \operatorname{ess\,inf}_{\alpha \in \mathbb{A}} \xi^\alpha, \qquad d\mathbb{P}\text{-}a.s.$$

Then,

$$Y_t = Y_t^{\hat{\alpha}} = \operatorname{ess\,inf}_{\alpha \in \mathbb{A}} Y_t^\alpha, \quad 0 \le t \le T, \ \mathbb{P}\text{-}a.s.$$

Proof. For each $\alpha \in \mathbb{A}$ we have $\xi \le \xi^\alpha$ a.s. and $\Psi(t, Y_t, Z_t) \le \Psi^\alpha(t, Y_t, Z_t)$ a.s. for a.e. t, so the comparison theorem for solutions of BSDEs implies that $Y_t \le Y_t^\alpha$ for $0 \le t \le T$, $\mathbb{P}$-a.s. Consequently,

$$Y_t \le \operatorname{ess\,inf}_{\alpha \in \mathbb{A}} Y_t^\alpha = Y_t^{\hat{\alpha}}, \quad 0 \le t \le T, \ \mathbb{P}\text{-}a.s.,$$

Now, since $(\mathbf{Y}, \mathbf{Z})$ solves $BSDE(\Psi, \xi)$, we have

$$Y_t = \xi + \int_t^T \Psi(s, Y_s, Z_s)ds - \int_t^T Z_s dW_s$$
$$= \xi^{\hat{\alpha}} + \int_t^T \Psi(\hat{\alpha}, Y_s, Z_s)ds - \int_t^T Z_s dW_s$$

so that $(\mathbf{Y}, \mathbf{Z})$ solves $BSDE(\Psi^{\hat{\alpha}}, \xi^{\hat{\alpha}})$, and by uniqueness, $Y_t = Y^{\hat{\alpha}}$. We conclude noticing that

$$\operatorname{ess\,inf}_{\alpha \in \mathbb{A}} Y_t^\alpha \le Y_t^{\hat{\alpha}} = Y_t \le \operatorname{ess\,inf}_{\alpha \in \mathbb{A}} Y_t^\alpha,$$

which is the desired result. □

Let us emphasize the way we want to (and already did) use this result. Think of $\mathbb{A}$ as the set of admissible controls of a stochastic control problem, and let us assume that for each $\alpha \in \mathbb{A}$, we are able to prove the representation $Y_0^\alpha = J(\alpha)$ of the value of the objective function for the control α as the solution $\mathbf{Y}^\alpha$ of a BSDE. Then, if we can find an element $\hat{\alpha} \in \mathbb{A}$ minimizing the driver and the terminal condition, this element $\hat{\alpha}$ of $\mathbb{A}$ is guaranteed to be an optimal control, and the solution of the corresponding BSDE gives a representation of the value function of the problem when computed along the optimal path. Notice that we not only get the identification and representation of the value function over the interval $[0, T]$, but we also get it over the interval $[t, T]$ for every $t \in [0, T]$.

4.1.3 ▪ Stochastic HJB Equations

The discussion of this subsection is motivated by the analysis of mean-field games with a common noise. While we may not be able to address these game models in Chapter 6 because of their high level of technicality, we thought it was appropriate to present this generalized form of the probabilistic representation of the value function of a control problem in a random environment in this section. As a model for the source of random shocks due

4.1. BSDEs and Stochastic Control

to the random environment, we consider an $\mathbb{R}^{d_0}$-valued Brownian motion $\mathbf{W}^0 = (W_t^0)_{t\geq 0}$ independent of $\mathbf{W} = (W_t)_{t\geq 0}$, together with its natural filtration $\mathbb{F}^0 = (\mathcal{F}_t^0)_{t\geq 0}$, as well as the joint filtration $\mathbb{F}^* = (\mathcal{F}_t^*)_{t\geq 0}$ generated by both Brownian motions. Furthermore, we assume that for each fixed $(x, \alpha) \in \mathbb{R}^d \times A$, the coefficients $b(t, \omega, x, \alpha)$, $\sigma(t, \omega, x, \alpha)$, and $f(t, \omega, x, \alpha)$ are progressively measurable processes for the filtration $\mathbb{F}^0$, and for each $x \in \mathbb{R}^d$, $g(x)$ is $\mathcal{F}_T^0$-measurable. In this subsection, the set $\mathbb{A}$ of admissible controls is chosen to be the set $\mathbb{H}^2(A)$ of all A-valued processes $\boldsymbol{\alpha} = (\alpha_t)_{0\leq t \leq T}$ which are $\mathbb{F}^*$-progressively measurable and satisfy

$$\mathbb{E}\int_0^T |\alpha_t|^2 dt < \infty.$$

Finally, for each $(t, x) \in [0, T] \times \mathbb{R}^d$, we let $\mathbf{X}^{t,x} = (X_s^{t,x})_{t\leq s \leq T}$ be the solution of the stochastic differential equation

$$dX_s = b(s, X_s, \alpha_s)ds + \sigma(s, X_s, \alpha_s)dW_s + \zeta(s, X_s, \alpha_s)dW_s^0, \qquad X_t = x. \qquad (4.2)$$

With this notation, we define the (conditional) cost

$$J_{t,x}(\boldsymbol{\alpha}) = \mathbb{E}\left[\int_t^T f(s, X_s^{t,x}, \alpha_s)ds + g(X_T^{t,x}) \Big| \mathcal{F}_t^0\right] \qquad (4.3)$$

and the (conditional) value function

$$V(t, x) = \text{ess inf}_{\boldsymbol{\alpha}\in\mathbb{A}} J_{t,x}(\boldsymbol{\alpha}). \qquad (4.4)$$

We shall drop the superscript and write X_s for $X_s^{t,x}$ when no confusion is possible. Under some regularity assumptions, we show that for each $x \in \mathbb{R}^d$, $(V(t, x))_{0\leq t\leq T}$ is an $\mathbb{F}^0$-semimartingale and that its Itô decomposition shows that it is the solution of a form of the stochastic Hamilton–Jacobi–Bellman equation. Because of the special form of the state dynamics (4.2), we introduce the operator symbol

$$L(t, x, y, z, z^0, \alpha) = b(t, x, \alpha) \cdot y + \frac{1}{2}\text{trace}([\sigma\sigma^\dagger(t, x, \alpha) + \zeta\zeta^\dagger(t, x, \alpha)] \cdot z)$$
$$+ \zeta(t, x, \alpha) \cdot z^0 + f(t, x, \alpha)$$

and its minimized form

$$\hat{L}(t, x, y, z, z^0) = \inf_{\alpha \in A} L(t, x, y, z, z^0, \alpha). \qquad (4.5)$$

Assuming that the value function is smooth enough, we can apply a generalization of the dynamic programming principle to the present setup of conditional value functions to show that $V(t, x)$ should satisfy a form of the stochastic HJB equation as given by a parametric family of BSDEs. More precisely, we expect V to satisfy

$$V(t, x) = g(x) + \int_t^T \hat{L}(s, x, \partial_x V(s, x), \partial_{xx}^2 V(s, x), \partial_x \tilde{V}(s, x))ds + \int_t^T V(s, x)dW_s^0. \qquad (4.6)$$

In analogy with the classical case, we prove the following verification theorem.

Theorem 4.6. *We assume that the random fields* $(\varphi_t(x))_{(t,x)\in[0,T]\times\mathbb{R}^d}$ *and* $(\phi_t(x))_{(t,x)\in[0,T]\times\mathbb{R}^d}$ *satisfy the following.*

1. *With probability one, for each $t \in [0, T]$, the function $\mathbb{R}^d \ni x \hookrightarrow (\varphi_t(x), \phi_t(x)) \in \mathbb{R} \times \mathbb{R}^m$ is twice continuously differentiable.*
2. *The processes $(\varphi_t(x), \phi_t(x))_{0 \leq t \leq T}$ and $(\partial_x \varphi_t(x), \partial_{xx}^2 \varphi_t(x), \partial_x \phi_t(x))_{0 \leq t \leq T}$ are adapted and continuous for each fixed $x \in \mathbb{R}^d$.*
3. *Almost surely, the function $[0, T] \times \mathbb{R}^d \ni (t, x) \hookrightarrow \hat{L}(t, x, \partial_x \varphi_t(x), \partial_{xx}^2 \varphi(s, x), \partial_x \phi_t(x))$ is smooth.*
4. *$(\varphi_t(x), \phi_t(x))$ is a solution of the stochastic HJB equation (4.6) in the sense that for each $(t, x) \in [0, T] \times \mathbb{R}^d$, we have*

$$\varphi_t(x) = g(x) + \int_t^T \hat{L}(s, x, \partial_x \varphi_s(x), \partial_{xx}^2 \varphi_s(x), \partial_x \phi_s(x))ds + \int_t^T \phi_s(x)dW_s^0. \quad (4.7)$$

5. *There exists a measurable function $[0, T] \times \mathbb{R}^d \ni (t, x) \hookrightarrow \alpha_t(x) \in A$ satisfying*

$$\begin{aligned}&\hat{L}(t, x, \partial_x \varphi_t(x), \partial_{xx}^2 \varphi(s, x), \partial_x \phi_t(x)) \\ &= L(t, x, \partial_x \varphi_t(x), \partial_{xx}^2 \varphi_t(x), \partial_x \phi_t(x), \alpha_t(x))\end{aligned} \quad (4.8)$$

for every $(t, x) \in [0, T] \times \mathbb{R}^d$, regular enough so that the state equation

$$dX_s = b(s, X_s, \alpha_s(X_s))ds + \sigma(s, X_s, \alpha_s(X_s))dW_s + \zeta(s, X_s, \alpha_s(X_s))dW_s^0 \quad (4.9)$$

with initial condition $X_t = x$, is well-posed (i.e., has a unique strong solution for each $(t, x) \in [0, T] \times \mathbb{R}^d$).

Then, $(\varphi_t(x), \phi_t(x))$ coincides with $(V(t, x), \tilde{V}(t, x))$, where $V(t, x)$ is the (conditional) value function defined in (4.4), and for any initial condition (t, x), the control $\hat{\alpha}_s = \alpha_s(X_s)$ minimizes the cost (4.3).

Proof. Let $\beta \in \mathbb{A}$ be an admissible control and let us denote by $\mathbf{X}^\beta$ the corresponding controlled state, namely the solution of the state equation

$$dX_s^\beta = b(s, X_s^\beta, \beta_s)ds + \sigma(s, X_s^\beta, \beta_s)dW_s + \zeta(s, X_s^\beta, \beta_s)dW_s^0, \qquad X_t = x.$$

Applying the Itô–Wentzell formula to $\varphi_s(X_s^\beta)$ over the interval $[t, T]$, and using (4.7), we get

$$\begin{aligned}\varphi_T(X_T^\beta) = \varphi_t(x) + \int_t^T &\Big[\partial_t \varphi_s(X_s^\beta) + b(s, X_s^\beta, \beta_s)\partial_x \varphi_s(X_s^\beta) \\
&+ \partial_x \varphi_s(X_s^\beta)\sigma(s, X_s^\beta, \alpha_s(X_s^\beta))dW_s + \partial_x \varphi_s(X_s^\beta)\zeta(s, X_s^\beta, \alpha_s(X_s^\beta))dW_s^0 \\
&+ \frac{1}{2}\mathrm{trace}([\sigma\sigma^\dagger(s, X_s^\beta, \beta_s) + \zeta\zeta^\dagger(s, X_s^\beta, \beta_s)]\partial_{xx}^2 \varphi_s(X_s^\beta))\Big]ds \\
&- \hat{L}(s, X_s^\beta, \partial_x \varphi_s(X_s^\beta), \partial_{xx}^2 \varphi_s(X_s^\beta), \partial_x \phi_s(X_s^\beta))ds - \phi_s(X_s^\beta)dW_s^0.\end{aligned}$$

So, taking conditional expectations of both sides, we get

$$\begin{aligned}\mathbb{E}[\varphi_T(X_T^\beta)|\mathcal{F}_t^0] = \varphi_t(x) + \mathbb{E}\Big[\int_t^T &\Big[\partial_t \varphi_s(X_s^\beta) + b(s, X_s^\beta, \beta_s)\partial_x \varphi_s(X_s^\beta) \\
&+ \frac{1}{2}\mathrm{trace}([\sigma\sigma^\dagger(s, X_s^\beta, \beta_s) + \zeta\zeta^\dagger(s, X_s^\beta, \beta_s)]\partial_{xx}^2 \varphi_s(X_s^\beta))\Big]ds \\
&- \hat{L}(s, X_s^\beta, \partial_x \varphi_s(X_s^\beta), \partial_{xx}^2 \varphi_s(X_s^\beta), \partial_x \phi_s(X_s^\beta))\Big]ds\Big|\mathcal{F}_t^0\Big],\end{aligned}$$

which gives

$$\varphi_t(x) \leq \mathbb{E}\left[\int_t^T f(s, X_s^\beta, \beta_s)ds + g(X_T^\beta)\bigg|\mathcal{F}_t^0\right]$$
$$= J_{t,x}(\boldsymbol{\beta}).$$

On the other hand, because of (4.8), this inequality becomes an equality if we use the control $\hat{\alpha}$ so that

$$\varphi_t(x) = \inf_{\boldsymbol{\alpha}\in\mathbb{A}} J_{t,x}(\boldsymbol{\alpha}) = V(t,x).$$

By uniqueness of the Itô decomposition of a semimartingale, we get $\phi_t(x) = \tilde{V}(t,x)$, and the last claim of the theorem follows from

$$\mathbb{E}\left[\int_t^T f(s, X_s^\beta, \beta_s)ds + g(X_T^\beta)\right] = \mathbb{E}\left[\mathbb{E}\left[\int_t^T f(s, X_s^\beta, \beta_s)ds + g(X_T^\beta)\bigg|\mathcal{F}_t^0\right]\right]. \quad \square$$

4.2 ▪ Pontryagin Stochastic Maximum Principle

From now on, we assume that the coefficients b, σ, and f are, for each $(t,\omega) \in [0,T] \times \Omega$, continuously differentiable with respect to the variables $(x,\alpha) \in \mathbb{R}^d \times A$ with Lipschitz continuous and bounded partial derivatives. Similarly, we assume that the function g giving the terminal cost is, for each $\omega \in \Omega$, continuously differentiable with respect to $x \in \mathbb{R}^d$, with Lipschitz continuous and bounded partial derivatives. In particular, the Hamiltonian of the system is differentiable with respect to the variables $(x,\alpha) \in \mathbb{R}^d \times A$. The results presented in the first two subsections below identify new connections between the optimal control of stochastic differential equations and BSDEs. They are usually referred to as the *stochastic maximum principle*. They provide a generalization to the stochastic framework of the classical Pontryagin maximum principle of deterministic control theory.

From now on, we assume that the set A of actions is a convex subset of $\mathbb{R}^k$ and that the conditions for a control process $\boldsymbol{\alpha}$ to be admissible are stable under convex combinations; in other words, we assume that the set $\mathbb{A}$ of admissible controls is convex. In this case, we are able to compute the Gâteaux derivative of the objective functional J as a function of $\boldsymbol{\alpha}$. More general forms of the Pontryagin maximum principle can be derived for nonconvex admissible sets of controls, but because of their much higher level of technicality, we refrain from discussing them here.

4.2.1 ▪ Necessary Conditions for Optimality

We fix an admissible control $\boldsymbol{\alpha} \in \mathbb{A}$ and we denote by $\mathbf{X} = \mathbf{X}^{\boldsymbol{\alpha}}$ the corresponding controlled state process. Next, we consider $\boldsymbol{\beta}$ as the direction in which we are computing the Gâteaux derivative of J. One can think of $\boldsymbol{\beta} = \boldsymbol{\alpha}' - \boldsymbol{\alpha}$ for some other admissible control $\boldsymbol{\alpha}'$. Remember that we are assuming that $\mathbb{A}$ is convex. For each $\epsilon > 0$ small enough, we consider the admissible control $\boldsymbol{\alpha}^\epsilon$ defined by $\alpha_t^\epsilon = \alpha_t + \epsilon\beta_t$, and the corresponding controlled state $\mathbf{X}^\epsilon = \mathbf{X}^{\boldsymbol{\alpha}^\epsilon}$.

The Variation Process

We then define $\mathbf{V}$ as the solution of the linear SDE (with random coefficients)

$$dV_t = [\partial_x b(t, X_t, \alpha_t)V_t + \partial_\alpha b(t, X_t, \alpha_t)\beta_t]dt + [\partial_x \sigma(t, X_t, \alpha_t)V_t + \partial_\alpha \sigma(t, X_t, \alpha_t)\beta_t]dW_t$$
(4.10)

with initial condition $V_0 = 0$. Note that $\partial_x b$ is in $\mathbb{R}^{d \times d}$, $\partial_\alpha b$, is in $\mathbb{R}^{d \times k}$, $\partial_x \sigma$ is in $\mathbb{R}^{d \times m \times d}$, and $\partial_\alpha \sigma$ is in $\mathbb{R}^{d \times m \times k}$. Finally, V_t is in $\mathbb{R}^d$ like X_t. Because of our assumption on the boundedness of the partial derivatives of the coefficients, Theorem 1.2 gives existence and uniqueness of **V**, as well as the fact that it satisfies

$$\mathbb{E} \sup_{0 \leq t \leq T} |V_t|^p < \infty \tag{4.11}$$

for every finite $p \geq 1$. The following result says that at least in some L^2 sense, V_t is the derivative of the controlled state when we vary the control in the direction β.

Lemma 4.7. *We have*

$$\lim_{\epsilon \searrow 0} \mathbb{E} \sup_{0 \leq t \leq T} \left| \frac{X_t^\epsilon - X_t}{\epsilon} - V_t \right|^2 = 0. \tag{4.12}$$

Proof. For the purpose of this proof we set

$$V_t^\epsilon = \frac{X_t^\epsilon - X_t}{\epsilon} - V_t.$$

Notice that $V_0^\epsilon = 0$ and that

$$dV_t^\epsilon = \frac{1}{\epsilon} \bigg[b(t, X_t + \epsilon(V_t^\epsilon + V_t), \alpha_t + \epsilon \beta_t) - b(t, X_t, \alpha_t)$$
$$- \epsilon \partial_x b(t, X_t, \alpha_t) V_t - \epsilon \partial_\alpha b(t, X_t, \alpha_t) \beta_t \bigg] dt$$
$$+ \frac{1}{\epsilon} \bigg[\sigma(t, X_t + \epsilon(V_t^\epsilon + V_t), \alpha_t + \epsilon \beta_t) - \sigma(t, X_t, \alpha_t)$$
$$- \epsilon \partial_x \sigma(t, X_t, \alpha_t) V_t - \epsilon \partial_\alpha \sigma(t, X_t, \alpha_t) \beta_t \bigg] dW_t. \tag{4.13}$$

Now for each $t \in [0, T]$ and each $\epsilon > 0$, we have

$$\frac{1}{\epsilon} [b(t, X_t + \epsilon(V_t^\epsilon + V_t), \alpha_t + \epsilon \beta_t) - b(t, X_t, \alpha_t)]$$
$$= \int_0^1 \partial_x b(t, X_t + \lambda \epsilon (V_t^\epsilon + V_t), \alpha_t + \lambda \epsilon \beta_t)(V_t^\epsilon + V_t) d\lambda$$
$$+ \int_0^1 \partial_\alpha b(t, X_t + \lambda \epsilon (V_t^\epsilon + V_t), \alpha_t + \lambda \epsilon \beta_t) \beta_t d\lambda. \tag{4.14}$$

In order to simplify the notation we set $X_t^{\lambda, \epsilon} = X_t + \lambda \epsilon (V_t^\epsilon + V_t)$ and $\alpha_t^{\lambda, \epsilon} = \alpha_t + \lambda \epsilon \beta_t$. If we compute the "$dt$"-term in (4.13) by plugging (4.14) into (4.13), we get

$$\frac{1}{\epsilon} [b(t, X_t + \epsilon(V_t^\epsilon + V_t), \alpha_t + \epsilon \beta_t) - b(t, X_t, \alpha_t)$$
$$- \epsilon \partial_x b(t, X_t, \alpha_t) V_t - \epsilon \partial_\alpha b(t, X_t, \alpha_t) \beta_t]$$
$$= \int_0^1 \partial_x b(t, X_t^{\lambda, \epsilon}, \alpha_t^{\lambda, \epsilon}) V_t^\epsilon d\lambda + \int_0^1 [\partial_x b(t, X_t^{\lambda, \epsilon}, \alpha_t^{\lambda, \epsilon}) - \partial_x b(t, X_t, \alpha_t)] V_t d\lambda$$
$$+ \int_0^1 [\partial_\alpha b(t, X_t^{\lambda, \epsilon}, \alpha_t^{\lambda, \epsilon}) - \partial_\alpha b(t, X_t, \alpha_t)] \beta_t d\lambda. \tag{4.15}$$

4.2. Pontryagin Stochastic Maximum Principle

The last two terms appearing in the right-hand side of (4.15) converge to 0 in $L^2([0,T] \times \Omega)$ as $\epsilon \searrow 0$. Indeed, if we denote by I_t the first of these two terms, we have

$$\mathbb{E}\int_0^T |I_t|^2 dt = \mathbb{E}\int_0^T \left|\int_0^1 [\partial_x b(t, X_t^{\lambda,\epsilon}, \alpha_t^{\lambda,\epsilon}) - \partial_x b(t, X_t, \alpha_t)] V_t d\lambda\right|^2 dt$$

$$\leq \mathbb{E}\int_0^T \int_0^1 |\partial_x b(t, X_t^{\lambda,\epsilon}, \alpha_t^{\lambda,\epsilon}) - \partial_x b(t, X_t, \alpha_t)|^2 |V_t|^2 d\lambda dt$$

$$\leq c\mathbb{E}\int_0^T \int_0^1 (\lambda\epsilon)^2 (|V_t^\epsilon + V_t|^2 + |\beta_t|^2)|V_t|^2 d\lambda dt$$

$$\leq c\left(\int_0^T \int_0^1 \mathbb{E}|\lambda\epsilon(V_t^\epsilon + V_t)|^4 d\lambda dt\right)^{1/2} \left(\int_0^T \mathbb{E}|V_t|^4 dt\right)^{1/2}$$

$$+ c\left(\int_0^T \int_0^1 \mathbb{E}|\lambda\epsilon\beta_t|^4 d\lambda dt\right)^{1/2} \left(\int_0^T \mathbb{E}|V_t|^4 dt\right)^{1/2},$$

which clearly converges to 0 as $\epsilon \searrow 0$ since the expectations are finite. Now the same argument can be used to show that the second term also goes to 0 as $\epsilon \searrow 0$. Next, we treat the diffusion part of (4.13) in the same way using Jensen's inequality, except that we have to use the Burkholder–Davis–Gundy inequality to control the quadratic variation of the stochastic integrals. Consequently, going back to (4.13), we see that

$$\mathbb{E}\sup_{0\leq t\leq T} |V_t^\epsilon|^2 dt \leq c\left(\int_0^T \mathbb{E}\sup_{0\leq s\leq t} |V_s^\epsilon|^2 dt + \int_0^T \sup_{0\leq s\leq t} |(\mathbb{E}V_s^\epsilon)|^2 dt\right) + \delta_\epsilon$$

$$\leq c\int_0^T \mathbb{E}\sup_{0\leq s\leq t} |V_s^\epsilon|^2 dt + \delta_\epsilon,$$

where, as usual, $c > 0$ is a generic constant and $\lim_{\epsilon \searrow 0} \delta_\epsilon = 0$. Finally, we get the desired result applying Gronwall's inequality. □

Gâteaux Derivative of the Cost Functional

Lemma 4.8. *The function* $\alpha \hookrightarrow J(\alpha)$ *is Gâteaux differentiable and*

$$\frac{d}{d\epsilon}J(\alpha + \epsilon\beta)\Big|_{\epsilon=0}$$
$$= \mathbb{E}\left[\int_0^T [\partial_x f(t, X_t, \alpha_t) V_t + \partial_\alpha f(t, X_t, \alpha_t)\beta_t] dt + \partial_x g(X_T) V_T\right]. \quad (4.16)$$

Proof. We use freely the notation introduced in the proof of the previous lemma:

$$\frac{d}{d\epsilon}J(\alpha + \epsilon\beta)\Big|_{\epsilon=0} = \lim_{\epsilon \searrow 0} \frac{1}{\epsilon}\mathbb{E}\int_0^T [f(t, X_t + \epsilon(V_t^\epsilon + V_t), \alpha_t + \epsilon\beta_t) - f(t, X_t, \alpha_t)] dt$$

$$+ \lim_{\epsilon \searrow 0} \frac{1}{\epsilon}\mathbb{E}[g(X_T + \epsilon(V_T^\epsilon + V_T)) - g(X_T)].$$

Treating the two limits separately, we get

$$\lim_{\epsilon \searrow 0} \frac{1}{\epsilon} \mathbb{E} \int_0^T [f(t, X_t + \epsilon(V_t^\epsilon + V_t), \alpha_t + \epsilon \beta_t) - f(t, X_t, \alpha_t)] dt$$

$$= \lim_{\epsilon \searrow 0} \frac{1}{\epsilon} \mathbb{E} \int_0^T \int_0^1 \frac{d}{d\lambda} f(t, X_t + \lambda\epsilon(V_t^\epsilon + V_t), \alpha_t + \lambda\epsilon\beta_t) d\lambda dt$$

$$= \lim_{\epsilon \searrow 0} \mathbb{E} \int_0^T \int_0^1 [\partial_x f(t, X_t + \lambda\epsilon(V_t^\epsilon + V_t), \alpha_t + \lambda\epsilon\beta_t)(V_t^\epsilon + V_t)$$
$$+ \partial_\alpha f(t, X_t + \lambda\epsilon(V_t^\epsilon + V_t), \alpha_t + \lambda\epsilon\beta_t)\beta_t] d\lambda dt$$

$$= \mathbb{E} \int_0^T [\partial_x f(t, X_t, \alpha_t) V_t + \partial_\alpha f(t, X_t, \alpha_t) \beta_t] dt$$

using Lebesgue's dominated convergence theorem and the result of the previous lemma. Similarly,

$$\lim_{\epsilon \searrow 0} \frac{1}{\epsilon} \mathbb{E}[g(X_T + \epsilon(V_T^\epsilon + V_T)) - g(X_T)]$$

$$= \lim_{\epsilon \searrow 0} \frac{1}{\epsilon} \mathbb{E} \int_0^1 \frac{d}{d\lambda} g(X_T + \lambda\epsilon(V_T^\epsilon + V_T)) d\lambda$$

$$= \lim_{\epsilon \searrow 0} \mathbb{E} \int_0^1 \partial_x g(X_T + \lambda\epsilon(V_T^\epsilon + V_T))(V_T^\epsilon + V_T) d\lambda$$

$$= \mathbb{E} \partial_x g(X_T) V_T,$$

which completes the proof. □

Duality: The Adjoint Equations and the Adjoint Processes

Definition 4.9. *Let α be an admissible control and $\mathbf{X}^\alpha$ be the corresponding controlled state process. We call adjoint processes associated with α any solution $(\mathbf{Y}, \mathbf{Z})$ of the BSDE*

$$dY_t = -\partial_x H(t, X_t, Y_t, Z_t, \alpha_t) dt + Z_t dW_t, \qquad Y_T = \partial_x g(X_T). \tag{4.17}$$

This equation is called the adjoint equation associated with the admissible control α.

Notice that $Y_t \in \mathbb{R}^d$ like X_t, and that $Z_t \in \mathbb{R}^{d \times m}$. Notice also that the adjoint equation can be rewritten as

$$dY_t = -[\partial_x b(t, X_t, \alpha_t) Y_t + \partial_x \sigma(t, X_t, \alpha_t) Z_t + \partial_x f(t, X_t, \alpha_t)] dt + Z_t dW_t \tag{4.18}$$

with the same terminal condition, and which, given α and $\mathbf{X}$, and given the boundedness assumption on the partial derivatives of the coefficients, is a linear BSDE for which existence and uniqueness of a solution are guaranteed by Theorem 2.2. Moreover, this solution $(\mathbf{Y}, \mathbf{Z})$ satisfies

$$\mathbb{E} \sup_{0 \leq t \leq T} |Y_t|^2 + \mathbb{E} \int_0^T |Z_t|^2 dt < \infty.$$

The actual duality relationship is given as follows.

Lemma 4.10. *It holds that*

$$\mathbb{E}[Y_T^\dagger V_T] = \mathbb{E} \int_0^T [Y_t^\dagger \partial_\alpha b(t, X_t, \alpha_t) \beta_t - V_t^\dagger \partial_x f(t, X_t, \alpha_t)$$
$$+ \text{trace}(Z_t^\dagger \partial_\alpha \sigma(t, X_t, \alpha_t) \beta_t)] dt. \tag{4.19}$$

4.2. Pontryagin Stochastic Maximum Principle

Proof. We switch from the inner product notation to the matrix/vector product notation to be able to check more easily the consistency of our computations. Using the definitions (4.10) of the variation process **V** and (4.18) of the adjoint process **Y**, integration by parts gives

$$Y_T^\dagger V_T = Y_0^\dagger V_0 + \int_0^T Y_t^\dagger dV_t + \int_0^T V_t^\dagger dY_t + \int_0^T \text{trace}[dY_t \otimes dV_t]$$

$$= \int_0^T [Y_t^\dagger \partial_x b(t, X_t, \alpha_t) V_t + Y_t^\dagger \partial_\alpha b(t, X_t, \alpha_t)\beta_t - Y_t^\dagger \partial_x b(t, X_t, \alpha_t) V_t$$
$$- \text{trace}[Z_t^\dagger \partial_x \sigma(t, X_t, \alpha_t) V_t] - V_t^\dagger \partial_x f(t, X_t, \alpha_t)$$
$$+ \text{trace}(Z_t^\dagger \partial_x \sigma(t, X_t, \alpha_t) V_t + Z_t \partial_\alpha \sigma(t, X_t, \alpha_t)\beta_t)]dt + M_t,$$

where $(M_t)_{0 \leq t \leq T}$ is a mean zero square integrable martingale. By taking expectations on both sides, we get the desired equality (4.62). □

The duality relation provides an expression of the Gâteaux derivative of the cost functional in terms of the Hamiltonian of the system. Indeed, we have the following.

Corollary 4.11.

$$\frac{d}{d\epsilon}J(\boldsymbol{\alpha} + \epsilon\boldsymbol{\beta})\Big|_{\epsilon=0} = \mathbb{E}\int_0^T \partial_\alpha H(t, X_t, Y_t, Z_t, \alpha_t) \cdot \beta_t dt.$$

Proof. Since $Y_T = \partial_x g(X_T)$, replacing $\mathbb{E}\partial_x g(X_T) V_T$ in the expression of the Gâteaux derivative found in Lemma 4.8 by the expression of $\mathbb{E}Y_Y V_T$ found in Lemma 4.10 gives the desired result. □

Pontryagin Maximum Principle: Necessary Condition

The main result of this subsection is the following theorem.

Theorem 4.12. *Under the above assumptions, if the admissible control $\boldsymbol{\alpha} \in \mathbb{A}$ is optimal, **X** is the associated (optimally) controlled state, and $(\mathbf{Y}, \mathbf{Z})$ are the associated adjoint processes solving the adjoint equation (4.17):*

$$\forall \alpha \in A, \quad \partial_\alpha H(t, X_t, Y_t, Z_t, \alpha_t) \cdot (\alpha - \alpha_t) \geq 0 \quad a.e. \text{ in } t \in [0, T], \text{ a.s.} \quad (4.20)$$

Remark 4.13. *The above conclusion essentially means that*

$$\forall \alpha \in A, \quad H(t, X_t, Y_t, Z_t, \alpha_t) \leq H(t, X_t, Y_t, Z_t, \alpha) \quad a.e. \text{ in } t \in [0, T], \text{ a.s.,} \quad (4.21)$$

which can be derived from (4.20) under mild regularity conditions.

Proof. Since A is convex, given $\boldsymbol{\beta} \in \mathbb{A}$ we can choose perturbations $\alpha_t^\epsilon = \alpha_t + \epsilon(\beta_t - \alpha_t)$ still in $\mathbb{A}$ for $0 \leq \epsilon \leq 1$. Since $\boldsymbol{\alpha}$ is optimal, we have the inequality

$$\frac{d}{d\epsilon}J(\boldsymbol{\alpha} + \epsilon(\boldsymbol{\beta} - \boldsymbol{\alpha}))\Big|_{\epsilon=0} = \mathbb{E}\int_0^T \partial_\alpha H(t, X_t, Y_t, Z_t, \alpha_t) \cdot (\beta_t - \alpha_t) dt \geq 0.$$

From this we conclude that

$$\partial_\alpha H(t, X_t, Y_t, Z_t, \alpha_t) \cdot (\beta - \alpha_t) \geq 0 \quad \text{for } dt \otimes d\mathbb{P}\text{-a.e.}$$

for all $\beta \in A$. □

4.2.2 • A Sufficient Condition for Optimality

The necessary conditions of optimality identified by Pontryagin's maximum principle can be turned into a sufficient condition for optimality under some technical assumptions.

We assume that the admissible control $\alpha \in \mathbb{A}$ is fixed, and as before, we denote by $\mathbf{X} = \mathbf{X}^\alpha$ the corresponding controlled state process, and by $(\mathbf{Y}, \mathbf{Z})$ the adjoint processes.

Theorem 4.14. *In addition to the assumptions in force in this section, let us assume that*

1. *the function g giving the terminal cost is convex;*
2. *for each $t \in [0,T]$, the function $(x,a) \hookrightarrow H(t,x,Y_t,Z_t,\alpha)$ is convex $\mathbb{P}$-almost surely.*

Then if
$$H(t, X_t, Y_t, Z_t, \alpha_t) = \inf_{\alpha \in A} H(t, X_t, Y_t, Z_t, \alpha), \quad a.s., \tag{4.22}$$

then α is an optimal control, i.e.,
$$J(\alpha) = \inf_{\alpha' \in \mathbb{A}} J(\alpha'). \tag{4.23}$$

Proof. Let $\alpha' \in \mathbb{A}$ be a generic admissible control, and let us denote by $\mathbf{X}' = \mathbf{X}^{\alpha'}$ the associated controlled state process. g being convex, we have $g(y) - g(x) \leq (y-x)\partial_x g(y)$ so that

$$\begin{aligned}
&\mathbb{E}[g(X_T) - g(X'_T)] \\
&\leq \mathbb{E}[\partial_x g(X_T) \cdot (X_T - X'_T)] \\
&= \mathbb{E}[Y_T \cdot (X_T - X'_T)] \\
&= \mathbb{E}\Big[\int_0^T (X_t - X'_t)^\dagger dY_t + \int_0^T Y_t^\dagger [dX_t - dX'_t] \\
&\qquad + \int_0^T \operatorname{trace}[(\sigma(t, X_t, \alpha_t) - \sigma(t, X'_t, \alpha'_t))^\dagger Z_t] dt \Big] \\
&= -\mathbb{E}\int_0^T (X_t - X'_t)^\dagger \partial_x H(t, X_t, Y_t, Z_t, \alpha_t) dt \\
&\qquad + \mathbb{E}\int_0^T Y_t^\dagger [b(t, X_t, \alpha_t) - b(t, X'_t, \alpha'_t)] dt \\
&\qquad + \mathbb{E}\int_0^T \operatorname{trace}[(\sigma(t, X_t \alpha_t) - \sigma(t, X'_t, \alpha'_t))^\dagger Z_t] dt,
\end{aligned}$$

where we used integration by parts and the fact that the stochastic integrals are bona fide martingales, and consequently have expectations 0. On the other hand,

$$\begin{aligned}
&\mathbb{E}\int_0^T [f(t, X_t, \alpha_t) - f(t, X'_t, \alpha'_t)] dt \\
&= \mathbb{E}\int_0^T [H(t, X_t, Y_t, Z_t, \alpha_t) - H(t, X'_t, Y_t, Z_t, \alpha'_t)] dt \\
&\quad - \mathbb{E}\int_0^T Y_t [b(t, X_t, \alpha_t) - b(t, X'_t, \alpha'_t)] dt \\
&\quad - \mathbb{E}\int_0^T \operatorname{trace}([\sigma(t, X_t, \alpha_t) - \sigma(t, X'_t, \alpha'_t)]^\dagger Z_t) dt
\end{aligned}$$

4.2. Pontryagin Stochastic Maximum Principle

by definition of the Hamiltonian. Adding the two together we get

$$J(\alpha) - J(\alpha') = \mathbb{E}[g(X_T) - g(X'_T)] + \mathbb{E}\int_0^T [f(t, X_t, \alpha_t) - f(t, X'_t, \alpha'_t)]dt$$

$$\leq \mathbb{E}\int_0^T [H(t, X_t, Y_t, Z_t, \alpha_t) - H(t, X'_t, Y_t, Z_t, \alpha'_t)$$
$$- (X_t - X'_t)^\dagger \partial_x H(t, X_t, Y_t, Z_t, \alpha_t)]dt,$$

which is negative because of (4.22) and the convexity assumption. □

Remark 4.15. *If the minimized Hamiltonian H^* is defined by*

$$H^*(t, x, y, z) = \inf_{\alpha \in A} H(t, x, y, z, \alpha), \tag{4.24}$$

then the necessary part of the Pontryagin stochastic maximum principle tells us that if α is optimal, then $H^*(t, X_t, Y_t, Z_t) = H(t, X_t, Y_t, Z_t, \alpha_t)$. The term "Hamiltonian" comes from the fact that the equation giving the dynamics of the state of the system can be written in the form

$$dX_t = \partial_y H(t, X_t, Y_t, Z_t, \alpha_t)dt + \partial_z H(t, X_t, Y_t, Z_t, \alpha_t)dW_t,$$

the combination of this equation and the adjoint equation being a form familiar to aficionados of Hamiltonian mechanics.

Connection with the Dynamic Programming Principle and the HJB Equation

Proposition 4.16. *Let us assume that the HJB value function v is $C^{1,3}$, i.e., once continuously differentiable in t and three times continuously differentiable in x with bounded derivatives of all orders, and that there exists an optimal control α^*. If we denote by $\mathbf{X}^* = \mathbf{X}^{\alpha^*}$ the corresponding controlled state process, then the couple of processes $(\mathbf{Y}^*, \mathbf{Z}^*)$ defined by*

$$Y_t^* = \partial_x v(t, X_t^*) \quad \text{and} \quad Z_t^* = \partial_{xx}^2 v(t, X_t^*)\sigma(t, X_t^*, \alpha_t^*) \tag{4.25}$$

is a couple of adjoint processes solving the adjoint equations, and

$$H(t, X_t^*, \partial_x v(t, X_t^*), \partial_{xx}^2 v(t, X_t^*)\sigma(t, X_t^*, \alpha_t^*), \alpha_t^*) \tag{4.26}$$
$$= \inf_{\alpha \in A} H(t, X_t^*, \partial_x v(t, X_t^*), \partial_{xx}^2 v(t, X_t^*)\sigma(t, X_t^*, \alpha_t^*), \alpha),$$

$\mathbb{P}$-*almost surely for almost every $t \in [0, T]$.*

Proof. The fact that the processes defined in (4.25) solve the adjoint equation is a simple consequence of Itô's formula, and the claim (4.26) is a mere restatement of the necessary part of the Pontryagin maximum principle. □

4.2.3 ▪ Return to Utility of Terminal Wealth Maximization

We revisit the optimization problem of the expected utility of terminal wealth with the Pontryagin stochastic maximum principle in hand. Using the same notation as in Subsection 3.5.2, the wealth dynamics are given by

$$dX_t = X_t[\alpha_t \mu + (1 - \alpha_t)r]dt + \alpha_t X_t \sigma dW_t, \tag{4.27}$$

and the admissible controls $\alpha = (\alpha_t)_t$ giving the proportion of wealth invested in the stock are the progressively measurable processes such that $\int_0^T |\alpha_t|^2 dt < \infty$. The Hamiltonian takes the form

$$H(t, x, y, z, \alpha) = x[\alpha(\mu - r) + r]y + \alpha x \sigma z,$$

which is a linear function of α. So unless $x[(\mu - r)y + \sigma z] = 0$, the fact that we allow α to roam around the entire real line implies that the minimized Hamiltonian is identically $-\infty$! So we cannot expect to find a function $(t, x, y, z) \hookrightarrow \hat{\alpha}(t, x, y, z) \in \mathbb{R}$ such that

$$H^*(t, x, y, z) = \inf_{\alpha \in \mathbb{R}} H(t, x, y, z, \alpha) > -\infty$$

for all (t, x, y, z). This unfortunate conclusion says that we cannot expect to use the sufficient part of the Pontryagin stochastic maximum principle in the present situation. However, it is still possible to use the necessary part and gain some interesting information about the system in equilibrium (i.e., at the optimum).

The Case of the CRRA Utility

If we restrict ourselves to the CRRA utility function and use the admissible control α^* given by the constant $\alpha_t^* \equiv (\mu - r)/\gamma \sigma^2$ which was found to be optimal, we note that the corresponding optimal wealth dynamics are given by

$$dX_t^* = X_t^*[\alpha^*(\mu - r) + r]dt + \alpha^* X_t^* \sigma dW_t.$$

This shows that the optimal wealth is a geometric Brownian motion with a constant rate of growth greater than the interest rate. The corresponding adjoint equation reads

$$dY_t^* = -\left[[\alpha^*(\mu - r) + r]Y_t^* + \alpha^* \sigma Z_t^*\right]dt + Z_t^* dW_t$$

with terminal condition $Y_T^* = X_T^{-\gamma}$ which is unbounded, but whose moments are all finite. This BSDE is an affine BSDE with constant coefficients which can be solved explicitly (recall the first examples of BSDEs discussed in Chapter 2). In particular,

$$Y_t^* = \mathbb{E}\left[\Gamma_t^T X_T^* | \mathcal{F}_t\right]$$

with

$$\Gamma_t^T = \exp[(u - t)[\alpha^*(\mu - r) + r] + \alpha^* \sigma(W_u - W_t) - \alpha^{*2}\sigma^2(u - t)^2/2].$$

Moreover, since we have $H(t, X_t^*, Y_t^*, Z_t^*, \alpha^*) > -\infty$, if we use the necessary part of the stochastic maximum principle, we learn that

$$H(t, X_t^*, Y_t^*, Z_t^*, \alpha^*) = \inf_{\alpha \in \mathbb{R}} H(t, X_t^*, Y_t^*, Z_t^*, \alpha),$$

which is possible only if $X_t^*[(\mu - r)Y_t^* + \sigma Z_t^*] = 0$. Since $X_t^* \neq 0$ almost surely, this shows that we must have $Z_t^* = -(\mu - r)Y_t^*/\sigma$ almost surely.

The Case of the log Utility

If we use the notation $\tilde{X}_t = -\log X_t$, the problem is to minimize $\mathbb{E}[\tilde{X}_T]$ where the dynamics of the controlled state are given by Itô's formula,

$$d\tilde{X}_t = \left[\frac{\sigma^2}{2}\alpha_t^2 - (\mu - r)\alpha_t - r\right]dt - \alpha_t \sigma dW_t.$$

4.2. Pontryagin Stochastic Maximum Principle

The Hamiltonian H is given by:

$$H(t,x,y,z,\alpha) = \left[\frac{\sigma^2\alpha^2}{2} - (\mu-r)\alpha - r\right]y - \alpha\sigma z$$

and its minimization gives:

$$H^*(t,x,y,z) = -\infty \quad \text{if} \quad y < 0 \quad \text{or} \quad (y = 0 \text{ and } z \neq 0).$$

When H^* is finite, the minimum is attaind for

$$\alpha^* = \frac{\mu-r}{\sigma^2} + \frac{z}{\sigma y}$$

which is unforunately not constant.

The Case of a General Utility Function

We now consider the maximization of $\mathbb{E}[U(X_T)]$ for a general utility function U. The Hamiltonian is the same as above and we use the same notation α^* for it minimizer. As we noticed above when we revisited Merton's solution in the case of the CRRA utility function, the necessary part of the stochastic maximum principle implies that the adjoint processes must satisfy $Z_t^* = -(\mu-r)Y_t^*$, and consequently, if we set $\lambda = (\mu-r)/\sigma$, we have

$$Y_t^* = U'(X_T^*) - \lambda \int_t^T Y_s^* dW_s,$$

whose solution is $Y_t^* = Y_0^* e^{\lambda W_t - t\lambda^2/2}$. Consequently, we need to find Y_0^* so that $U(X_T) = Y_0^* e^{\lambda W_T - T\lambda^2/2}$. Notice that, once Y_0^* is found, the optimal terminal wealth is given by

$$X_T^* = (U')^{-1}(Y_T^*) = (U')^{-1}(Y_0^* e^{\lambda W_T - T\lambda^2/2}).$$

Remark 4.17. *The above argument is borrowed from the classical duality approach to utility maximization. It can be repeated mutatis mutandis when the rate of growth μ and the volatility σ are bounded adapted stochastic processes, as long as $\sigma_t > \underline{\sigma}$ for some strictly positive constant $\underline{\sigma}$.*

4.2.4 ▪ A Simple Example of a Constrained Optimization

Let us assume that γ, λ, and σ are positive constants, and let us consider the dynamics

$$\begin{cases} dX_t^0 = \gamma\alpha_t dt + \sigma dW_t, & X_0^0 = x_0, \\ dX_t^1 = \alpha_t dt, & X_0^1 = x, \end{cases} \quad (4.28)$$

for the two-dimensional controlled state $\boldsymbol{X}_t = [X_t^0, X_t^1]^\dagger$, where $\mathbf{W} = (W_t)_{0 \leq t \leq T}$ is a one-dimensional Wiener process and the set $\mathbb{A}$ of admissible controls $\boldsymbol{\alpha} = (\alpha_t)_{0 \leq t \leq T}$ is chosen to be the space $\mathbb{H}^2$ of square integrable adapted real-valued processes. X_t^0 should be interpreted as the *midprice* (midpoint between the best bid and ask prices) of some instrument, and X_t^1 the position of an investor trading at speed α_t. The drift term in the first equation accounts for the permanent impact that the speed of trading has on the midprice. Starting from an initial position $X_0^1 = x$, our (risk-neutral) trader wants to minimize the cost of trading over the period $[0, T]$ as given by

$$J(\alpha) = \mathbb{E}\left[\int_0^T (X_t^0 + \lambda \alpha_t)\alpha_t dt\right].$$

Here, the term $\lambda \alpha_t$ represents the temporary impact and $X_t^0 + \lambda \alpha_t$ the price at which the transaction actually takes place at time t. The purpose of this subsection is to compare the impact of adding a constraint to this optimal control problem. For the sake of the discussion, we shall solve the problem with and without adding the requirement that the trader should end *clean hand*, in the sense that his terminal position should be zero, i.e., $X_T^1 = 0$ and he does not carry any inventory at the end of the trading period.

Notice that

$$\int_0^T (X_t^0 + \lambda \alpha_t)\alpha_t dt = \int_0^T X_t^0 dX_t^1 + \lambda \int_0^T \alpha_t^2 dt$$

$$= X_T^1 X_T^0 - X_0^1 X_0^0 - \sigma \int_0^T X_t^1 dW_t - \gamma \int_0^T X_t^1 \alpha_t dt + \lambda \int_0^T \alpha_t^2 dt$$

$$= X_T^1 X_T^0 - xx_0 - \sigma \int_0^T X_t^1 dW_t - \frac{\gamma}{2}[(X_T^1)^2 - x^2] + \lambda \int_0^T \alpha_t^2 dt$$

so that

$$J(\alpha) = \mathbb{E} X_T^1 \left(X_T^0 - \frac{\gamma}{2} X_T^1\right) + x\left(\frac{\gamma}{2} x - x_0\right) + \lambda \mathbb{E}\left[\int_0^T \alpha_t^2 dt\right], \tag{4.29}$$

since the stochastic integral is a true martingale (as opposed to a local martingale) because $\mathbf{X}^1 \in \mathbb{H}^2$. Indeed, Jensen's inequality implies

$$\int_0^T (X_t^1 - X_0^1)^2 dt = \int_0^T \left(\int_0^t \alpha_s ds\right)^2 dt \leq \int_0^T t\left(\int_0^t \alpha_s^2 ds\right) dt \leq \frac{T^2}{2} \int_0^T \alpha_s^2 ds.$$

We first consider **case I** for which the terminal position X_T^1 is unconstrained and the set of admissible controls is the whole $\mathbb{H}^2$. In this case, the reduced Hamiltonian reads

$$\tilde{H}(t, x^0, x^1, y^0, y^1, \alpha) = \gamma \alpha y^0 + \alpha y^1 + (x^0 + \lambda \alpha)\alpha.$$

It attains its minimum for $\hat{\alpha} = \hat{\alpha}(x^0, y^0, y^1) = -(\gamma y^0 + y^1 + x^0)/(2\lambda)$. So the solution of the optimal control problem hinges on the solution of the FBSDE

$$\begin{cases} dX_t^0 = -\frac{\gamma}{2\lambda}(\gamma Y_t^0 + Y_t^1 + X_t^0)dt + \sigma dW_t, & X_0^0 = x_0, \\ dX_t^1 = -\frac{1}{2\lambda}(\gamma Y_t^0 + Y_t^1 + X_t^0)dt, & X_0^1 = x, \\ dY_t^0 = \frac{1}{2\lambda}(\gamma Y_t^0 + Y_t^1 + X_t^0)dt + Z_t^0 dW_t, & Y_T^0 = 0, \\ dY_t^1 = Z_t^1 dW_t, & Y_T^1 = 0, \end{cases} \tag{4.30}$$

from which we see that $Y_t^1 \equiv 0$, which reduces the FBSDE to

$$\begin{cases} dX_t^0 = -\frac{\gamma}{2\lambda}(\gamma Y_t^0 + X_t^0)dt + \sigma dW_t, & X_0^0 = x_0, \\ dY_t^0 = \frac{1}{2\lambda}(\gamma Y_t^0 + X_t^0)dt + Z_t^0 dW_t, & Y_T^0 = 0, \end{cases} \tag{4.31}$$

4.2. Pontryagin Stochastic Maximum Principle

since X_t^1 can be determined afterwards by computing $X_t^1 = x - \frac{1}{2\lambda}\int_0^t (\gamma Y_s^0 + X_s^0)ds$. FBSDE (4.31) is linear and its solution can be reduced to the solution of an ordinary Riccati equation by setting $Y_t^0 = a_t X_t^0$ for some deterministic function $t \hookrightarrow a_t$. This equation reads

$$2\lambda \dot{a}_t = \gamma^2(a_t)^2 + 2\gamma a_t + 1, \qquad a_T = 0,$$

whose solution is

$$a_t = -\frac{t-T}{\gamma(t-T) - 2\lambda}.$$

As a result, the midprice $\mathbf{X}^0$ is a Gaussian Ornstein–Uhlenbeck process solving the SDE

$$dX_t^0 = -\frac{\gamma}{2\lambda}(1 + \gamma a_t)X_t^0 dt + \sigma dW_t, \qquad X_0^0 = x_0, \qquad (4.32)$$

and the corresponding optimal control

$$\hat{\alpha}_t = -\frac{1}{2\lambda}(1 + \gamma a_t)X_t^0$$

is also a Gaussian process. Plugging these expressions into the formula for the cost J we get the optimal cost

$$\hat{J} = J(\hat{\underline{\alpha}}) = \mathbb{E}\left[\int_0^T (X_t^0 + \lambda \hat{\alpha}_t)\hat{\alpha}_t dt\right]$$

$$= \frac{1}{4\lambda}\int_0^T (\gamma^2(a_t)^2 - 1)\mathbb{E}[(X_t^0)^2]dt.$$

Since equation (4.32) can be solved explicitly,

$$X_t^0 = x_0 \exp\left[-\int_0^t \frac{\gamma}{2\lambda}(1+\gamma a_s)ds\right] + \sigma \int_0^t \exp\left[-\frac{\gamma}{2\lambda}(1+\gamma a_u)du\right]dW_s,$$

we get

$$\mathbb{E}[(X_t^0)^2] = x_0^2 \exp\left[-\int_0^t \frac{\gamma}{\lambda}(1+\gamma a_s)ds\right] + \sigma^2 \int_0^t \exp\left[-\frac{\gamma}{\lambda}(1+\gamma a_u)du\right],$$

proving that $\hat{J}$ depends upon the initial value x_0 of the midprice (and γ, λ, and σ) but not on the initial holding x!

We can also consider **case II** for which we impose the restriction $\alpha_t \leq 0$ (the trader is only allowed to sell), still imposing no restriction on the terminal position X_T^1. In this case, the reduced Hamiltonian should be minimized over $\alpha \in (-\infty, 0]$ and the minimizer is now $\hat{\alpha} = \hat{\alpha}(x^0, y^0, y^1) = -(\gamma y^0 + y^1 + x^0)^+/(2\lambda)$, and if we use the notation $D = \{(x^0, y^0, y^1); \gamma y^0 + y^1 + x^0 > 0\}$, then the FBSDE becomes

$$\begin{cases} dX_t^0 = -\frac{\gamma}{2\lambda}(\gamma Y_t^0 + Y_t^1 + X_t^0)\mathbf{1}_D(X_t^0, Y^0, t, Y_t^1)dt + \sigma dW_t, & X_0^0 = x_0, \\ dX_t^1 = -\frac{1}{2\lambda}(\gamma Y_t^0 + Y_t^1 + X_t^0)\mathbf{1}_D(X_t^0, Y^0, t, Y_t^1)dt, & X_0^1 = x, \\ dY_t^0 = -\frac{1}{2\lambda}(\gamma Y_t^0 + Y_t^1 + X_t^0)dt + Z_t^0 dW_t, & Y_T^0 = 0, \\ dY_t^1 = Z_t^1 dW_t, & Y_T^1 = 0. \end{cases} \qquad (4.33)$$

Finally, we consider **case III** for which we impose the restriction $X_T^1 = 0$ for a control to be admissible. Let us assume momentarily that $\boldsymbol{\alpha}$ is a deterministic function of time which satisfies $\int_0^T \alpha_s^2 ds < \infty$ and $\int_0^T \alpha_s ds = 0$, and which minimizes J over all such deterministic functions. Now, if $\beta \in \mathbb{H}^2$ is such that $X_t^1 = x + \int_0^t \beta_s ds$ satisfies $X_T^1 = 0$ almost surely, then if we use (4.29) and the notation $\overline{\beta}_t$ for the expectation $\mathbb{E}[\beta_t]$, we have

$$J(\beta) = x\left(\frac{\gamma}{2}x - x_0\right) + \lambda \mathbb{E}\left[\int_0^T \beta_t^2 dt\right]$$

$$\geq -xx_0 + \lambda \int_0^T \overline{\beta}_t^2 dt$$

$$= J(\overline{\beta}) \geq J(\alpha),$$

which shows that α also minimizes J over all the square integrable adapted control strategies satisfying the terminal constraint. We see from (4.29) that $\alpha \hookrightarrow J(\alpha)$ is strictly convex so that its minimum, if it exists, is unique, and it can be found by minimizing J over the deterministic function of time which satisfies $\int_0^T \alpha_s^2 ds < \infty$ and $\int_0^T \alpha_s ds = 0$. This is a classical (nonrandom) calculus of variations problem and one easily finds that the constant function $\alpha_t \equiv -x/(2T)$ gives the optimal rate of trading in the constrained case.

4.3 ▪ Linear-Quadratic (LQ) Models

In this section we consider the case of linear-quadratic (LQ) models. They are often used because they offer an important class of solvable models.

4.3.1 ▪ The Model

The dynamics of the controlled state are linear in the sense that

$$b(t,x,\alpha) = A_t x + B_t \alpha + \beta_t \quad \text{and} \quad \sigma(t,x,\alpha) = C_t x + D_t \alpha + \sigma_t,$$

where $\mathbf{A} = (A_t)_{0\leq t\leq T}$, $\mathbf{B} = (B_t)_{0\leq t\leq T}$, $\boldsymbol{\beta} = (\beta_t)_{0\leq t\leq T}$, $\mathbf{C} = (C_t)_{0\leq t\leq T}$, $\mathbf{D} = (D_t)_{0\leq t\leq T}$, and $\boldsymbol{\sigma} = (\sigma_t)_{0\leq t\leq T}$ are bounded progressively measurable processes in $\mathbb{R}^{d\times d}$, $\mathbb{R}^{d\times k}$, $\mathbb{R}^d$, $\mathbb{R}^{d\times m\times d}$, $\mathbb{R}^{d\times m\times k}$, and $\mathbb{R}^{d\times m}$, respectively, and as always we assume that $\mathbf{W} = (W_t)_{0\leq t\leq T}$ is a Wiener process in $\mathbb{R}^m$. So the controlled state dynamics are of the form

$$dX_t = [A_t X_t + B_t \alpha_t + \beta_t]dt + [C_t x + D_t \alpha_t + \sigma_t]dW_t, \qquad X_0 = x. \quad (4.34)$$

In this section, we assume that the set $\mathbb{A}$ of admissible controls is the space $\mathbb{H}^{2,k}$, and as before, for each admissible control α we denote by $\mathbf{X} = \mathbf{X}^\alpha$ the corresponding controlled state process, and the associated cost is given by

$$J(\alpha) = \mathbb{E}\left[\int_0^T f(t, X_t, \alpha_t)dt + g(X_T)\right],$$

where in this section we assume that the running and terminal cost functionals are quadratic in the state and control variables. To be specific, we assume that they are of the form

$$f(t,x,\alpha) = \frac{1}{2}x^\dagger L_t x + \frac{1}{2}\alpha^\dagger M_t \alpha \quad \text{and} \quad g(x) = \frac{1}{2}x^\dagger N x,$$

where $\mathbf{L} = (L_t)_{0\leq t\leq T}$ and $\mathbf{M} = (M_t)_{0\leq t\leq T}$ are bounded progressively measurable processes in the subspaces of $\mathbb{R}^{d\times d}$ and $\mathbb{R}^{k\times k}$ of symmetric positive definite matrices, and

4.3. Linear-Quadratic (LQ) Models

N is a bounded random symmetric positive definite matrix in $\mathbb{R}^{d\times d}$. We also assume the existence of a positive (deterministic) constant $c > 0$ such that for all $t \in [0, T]$,

$$c \leq L_t, \quad c \leq M_t \leq c^{-1}, \quad \text{and} \quad N \geq c,$$

$\mathbb{P}$-almost surely, where these inequalities are understood in the sense of inequalities between quadratic forms. Under these conditions, the Hamiltonian of the system reads

$$H(t,x,y,z,\alpha) = y \cdot [A_t x + B_t \alpha + \beta_t] + z \cdot [C_t x + D_t \alpha + \sigma_t] + \frac{1}{2}x^\dagger L_t x + \frac{1}{2}\alpha^\dagger M_t \alpha. \quad (4.35)$$

4.3.2 ▪ The Weak Formulation Approach

In order to apply ourselves to the setup of the weak formulation described in Section 4.1, we need to assume that $d = m$, and that the volatility is uncontrolled (i.e., $D_t = 0$) and independent of the state variable (i.e., $C_t = 0$), if one wants to satisfy the invertibility condition. In other words, we assume that the processes $\mathbf{C} = (C_t)_{0 \leq t \leq T}$ and $\mathbf{D} = (D_t)_{0 \leq t \leq T}$ are zero. Moreover, we also assume that the process $\boldsymbol{\sigma} = (\sigma_t)_{0 \leq t \leq T}$ is bounded and progressively measurable in $\mathbb{R}^{d\times d}$, and that the matrices σ_t are symmetric and positive definite with a uniform lower bound $\sigma_t \geq \lambda > 0$ independent of t and the random scenario. As explained earlier, the symmetry is only assumed for notational convenience, so that we do not have to carry all the adjoint volatility matrices. As always, we assume that $\mathbf{W} = (W_t)_{0 \leq t \leq T}$ is a Wiener process in $\mathbb{R}^d$ (recall that we assume $d = m$). So the controlled state dynamics are of the form

$$dX_t = [A_t X_t + B_t \alpha_t + \beta_t]dt + \sigma_t dW_t, \quad X_0 = x. \quad (4.36)$$

The reduced Hamiltonian reads

$$\tilde{H}(t,x,y,\alpha) = y \cdot [A_t x + B_t \alpha + \beta_t] + \frac{1}{2}x^\dagger L_t x + \frac{1}{2}\alpha^\dagger M_t \alpha$$

so that

$$\partial_\alpha \tilde{H} = 0 \quad \Leftrightarrow \quad y^\dagger B_t + \alpha^\dagger M_t = 0,$$

which is minimized by

$$\hat{\alpha}(t,x,y) = -M_t^{-1} B_t^\dagger y. \quad (4.37)$$

The BSDE identified in Proposition 4.1 reads

$$\begin{aligned}
d\hat{Y}_t &= -\tilde{H}(t, X_t^0, -\sigma_t^{-1}\hat{Z}_t, \hat{\alpha}(t, X_t^0, -\sigma_t^{-1}\hat{Z}_t))dt - \hat{Z}_t dW_t, \\
&= \left([\sigma_t^{-1}\hat{Z}_t] \cdot [A_t X_t^0 + B_t M_t^{-1} B_t^\dagger \sigma_t^{-1}\hat{Z}_t + \beta_t] - \frac{1}{2}X_t^{0\dagger} L_t X_t^0 \right. \\
&\qquad \left. - \frac{1}{2}\hat{Z}_t^\dagger[\sigma_t^{-1} B_t M_t^{-1} B_t^\dagger \sigma_t^{-1}]\hat{Z}_t\right)dt - \hat{Z}_t dW_t, \\
&= \left(-\frac{1}{2}X_t^{0\dagger} L_t X_t^0 + \hat{Z}_t^\dagger \sigma_t^{-1}[A_t X_t^0 + \beta_t] \right. \\
&\qquad \left. + \frac{1}{2}\hat{Z}_t[\sigma_t^{-1} B_t M_t^{-1} B_t^\dagger \sigma_t^{-1}]\hat{Z}_t\right)dt - \hat{Z}_t dW_t,
\end{aligned}$$

where we substituted the minimizer $\hat{\alpha}$ identified in (4.37). The driver of this BSDE is quadratic in z, and even though the variable y does not appear in the driver, this equation will be seldom solvable. In fact, even the Kobylanski existence theorem for quadratic BSDEs cannot be used in general because the terminal condition $\hat{Y}_T = \frac{1}{2}X_T^{0\dagger} N X_T^0$ is quadratic in X_T^0, and as a result is unlikely to be a bounded random variable.

4.3.3 ▪ The Stochastic Maximum Approach

Having mostly failed to use the first prong of the probabilistic approach in the solution of control problems, we now return to the general setting of the section, and try to use the Pontryagin stochastic maximum principle approach. We compute the partial derivative of the Hamiltonian (4.35) with respect to the state variable in order to state the adjoint equations. We get

$$\partial_x H(t, x, y, z, \alpha) = A_t^\dagger y + z \cdot C_t^{(1,2)} + L_t x,$$

where $z \cdot C_t^{(1,2)}$ means that the inner product is computed as a function of the third dimension of C_t, which is not contracted. For a given admissible control $\alpha \in \mathbb{A}$, if $\boldsymbol{X}$ denotes the corresponding controlled state process, the adjoint equation is the BSDE

$$dY_t = -\Psi(t, Y_t, Z_t)dt + Z_t dW_t$$

with terminal value $Y_T = NX_T$ and driver $\Psi(t, y, z) = y^\dagger A_t + z \cdot C_t^{(1,2)} + L_t X_t$, which is affine in the variables y and z. Because of the bounded hypothesis on N and $\mathbf{L}$, together with the $\mathbb{S}^{2,d}$ bound on the solution X_t of the state dynamics SDE, the conditions of existence and uniqueness of Theorem 2.2 are satisfied and the adjoint processes always exist. However, both the necessary and the sufficient parts of the stochastic maximum principle suggest that we look for an optimal control by minimizing the Hamiltonian. In other words, for each fixed (t, x, y, z) we look for a control $\hat{\alpha}(t, x, y, z) \in A$ satisfying

$$\hat{\alpha}(t, x, y, z) \in \arg\inf_{\alpha \in A} H(t, x, y, z, \alpha).$$

In the present LQ model, the first-order condition reads

$$\partial_\alpha H = 0 \quad \Leftrightarrow \quad y^\dagger B_t + z \cdot D_t^{(1,2)} + \alpha^\dagger M_t = 0$$

so that

$$\hat{\alpha}(t, x, y, z) = -M_t^{-1}[B_t^\dagger y + \mathrm{trace}(z \cdot D_t^{(1,2)\dagger})]$$

satisfies

$$H(t, x, y, z, \hat{\alpha}(t, x, y, z)) = H^*(t, x, y, z) = \inf_{\alpha \in A} H(t, x, y, z, \alpha)$$

for all t, x, y, and z. Recall that a function $\hat{\alpha}$ satisfying this property is said to satisfy Isaacs' condition. While the BSDE to be solved in order to construct the adjoint processes is decoupled from the forward dynamics, we now try to construct both the forward controlled state $\boldsymbol{X}$ and the backward (adjoint) processes $(\boldsymbol{Y}, \boldsymbol{Z})$ when the admissible control process is of the form $\hat{\alpha}_t = \hat{\alpha}(t, X_t, Y_t, Z_t)$ for the minimizer function $\hat{\alpha}$. The forward SDE and the adjoint equations are now fully coupled and we need to solve a coupled FBSDE in order to solve our control problem. While the necessary part of the stochastic maximum principle only requires that the Hamiltonian be minimized along the optimal path, our choice of control $\hat{\alpha}_t$ will guarantee that the Hamiltonian is minimized everywhere. Consequently, if we can solve the resulting coupled FBSDE, the sufficient part of the stochastic maximum principle will guarantee that we have actually solved the control problem. In the present situation, this FBSDE reads

$$\begin{cases} dX_t = [A_t X_t - B_t M_t^{-1} B_t^\dagger Y_t - \beta_t M_t^{-1} \mathrm{trace}(Z \cdot D_t^{(1,2)\dagger}) + \beta_t] dt \\ \qquad + [C_t X_t - D_t M_t^{-1} B_t^\dagger Y_t - D_t M_t^{-1} \mathrm{trace}(Z \cdot D_t^{(1,2)\dagger}) + \sigma_t] dW_t, \quad X_0 = x, \\ dY_t = -[L_t X_t + Y_t^\dagger A_t + Z_t \cdot C_t^{(1,2)}] dt + Z_t dW_t, \quad Y_T = NX_T. \end{cases}$$

(4.38)

4.3. Linear-Quadratic (LQ) Models

For the sake of comparison with the HJB approach, we consider the particular case of uncontrolled volatility for which $D_t = 0$. In this case, the FBSDE becomes

$$\begin{cases} dX_t = [A_t X_t - B_t M_t^{-1} B_t^\dagger Y_t + \beta_t]dt + [C_t X_t + \sigma_t]dW_t, & X_0 = x, \\ dY_t = -[L_t X_t + A_t^\dagger Y_t + Z_t \cdot C_t^{(1,2)}]dt + Z_t dW_t, & Y_T = NX_T. \end{cases} \quad (4.39)$$

This is an affine FBSDE of the type considered in Section 2.6. Recall that our preferred way to solve this class of FBSDE is to use the fact that the decoupling field is expected to be an affine function of the state variable, and to search for the coefficients of this affine function by identifying the ODEs that these coefficients have to satisfy.

Remark 4.18. *Recall that the decoupling field of the FBSDE (4.39) (sometimes called the FBSDE value function) is expected to be the derivative with respect to the state variable x of the value function of the optimization problem. For this reason, the Riccati equation whose solution determines the affine form of the decoupling field, should be different from the Riccati equation whose solution determines the quadratic form of the HJB value function, even though simple arithmetic allows us to derive one from the other!*

But since Section 2.6 was mostly focused on the reduction of the problem to a nonlinear ODE of the Riccati type, we here show that it is possible, at least under some specific assumptions, to identify conditions on the coefficients which could allow the use of the monotonicity arguments of Theorem 2.20. For the sake of simplicity, we restrict ourselves to the case $d = p$ and we choose $G = I_d$. The second condition is satisfied in the present situation since

$$\langle g(x) - g(\tilde{x}), x - \tilde{x} \rangle = (x - \tilde{x})^\dagger N(x - \tilde{x}) \geq c|x - tx|^2$$

by assumption on the lower boundedness of the spectrum of N. The first condition is more difficult to check. It reads

$$-[L_t(x - \tilde{x}) + A_t^\dagger(y - \tilde{y}) + (z - \tilde{z}) \cdot C_t^{(1,2)}]^\dagger (x - \tilde{x})$$
$$+ [A_t(x - \tilde{x}) - B_t M_t^{-1} B_t^\dagger(y - \tilde{y})]^\dagger (y - \tilde{y}) + [C_t(x - \tilde{x})]^\dagger (z - \tilde{z})$$
$$\leq -\beta_1 |x - \tilde{x}|^2 - \beta_2 |y - \tilde{y}|^2 + |z - \tilde{z}|^2,$$

which reduces to

$$-(x - \tilde{x})^\dagger L_t(x - \tilde{x}) - (y - \tilde{y})^\dagger B_t M_t^{-1} B_t^\dagger(y - \tilde{y}) + [C_t(x - \tilde{x})]^\dagger (z - \tilde{z})$$
$$\leq -\beta_1 |x - \tilde{x}|^2 - \beta_2 |y - \tilde{y}|^2 + |z - \tilde{z}|^2,$$

which is satisfied with β_1, the uniform lower bound on the spectrum of L_t, and β_2, the uniform lower bound on the spectrum of $B_t M_t^{-1} B_t^\dagger$.

4.3.4 ▪ Mean-Variance Portfolio Selection

One form of the mean-variance portfolio optimization problem is to minimize, for each fixed value $m = \mathbb{E}[X_T]$ of the expected terminal wealth, the variance of the terminal value of the portfolio, namely, to solve the constrained minimization problem and compute

$$J(m) = \inf_{\alpha \in \mathbb{A}, \, \mathbb{E}[X_T] = m} \text{var}(X_T).$$

A simple Lagrangian duality argument relates this value function to its dual defined by

$$\tilde{J}(\lambda) = \inf_{\alpha \in \mathbb{A}} \mathbb{E}[(X_T - \lambda)^2] \qquad (4.40)$$

for $\lambda \in \mathbb{R}$. Indeed, the duality relationships give

$$\tilde{J}(\lambda) = \inf_{m \in \mathbb{R}} [J(m) + (m - \lambda)^2]$$

and

$$J(m) = \sup_{\lambda \in \mathbb{R}} [\tilde{J}(\lambda) - (m - \lambda)^2]. \qquad (4.41)$$

Moreover, for any fixed level m of expected return, the optimal control is given by the Markovian control $\hat{\alpha}_{\lambda_m}$ given by formula (4.50), where λ_m is the argument of the supremum (4.41) as given by

$$\lambda_m = \frac{m - e^{\int_0^T (r_s - \rho_s) ds}}{1 - e^{-\int_0^T \rho_s ds}},$$

where ρ_t is defined below in (4.49).

So, the mean-variance optimal portfolio search reduces to the solution of the set of LQ control problems (4.40), and for each $\lambda \in \mathbb{R}$ we consider the problem of minimizing

$$J^\lambda(\boldsymbol{\alpha}) = \mathbb{E}[(X_T - \lambda)^2] \qquad (4.42)$$

under the dynamic constraint

$$dX_t = \left[X_t r_t dt + \sum_{i=1}^m \alpha_t^i [(\mu_t^i - r_t) dt + \sigma_t^i dW_t]\right]. \qquad (4.43)$$

The Hamiltonian reads

$$H(t, x, y, z, \alpha) = y[r_t x + \alpha \cdot (\mu - r_t \mathbf{1})] + z\alpha \cdot \sigma_t$$

so that, for each admissible control process $\boldsymbol{\alpha} \in \mathbb{A}$, the adjoint equation reads

$$dY_t = -r_t Y_t dt + Z_t dW_t, \qquad Y_T = 2(X_T - \lambda). \qquad (4.44)$$

The Hamiltonian being linear in the control variable α, we cannot expect to use the sufficient part of the stochastic maximum principle to prove the existence of an optimum. However, we can still use the necessary part of the principle. It says that if $\hat{\alpha}$ is an optimal control process, $\hat{\boldsymbol{X}}$ the corresponding controlled state, and $(\hat{\boldsymbol{Y}}, \hat{\boldsymbol{Z}})$ the corresponding adjoint processes, then the Hamiltonian should be minimized along the optimal trajectories so that

$$H^*(t, \hat{X}_t, \hat{Y}_t, \hat{Z}_t, \alpha) = \hat{Y}_t r_t \hat{X}_t + \alpha \cdot [(\mu - r_t \mathbf{1}) \hat{Y}_t + \hat{Z} \sigma_t]$$

should be minimized in α for $\hat{\alpha}_t$, which implies that

$$(\mu_t - r_t \mathbf{1}) \hat{Y}_t + \hat{Z} \sigma_t = 0, \qquad 0 \le t \le T, \text{ a.s.} \qquad (4.45)$$

Notice that the forward equation $dX_t = [X_t r_t + \alpha_t \cdot (\mu_t - r_t \mathbf{1})] dt + \alpha_t \cdot \sigma_t dW_t$ is affine, and as a result, the Hamiltonian is linear in the control variable α, which says that the control problem is singular. Even though the adjoint equation (4.44) is affine, we do not know its structure after the control process α is replaced by the optimal control $\hat{\alpha}$ if and when we can identify it, since we do not know how the latter depends upon $\hat{X}_t$, $\hat{Y}_t$, and $\hat{Z}_t$.

Nevertheless, we are going to behave as if the resulting FBSDE was known to be affine, and assume that its decoupling field is affine. In other words, we will look for a solution in the form $Y_t = \eta_t X_t + \chi_t$ for two deterministic smooth functions η and χ of the time variable t. This approach is reasonable because the adjoint equation (4.44) does not depend upon α_t, and because condition (4.45) identifies $\hat{Z}_t$ in terms of $\hat{Y}_t$. Computing the differential of Y_t from this ansatz, plugging in the adjoint equation (4.44), and identifying the bounded variation parts and the martingale parts, we get

$$\begin{cases} (\dot{\eta}_t + \eta_t r_t) X_t + \eta_t \alpha_t \cdot (\mu_t - r_t \mathbf{1}) + \dot{\chi}_t = -r_t \eta_t X_t - r_t \chi_t, \\ Z_t = \eta_t \alpha_t \cdot \sigma_t. \end{cases} \qquad (4.46)$$

Plugging this expression of Z_t into (4.45) allows us to determine the necessary form of the optimal control

$$\hat{\alpha}_t = -[\eta_t \sigma_t^\dagger \sigma_t]^{-1} (\mu_t - r_t \mathbf{1})(\eta_t X_t + \chi_t). \qquad (4.47)$$

Plugging this expression for the optimal control process into the first of the two equations of (4.46), and identifying the coefficients of X_t and the constant terms, we get

$$\begin{cases} \dot{\eta}_t + \eta_t (2r_t - \rho_t) = 0, \\ \dot{\chi}_t + (r_t - \rho_t) = 0, \end{cases} \qquad (4.48)$$

where we set

$$\rho_t = (\mu_t - r_t \mathbf{1})^\dagger [\sigma_t^\dagger \sigma_t]^{-1} (\mu_t - r_t \mathbf{1}). \qquad (4.49)$$

Recalling the terminal condition $\eta_T = 2$, the first of the two ODEs in (4.72) can be solved explicitly. We get

$$\eta_t = 2 e^{\int_t^T (2r_s - \rho_s) ds}.$$

The function η_t is independent of the multiplier λ. However, because of the terminal condition $\chi_T = -2\lambda$, the function χ_t depends upon λ, a fact which we emphasize by using a superscript λ. We get

$$\chi_t^\lambda = -2\lambda e^{\int_t^T (r_s - \rho_s) ds}.$$

Finally, we emphasize the fact that the optimal control can be given in closed loop feedback (Markovian) form using these two expressions for η_t and χ_t^λ:

$$\hat{\alpha}_\lambda(t, x) = -[\eta_t \sigma_t^\dagger \sigma_t]^{-1} (\mu_t - r_t \mathbf{1})(\eta_t x + \chi_t^\lambda). \qquad (4.50)$$

4.4 · Optimal Control of McKean–Vlasov Dynamics

In what follows, we assume that $\mathbf{W} = (W_t)_{0 \le t \le T}$ is an m-dimensional standard Wiener process defined on a probability space $(\Omega, \mathcal{F}, \mathbb{P})$ and $\mathbb{F} = (\mathcal{F}_t)_{0 \le t \le T}$ is its natural filtration. For each random variable/vector or stochastic process X, we denote by $\mathbb{P}_X$, or whenever possible by the alternative notation $\mathcal{L}(X)$, the law (also called the distribution) of X. The stochastic dynamics of interest in this section are given by a stochastic process $\mathbf{X} = (X_t)_{0 \le t \le T}$ satisfying a nonlinear stochastic differential equation of the form

$$dX_t = b(t, X_t, \mathcal{L}(X_t), \alpha_t) dt + \sigma(t, X_t, \mathcal{L}(X_t), \alpha_t) dW_t, \qquad 0 \le t \le T, \quad (4.51)$$

where the drift and volatility of the state X_t of the system are given by a deterministic function $(b, \sigma) : [0, T] \times \mathbb{R}^d \times \mathcal{P}_2(\mathbb{R}^d) \times A \hookrightarrow \mathbb{R}^d \times \mathbb{R}^{d \times m}$ and $\boldsymbol{\alpha} = (\alpha_t)_{0 \le t \le T}$ is a progressively measurable process with values in a measurable space $(A, \mathcal{A})$. Typically, A will be a Borel subset of a Euclidean space $\mathbb{R}^k$, and $\mathcal{A}$ the σ-field induced by the Borel σ-

field of this Euclidean space. Also, for each measurable space $(E, \mathcal{E})$, we use the notation $\mathcal{P}(E)$ for the space of probability measures on $(E, \mathcal{E})$, assuming that the σ-field $\mathcal{E}$ on which the measures are defined is understood. When E is a metric or a normed space (most often $\mathbb{R}^d$ in what follows), we denote by $\mathcal{P}_p(E)$ the subspace of $\mathcal{P}(E)$ of the probability measures which integrate the pth power of the distance to a fixed point (whose choice is irrelevant in the definition of $\mathcal{P}_p(E)$). We call the dynamics (4.51) *nonlinear*, not so much because the coefficients b and σ can be nonlinear functions of x, but because they depend not only on the value of the unknown process X_t at time t, but also on its marginal distribution $\mathcal{L}(X_t)$. We shall assume that the drift coefficient b and the volatility σ satisfy the following assumptions.

(A1) For each $x \in \mathbb{R}^d$ and $\mu \in \mathcal{P}_2(\mathbb{R}^d)$, the processes $(b(t, x, \mu))_{0 \leq t \leq T}$ and $(\sigma(t, x, \mu))_{0 \leq t \leq T}$ are in $\mathbb{H}^{2,d}$ and $\mathbb{H}^{2,dm}$, respectively;

(A2) $\exists c > 0, \forall t \in [0, T], \forall \alpha \in A, \forall x, x' \in \mathbb{R}^d, \forall \mu, \mu' \in \mathcal{P}_2(\mathbb{R}^d)$,

$$|b(t, x, \mu, \alpha) - b(t, x', \mu', \alpha)| + |\sigma(t, \omega, x, \mu) - \sigma(t, \omega, x', \mu')|$$
$$\leq c(|x - x'| + W^{(2)}(\mu, \mu')),$$

where $W^{(2)}(\mu, \mu'))$ denotes the Wasserstein distance introduced in Section 1.3. Notice that if X and X' are random variables of order 2, then by definition we have

$$W^{(2)}(\mathbb{P}_X, \mathbb{P}_{X'}) \leq [\mathbb{E}|X - X'|^2]^{1/2}.$$

The measurability, integrability, etc., properties required from a process α in order to belong to $\mathbb{A}$ (and be called admissible) may change from time to time, but for the time being we assume that $\mathbb{A}$ is the set of progressively measurable processes α satisfying

$$\mathbb{E} \int_0^T |\alpha_t|^2 dt < \infty \quad \text{and} \quad \mathbb{E} \int_0^T [|b(t, 0, \delta_0, \alpha_t)|^2 + |\sigma(t, 0, \delta_0, \alpha_t)|^2] dt < \infty.$$

Because of Theorem 1.7, together with the Lipschitz assumption **(A2)**, this definition guarantees that for any $\alpha \in \mathbb{A}$, there exists a unique solution $\mathbf{X} = \mathbf{X}^\alpha$ of (4.51) and that moreover this solution satisfies

$$\mathbb{E} \sup_{0 \leq t \leq T} |X_t|^p < \infty$$

for every $p > 1$. The stochastic optimization problem which we consider is to minimize the objective function

$$J(\boldsymbol{\alpha}) = \mathbb{E} \left\{ \int_0^T f(t, X_t, \mathcal{L}(X_t), \alpha_t) dt + g(X_T, \mathcal{L}(X_t)) \right\} \quad (4.52)$$

over the set $\mathbb{A}$ of admissible control processes $\boldsymbol{\alpha} = (\alpha_t)_{0 \leq t \leq T}$. We explained in Section 1.3 why the McKean–Vlasov dynamics posited in (4.51) are often called of *mean-field type*.

Our goal is to develop an analog of the probabilistic approach presented earlier via the generalization to the stochastic control setting of the Pontryagin maximum principle.

4.4.1 ▪ Differentiability and Convexity of Functions of Measures

There are many notions of differentiability for functions defined on spaces of measures, and recent progress in the theory of optimal transportation have put several forms in the limelight. See, for example, [115, 6] for exposés of these geometric approaches in textbook form. However, the approach to differentiability which we find convenient for the control of McKean–Vlasov dynamics is slightly different, even if the notion of differentiability ends up being equivalent in many situations. It is more of a functional analytic nature. It was introduced by P. L. Lions in his lectures at the *Collège de France*. This notion of differentiability is based on a specific *lifting* of functions $\mathcal{P}_2(\mathbb{R}^d) \ni \mu \hookrightarrow H(\mu)$ into functions $\tilde{H}$ defined on the Hilbert space $L^2(\tilde{\Omega})$ over some probability space $(\tilde{\Omega}, \tilde{\mathcal{F}}, \tilde{\mathbb{P}})$ by setting $\tilde{H}(X) = H(\mathcal{L}(X))$. Then, a function H is said to be differentiable at $\mu_0 \in \mathcal{P}_2(\mathbb{R}^d)$ if there exists a random variable X_0 with law μ_0, in other words, such that $\mathcal{L}(X_0) = \mu_0$, and such that the lifted function $\tilde{H}$ is Fréchet differentiable at X_0. It turns out that, whenever this is the case, this Fréchet derivative depends only upon the law μ_0 and not upon the particular random variable X_0 having distribution μ_0. See [27] for details. This Fréchet derivative is called the derivative of H at μ_0. By definition,

$$H(\nu) = H(\mu) + [D\tilde{H}](X) \cdot (Y - X) + o(\|Y - X\|)$$

whenever X and Y are random variables with distributions μ and ν, respectively. Notice that the random variables X and Y form a *coupling* of μ and ν, and that the term $o(\|X - Y\|)$ compares to the Wasserstein distance $W^{(2)}(\mu, \nu)$. The derivative $D_\mu H$ at $\mu \in \mathcal{P}_2(\mathbb{R}^d)$ is identified to the Fréchet derivative $D\tilde{H}$ of $\tilde{H}$ at any random variable $\tilde{X}$ on $(\tilde{\Omega}, \tilde{\mathcal{F}}, \tilde{\mathbb{P}})$ whose distribution is μ, and if we identify $L^2(\tilde{\Omega})$ and its dual, it is an element of $L^2(\tilde{\Omega})$. It is shown in [27] that this random variable is of the form $\tilde{h}(\tilde{X})$ for some deterministic measurable function $\tilde{h}$ defined μ-a.e. on $\mathbb{R}^d$ with values in $\mathbb{R}^d$. We will often identify the derivative $D_\mu H$ with this deterministic function. It is plain to see how this works when the function H is of the form

$$H(\mu) = \int h(x)\mu(dx) = \langle h, \mu \rangle$$

for some scalar function h differentiable on $\mathbb{R}^d$. Indeed, in this case $\tilde{H}(\tilde{X}) = \tilde{\mathbb{E}}h(\tilde{X})$ and $D\tilde{H}(\tilde{X}) \cdot \tilde{Y} = \tilde{\mathbb{E}}[h'(\tilde{X}) \cdot \tilde{Y}]$, and we may want to think of $D_\mu H(\mu)$ as the deterministic function h'. We will use this particular example several times in the the following, in particular to recover the Pontryagin principle originally derived in [7] for the scalar case.

We define a notion of convexity associated with this notion of differentiability. A function g on $\mathbb{R}^d \times \mathcal{P}_2(\mathbb{R}^d)$ which is continuously differentiable in its first variable, and differentiable in the above sense with respect to its second variable, is said to be convex if for every (x, μ) and (x', μ') in $\mathbb{R}^d \times \mathcal{P}_1(\mathbb{R}^d)$ we have

$$g(x', \mu') - g(x, \mu) - \partial_x g(x, \mu) \cdot (x' - x) - \tilde{\mathbb{E}}[\partial_\mu g(x, \tilde{X}) \cdot (\tilde{X}' - \tilde{X})] \geq 0 \quad (4.53)$$

whenever $\tilde{X}$ and $\tilde{X}'$ are square integrable random variables with distributions μ and μ', respectively.

4.4.2 ▪ The Hamiltonian and the Dual Equations

In full analogy with the classical case, the Hamiltonian of the stochastic optimization problem is defined as the function H given by

$$H(t, x, \mu, y, z, \alpha) = b(t, x, \mu, \alpha) \cdot y + \sigma(t, x, \mu, \alpha) \cdot z + f(t, x, \mu, \alpha). \quad (4.54)$$

Because we need to use derivatives of H with respect to its variable μ, we consider the lifting $\tilde{H}$ defined by

$$\tilde{H}(t, x, \tilde{X}, y, z, \alpha) = H(t, x, \mu, y, z, \alpha) \quad (4.55)$$

for any random variable $\tilde{X}$ with distribution μ, and we denote by $\partial_\mu \tilde{H}$ the Fréchet derivative of $\tilde{H}$ with respect to X whenever all the other variables t, x, y, z, and α are held fixed.

Definition 4.19. *Given an admissible control $\boldsymbol{\alpha} = (\alpha_t)_{0 \leq t \leq T} \in \mathbb{A}$, we denote by $\mathbf{X} = \mathbf{X}^\alpha$ the corresponding controlled state process, and we call adjoint processes any couple $(\mathbf{Y}, \mathbf{Z}) = (Y_t, Z_t)_{0 \leq t \leq T}$ of progressively measurable stochastic processes satisfying the equation (which we call the adjoint equation)*

$$\begin{cases} dY_t = -\partial_x H(t, X_t, \mathcal{L}(X_t), Y_t, Z_t, \alpha_t) dt + Z_t dW_t \\ \qquad\qquad - \tilde{\mathbb{E}}[\partial_\mu \tilde{H}(t, \tilde{X}_t, \mathbb{P}_{\tilde{X}_t}, \tilde{Y}_t, \tilde{Z}_t, \tilde{\alpha}_t)](X_t) dt, \\ Y_T = \partial_x g(X_T, \mathcal{L}(X_t)) + \tilde{\mathbb{E}}[\partial_\mu g(\tilde{X}_T, \mathcal{L}(\tilde{X}_t))(X_T)], \end{cases} \quad (4.56)$$

where $(\tilde{\alpha}, \tilde{X}, \tilde{Y}, \tilde{Z})$ is an independent copy of (α, X, Y, Z) and $\tilde{\mathbb{E}}$ denotes the expectation with respect to this independent copy.

We refer to Section 2.2 for the introduction and first systematic use of these independent copies in the analysis of the adjoint equation. Notice that, when b, σ, and g do not depend upon the marginal distributions of the process, the extra terms appearing in the adjoint equation and the terminal condition disappear and this equation coincides with the classical adjoint equation of stochastic control. In the present setup, the adjoint equation rewrites as

$$\begin{aligned} dY_t = &-[\partial_x b(t, X_t, \mathcal{L}(X_t), \alpha_t) Y_t + \partial_x \sigma(t, X_t, \mathcal{L}(X_t), \alpha_t) Z_t \\ &+ \partial_x f(t, X_t, \mathcal{L}(X_t), \alpha_t)] dt + Z_t dW_t \\ &- \tilde{\mathbb{E}}[\partial_\mu \tilde{b}(t, \tilde{X}_t, x, \tilde{Y}_t, \tilde{Z}_t, \tilde{\alpha}_t) + \partial_\mu \tilde{\sigma}(t, \tilde{X}_t, x, \tilde{Y}_t, \tilde{Z}_t, \tilde{\alpha}_t) \\ &+ \partial_\mu \tilde{f}(t, \tilde{X}_t, x, \tilde{Y}_t, \tilde{Z}_t, \tilde{\alpha}_t)]|_{x=X_t} dt \end{aligned}$$

with the terminal condition $Y_T = \partial_x g(X_T, \mathcal{L}(X_t)) + \tilde{\mathbb{E}}[\partial_\mu g(x, \tilde{X}_t)]|_{x=X_T}$. Notice that, given an admissible control $\boldsymbol{\alpha} \in \mathbf{A}$ and the corresponding controlled state process $\mathbf{X} = \mathbf{X}^\alpha$, despite the boundedness assumption on the partial derivatives of the coefficients, and despite the fact that the first part of the equation appears to be linear in the unknown processes Y_t and Z_t, existence and uniqueness of a solution $(\mathbf{Y}, \mathbf{Z})$ of the adjoint equation are not provided by standard results on BSDEs such as for example in Theorem 2.2, as the distributions of the solution processes (more precisely, their joint distributions with the control and state processes $\boldsymbol{\alpha}$ and $\mathbf{X}$) appear in the coefficients of the equation. However, a slight modification of the original existence and uniqueness result of Pardoux and Peng [102] gives a result perfectly tailored to our present needs. It is given in Theorem 2.9 in Section 2.2. It implies that, for each admissible $\boldsymbol{\alpha}$, there exists a unique couple of adjoint processes $(\mathbf{Y}, \mathbf{Z})$ solving the adjoint equation (4.17) and satisfying

$$\mathbb{E} \sup_{0 \leq t \leq T} |Y_t|^2 + \mathbb{E} \int_0^T |Z_t|^2 dt < \infty.$$

4.4.3 ▪ Special Cases

For the purpose of illustration, we consider in detail the special cases of interactions introduced in Section 1.3, and we provide the special forms of the adjoint equations so that, once we prove appropriate forms of the Pontryagin maximum principle which apply in these cases, we should be able to pursue the analysis of the optimization problems.

4.4. Optimal Control of McKean–Vlasov Dynamics

Scalar Interactions

In this subsection, we lay the foundations for the proofs of the results of [7], which will appear as corollaries of the general Pontryagin principle proven later in this section. In the case of scalar interactions, the dependence upon the probability measure of the coefficients of the dynamics of the state and the cost functions is through functions of scalar moments of the measure. More specifically, we assume that

$$b(t, x, \mu, \alpha) = \hat{b}(t, x, \langle \psi, \mu \rangle, \alpha), \qquad \sigma(t, x, \mu, \alpha) = \hat{\sigma}(t, x, \langle \phi, \mu \rangle, \alpha),$$

$$f(t, x, \mu, \alpha) = \hat{f}(t, x, \langle \gamma, \mu \rangle, \alpha), \qquad g(x, \mu) = \hat{g}(x, \langle \zeta, \mu \rangle)$$

for some scalar functions ψ, ϕ, γ, and ζ growing at most quadratically at ∞, and functions $\hat{b}$, $\hat{\sigma}$, and $\hat{f}$ defined on $[0, T] \times \mathbb{R}^d \times \mathbb{R} \times A$ with values in $\mathbb{R}^d$, $\mathbb{R}^{d \times m}$, and $\mathbb{R}$, respectively, and a real-valued function $\hat{g}$ defined on $\mathbb{R}^d \times \mathbb{R}$. As usual, we use the notation $\langle h, \mu \rangle$ to denote the integral of the function h with respect to the measure μ. The functions $\hat{b}$, $\hat{\sigma}$, $\hat{f}$, and $\hat{g}$ are similar to the functions b, σ, f, and g with the variable μ, which was a measure, replaced by a numeric variable, say x'. We shall assume that the functions $\hat{b}$, $\hat{\sigma}$, $\hat{f}$, and $\hat{g}$ are continuously differentiable in the variables x, x', and α with bounded partial derivatives, and in order for the sufficient condition for optimality which we prove below to be satisfied, we shall assume that the partial derivatives with respect to x' are nonnegative (in other words, that these functions are nondecreasing), that the Hamiltonian is convex in the variables (x, x', α), and that the scalar functions ψ, ϕ, γ, and ζ are continuously differentiable and convex. These are exactly the assumptions used in [7]. We have

$$H(t, x, \mu, y, z, \alpha) = \hat{b}(t, x, \langle \psi, \mu \rangle, \alpha) \cdot y + \hat{\sigma}(t, x, \langle \phi, \mu \rangle, \alpha) \cdot z + \hat{f}(t, x, \langle \gamma, \mu \rangle, \alpha)$$

and we proceed to derive the particular form taken by the adjoint equation in the present situation. We start with the terminal condition as it is easier to identify. According to (4.17), this terminal condition reads

$$Y_T = \partial_x g(X_T, \mathcal{L}(X_t)) + \tilde{\mathbb{E}}[\partial_\mu g(\tilde{X}_T, \mathcal{L}(\tilde{X}_T))(X_T)]$$

with $g(x, \mu) = \hat{g}(x, \langle \zeta, \mu \rangle)$. Given our definition of differentiability with respect to the variable μ, we saw that $\partial_\mu g(x, \mu)(\cdot)$ was a function defined μ-a.e., and in this particular case of scalar dependence, it was given by $\partial_{x'} \hat{g}(x, , \langle \zeta, \mu \rangle)\zeta'(\cdot)$. Consequently, the above terminal condition can be rewritten as

$$Y_T = \partial_x \hat{g}(X_T, \mathbb{E}[\zeta(X_T)]) + \tilde{\mathbb{E}}[\partial_{x'} \hat{g}(\tilde{X}_T, \mathbb{E}[\zeta(X_T)])]\zeta'(X_T),$$

which is exactly the terminal condition used in [7] once we remark that the "tildes" can be removed. Using the same argument to compute the derivative with respect to μ of the Hamiltonian appearing in our form (4.17) of the adjoint equation, we find that $\partial_\mu H(t, x, \mu, y, z, \alpha)$ can be identified with the $\mathbb{R}^d$-valued function defined μ-a.e. on $\mathbb{R}^d$ by

$$\partial_\mu H(t, x, \mu, y, z, \alpha) = y^\dagger \partial_{x'} \hat{b}(t, x, \langle \psi, \mu \rangle, \alpha)\psi'(\cdot) + \text{trace}[z^\dagger \partial_{x'} \hat{\sigma}(t, x, \langle \phi, \mu \rangle, \alpha)\phi'(\cdot)]$$
$$+ \partial_{x'} \hat{f}(t, x, \langle \gamma, \mu \rangle, \alpha)\gamma'(\cdot),$$

and the dynamic part of the adjoint equation (4.17) rewrites

$$dY_t = -[\partial_x \hat{b}(t, X_t, \mathbb{E}[\psi(X_t)], \alpha_t) \cdot Y_t + \partial_x \hat{\sigma}(t, X_t, \mathbb{E}[\phi(X_t)], \alpha_t) \cdot Z_t$$
$$+ \partial_x \hat{f}(t, X_t, \mathbb{E}[\gamma(X_t)], \alpha_t)]dt + Z_t dW_t$$
$$- [\partial_{x'} \mathbb{E}[\tilde{Y}_t^\dagger \hat{b}(t, \tilde{X}_t, \mathbb{E}[\psi(\tilde{X}_t)], \tilde{\alpha}_t)]\psi'(X_t)$$
$$+ \tilde{\mathbb{E}}[\tilde{Z}_t^\dagger \partial_{x'} \hat{\sigma}(t, \tilde{X}_t, \mathbb{E}[\phi(\tilde{X}_t)], \tilde{\alpha}_t)]\phi'(X_t)$$
$$+ \tilde{\mathbb{E}}[\partial_{x'} \hat{f}((t, \tilde{X}_t, \mathbb{E}[\gamma(\tilde{X}_t)], \tilde{\alpha}_t)]\gamma'(X_t)]dt,$$

which again, is exactly the adjoint equation used in [7] once we remove the "tildes".

Several examples of control of McKean–Vlasov dynamics with scalar interactions will be given in Subsection 4.4.6 below. We shall revisit the mean-variance portfolio optimization problem, and discuss the optimal control of linear-quadratic (LQ) McKean–Vlasov dynamics.

First-Order Interactions

In the case of first-order interactions, the dependence upon the probability measure is linear in the sense that the coefficients b, σ, f, and g are given in the form

$$b(t, x, \mu, \alpha) = \langle \hat{b}(t, x, \cdot, \alpha), \mu \rangle, \qquad \sigma(t, x, \mu, \alpha) = \langle \hat{\sigma}(t, x, \cdot, \alpha), \mu \rangle,$$

$$f(t, x, \mu, \alpha) = \langle \hat{f}(t, x, \cdot, \alpha), \mu \rangle, \qquad g(x, \mu) = \langle \hat{g}(x, \cdot), \mu \rangle$$

for some functions $\hat{b}$, $\hat{\sigma}$, and $\hat{f}$ defined on $[0, T] \times \mathbb{R}^d \times \mathbb{R}^d \times A$ with values in $\mathbb{R}^d$, $\mathbb{R}^{d \times m}$, and $\mathbb{R}$ respectively, and a real-valued function $\hat{g}$ defined on $\mathbb{R}^d \times \mathbb{R}^d$. The form of this dependence comes from the original derivation of the McKean–Vlasov equation as the limit of the dynamics of a large system of particles evolving according to a system of stochastic differential equations with *mean-field* interactions of the form

$$dX_t^i = \frac{1}{N} \sum_{j=1}^N \hat{b}(t, X_t^i, X_t^j)dt + \frac{1}{N} \sum_{j=1}^N \hat{\sigma}(t, X_t^i, X_t^j)dW_t, \qquad i = 1, \cdots, N, \ 0 \le t \le T,$$
(4.57)

In the present situation, the linearity in μ implies that $\partial_\mu g(x, \mu)(\cdot) = \partial_{x'} \hat{g}(x, \cdot)$, and similarly,

$$\partial_\mu H(t, x, \mu, y, z, \alpha) = y^\dagger \partial_{x'} \hat{b}(t, x, \cdot, \alpha) + \text{trace}[z^\dagger \partial_{x'} \hat{\sigma}(t, x, \cdot, \alpha)] + \partial_{x'} \hat{f}(t, x, \cdot, \alpha),$$

and the dynamic part of the adjoint equation (4.17) rewrites as

$$dY_t = -\tilde{\mathbb{E}}[\partial_x \hat{b}(t, X_t, \tilde{X}_t, \alpha_t) \cdot Y_t + \partial_x \hat{\sigma}(t, X_t, \tilde{X}_t, \alpha_t) \cdot Z_t + \partial_x \hat{f}(t, X_t, \tilde{X}_t, \alpha_t)$$
$$+ \partial_{x'} \hat{b}(t, \tilde{X}_t, X_t, \tilde{\alpha}_t) \cdot \tilde{Y}_t + \partial_{x'} \hat{\sigma}(t, \tilde{X}_t, X_t, \tilde{\alpha}_t) \cdot \tilde{Z}_t$$
$$+ \partial_{x'} \hat{f}(t, \tilde{X}_t, X_t, \tilde{\alpha}_t)]dt + Z_t dW_t$$
$$= -\tilde{\mathbb{E}}[\partial_x \hat{H}(t, X_t, \tilde{X}_t, Y_t, Z_t, \alpha_t) + \partial_{x'} \hat{H}(t, \tilde{X}_t, X_t, \tilde{Y}_t, \tilde{Z}_t, \tilde{\alpha}_t)]dt + Z_t dW_t$$

if we use the obvious notation

$$\hat{H}(t, x, x', y, z, \alpha) = \hat{b}(t, x, x', \alpha) \cdot y + \hat{\sigma}(t, x, x', \alpha) \cdot z + \hat{f}(t, x,, x', \alpha)$$

and the terminal condition

$$Y_T = \tilde{\mathbb{E}}[\partial_x \hat{g}(X_T, \tilde{X}_T) + \partial_{x'} \hat{g}(\tilde{X}_T, X_T)].$$

4.4.4 ▪ Pontryagin Principle for Optimality

In this section, we assume that the coefficients b, σ, and f are continuously differentiable with respect to (x, α), with uniformly bounded partial derivatives, and are differentiable with respect to the variable μ (in the sense given above) with a continuous and uniformly bounded derivative in L^2. We provide generalizations of the necessary form as well as the sufficient form of the Pontryagin stochastic maximum principle to the optimal control of McKean–Vlasov dynamics. Many of the proofs follow very closely the strategies used in the classical case, the original arguments being most often limited to the manipulation of the derivatives with respect to various probability distributions. For this reason, we do not repeat those parts of the proofs which apply mutatis mutandis to both cases.

A Necessary Condition

We assume that the set $\mathbb{A}$ of admissible controls is convex, we fix $\boldsymbol{\alpha} \in \mathbb{A}$, and as before, we denote by $\mathbf{X} = \mathbf{X}^{\boldsymbol{\alpha}}$ the corresponding controlled state, namely, the solution of (4.51) with given initial condition $X_0 = x$. Our first task is to compute the Gâteaux derivative of the cost functional J at $\boldsymbol{\alpha}$ in all directions. In order to do so, we choose another admissible control $\boldsymbol{\beta} \in \mathbb{A}$ and we compute the variation of J at $\boldsymbol{\alpha}$ in the direction of $\boldsymbol{\beta}$ (think of $\boldsymbol{\beta}$ as the difference between another element of $\mathbb{A}$ and $\boldsymbol{\alpha}$).

We define the variation process $\mathbf{V} = (V_t)_{0 \leq t \leq T}$ as the solution of the equation

$$dV_t = [\gamma_t V_t + \delta_t(\mathcal{L}((X_t, V_t))) + \eta_t]dt + [\tilde{\gamma}_t V_t + \tilde{\delta}_t(\mathcal{L}((X_t, V_t))) + \tilde{\eta}_t]dW_t, \quad (4.58)$$

where the coefficients $\gamma_t, \delta_t, \eta_t, \tilde{\gamma}_t, \tilde{\delta}_t$, and $\tilde{\eta}_t$ are defined as

$$\gamma_t = \partial_x b(t, X_t, \mathcal{L}(X_t), \alpha_t) \quad \text{and} \quad \tilde{\gamma}_t = \partial_x \sigma(t, X_t, \mathcal{L}(X_t), \alpha_t),$$
$$\eta_t = \partial_\alpha b(t, X_t, \mathcal{L}(X_t), \alpha_t)\beta_t \quad \text{and} \quad \tilde{\eta}_t = \partial_\alpha \sigma(t, X_t, \mathcal{L}(X_t), \alpha_t)\beta_t,$$

which are progressively bounded or square integrable processes in $\mathbb{R}^{d \times d}$, $\mathbb{R}^{d \times m \times d}$, $\mathbb{R}^d$, and $\mathbb{R}^{d \times m}$, respectively, and

$$\delta_t = \tilde{\mathbb{E}} \partial_\mu b(t, x, \mathcal{L}(X_t), \alpha)(\tilde{X}_t) \cdot \tilde{V}_t \big|_{\substack{x = X_t \\ \alpha = \alpha_t}} \quad \text{and}$$
$$\tilde{\delta}_t = \tilde{\mathbb{E}} \partial_\mu \sigma(t, x, \mathcal{L}(X_t), \alpha)(\tilde{X}_t) \cdot \tilde{V}_t \big|_{\substack{x = X_t \\ \alpha = \alpha_t}}, \quad (4.59)$$

where $(\tilde{X}_t, \tilde{V}_t)$ is an independent copy of (X_t, V_t). Notice that δ_t and $\tilde{\delta}_t$ are linear functions of the distribution $\mathcal{L}((X_t, V_t))$ of (X_t, V_t). Even though we are dealing with possibly random coefficients, the existence and uniqueness of the variation process is guaranteed by Proposition 2.1 of [72] applied to the couple $(\mathbf{X}, \mathbf{V})$ and the system formed by (4.51) and (4.10). Because of our assumption on the boundedness of the partial derivatives of the coefficients, $\mathbf{V}$ satisfies

$$\mathbb{E} \sup_{0 \leq t \leq T} |V_t|^p < \infty \quad (4.60)$$

for every finite $p \geq 1$.

Lemma 4.20. *For each $\epsilon > 0$ small enough, we consider the admissible control $\boldsymbol{\alpha}^\epsilon$ defined by $\alpha_t^\epsilon = \alpha_t + \epsilon \beta_t$, and the corresponding controlled state $\mathbf{X}^\epsilon = \mathbf{X}^{\boldsymbol{\alpha}^\epsilon}$. We have*

$$\lim_{\epsilon \searrow 0} \mathbb{E} \sup_{0 \leq t \leq T} \left| \frac{X_t^\epsilon - X_t}{\epsilon} - V_t \right|^2 = 0. \quad (4.61)$$

Proof. For the purpose of this proof we set $V_t^\epsilon = \epsilon^{-1}(X_t^\epsilon - X_t) - V_t$. Notice that $V_0^\epsilon = 0$ and that

$$dV_t^\epsilon = \left[\frac{1}{\epsilon}[b(t, X_t^\epsilon, \mathcal{L}(X_t^\epsilon), \alpha_t^\epsilon) - b(t, X_t, \mathcal{L}(X_t), \alpha_t)] - \partial_x b(t, X_t, \mathcal{L}(X_t), \alpha_t)V_t\right.$$
$$\left. - \partial_\alpha b(t, X_t, \mathcal{L}(X_t), \alpha_t)\beta_t - \tilde{\mathbb{E}}\partial_\mu b(t, x, \mathcal{L}(X_t), \alpha)(\tilde{X}_t) \cdot \tilde{V}_t\big|_{\substack{x=X_t \\ \alpha=\alpha_t}}\right]dt$$
$$+ \left[\frac{1}{\epsilon}[\sigma(t, X_t^\epsilon, \mathcal{L}(X_t^\epsilon), \alpha_t^\epsilon) - \sigma(t, X_t, \mathcal{L}(X_t), \alpha_t)] - \partial_x \sigma(t, X_t, \mathcal{L}(X_t), \alpha_t)V_t\right.$$
$$\left. - \partial_\alpha \sigma(t, X_t, \mathcal{L}(X_t), \alpha_t)\beta_t - \tilde{\mathbb{E}}\partial_\mu \sigma(t, x, \mathcal{L}(X_t), \alpha)(\tilde{X}_t) \cdot \tilde{V}_t\big|_{\substack{x=X_t \\ \alpha=\alpha_t}}\right]dW_t$$
$$= V_t^{\epsilon,1}dt + V_t^{\epsilon,2}dW_t.$$

Now for each $t \in [0, T]$ and each $\epsilon > 0$, we have

$$\frac{1}{\epsilon}[b(t, X_t^\epsilon, \mathcal{L}(X_t^\epsilon), \alpha_t^\epsilon) - b(t, X_t, \mathcal{L}(X_t), \alpha_t)]$$
$$= \int_0^1 \partial_x b(t, X_t + \lambda\epsilon(V_t^\epsilon + V_t), \mathcal{L}(X_t + \lambda\epsilon(V_t^\epsilon + V_t)), \alpha_t + \lambda\epsilon\beta_t)(V_t^\epsilon + V_t)d\lambda$$
$$+ \int_0^1 \tilde{\mathbb{E}}\partial_\mu b(t, X_t + \lambda\epsilon(V_t^\epsilon + V_t), \mathcal{L}(X_t + \lambda\epsilon(V_t^\epsilon + V_t)), \alpha_t + \lambda\epsilon\beta_t)(\tilde{X}_t + \lambda\epsilon(\tilde{V}_t^\epsilon + \tilde{V}_t))$$
$$\cdot (\tilde{V}_t^\epsilon + \tilde{V}_t)d\lambda$$
$$+ \int_0^1 \partial_\alpha b(t, X_t + \lambda\epsilon(V_t^\epsilon + V_t), \mathcal{L}(X_t + \lambda\epsilon(V_t^\epsilon + V_t)), \alpha_t + \lambda\epsilon\beta_t)\beta_t d\lambda.$$

In order to simplify the notation slightly, we set $X_t^{\lambda,\epsilon} = X_t + \lambda\epsilon(V_t^\epsilon + V_t)$ and $\alpha_t^{\lambda,\epsilon} = \alpha_t + \lambda\epsilon\beta_t$. If we compute the "$dt$"-term we get

$$V_t^{\epsilon,1} = \int_0^1 \partial_x b(t, X_t^{\lambda,\epsilon}, \mathcal{L}(X_t^{\lambda,\epsilon}), \alpha_t^{\lambda,\epsilon})V_t^\epsilon d\lambda$$
$$+ \int_0^1 \tilde{\mathbb{E}}\partial_\mu b(t, X_t^{\lambda,\epsilon}, \mathcal{L}(X_t^{\lambda,\epsilon}), \alpha_t^{\lambda,\epsilon})(\tilde{X}_t^{\lambda,\epsilon}) \cdot \tilde{V}_t^\epsilon d\lambda$$
$$+ \int_0^1 [\partial_x b(t, X_t^{\lambda,\epsilon}, \mathcal{L}(X_t^{\lambda,\epsilon}), \alpha_t^{\lambda,\epsilon}) - \partial_x b(t, X_t, \mathcal{L}(X_t), \alpha_t)]V_t d\lambda$$
$$+ \int_0^1 [\tilde{\mathbb{E}}\partial_\mu b(t, X_t^{\lambda,\epsilon}, \mathcal{L}(X_t^{\lambda,\epsilon}), \alpha_t^{\lambda,\epsilon}) - \partial_x b(t, X_t, \mathcal{L}(X_t), \alpha_t)](\tilde{X}_t^{\lambda,\epsilon}) \cdot \tilde{V}_t d\lambda$$
$$+ \int_0^1 [\partial_\alpha b(t, X_t^{\lambda,\epsilon}, \mathcal{L}(X_t^{\lambda,\epsilon}), \alpha_t^{\lambda,\epsilon}) - \partial_\alpha b(t, X_t, \mathcal{L}(X_t), \alpha_t)]\beta_t d\lambda$$
$$= \int_0^1 \partial_x b(t, X_t^{\lambda,\epsilon}, \mathcal{L}(X_t^{\lambda,\epsilon}), \alpha_t^{\lambda,\epsilon})V_t^\epsilon d\lambda$$
$$+ \int_0^1 \tilde{\mathbb{E}}\partial_\mu b(t, X_t^{\lambda,\epsilon}, \mathcal{L}(X_t^{\lambda,\epsilon}), \alpha_t^{\lambda,\epsilon})(\tilde{X}_t^{\lambda,\epsilon}) \cdot \tilde{V}_t^\epsilon d\lambda + I_t^{\epsilon,1} + I_t^{\epsilon,2} + I_t^{\epsilon,3}.$$

4.4. Optimal Control of McKean–Vlasov Dynamics

The last three terms of the above right-hand side converge to 0 in $L^2([0,T] \times \Omega)$ as $\epsilon \searrow 0$. Indeed,

$$\mathbb{E}\int_0^T |I_t^{\epsilon,1}|^2 dt = \mathbb{E}\int_0^T \left| \int_0^1 [\partial_x b(t, X_t^{\lambda,\epsilon}, \mathcal{L}(X_t^{\lambda,\epsilon}), \alpha_t^{\lambda,\epsilon}) \right.$$
$$\left. - \partial_x b(t, X_t, \mathcal{L}(X_t), \alpha_t)] V_t d\lambda \right|^2 dt$$

$$\leq \mathbb{E}\int_0^T \int_0^1 |\partial_x b(t, X_t^{\lambda,\epsilon}, \mathcal{L}(X_t^{\lambda,\epsilon}), \alpha_t^{\lambda,\epsilon})$$
$$- \partial_x b(t, X_t, \mathcal{L}(X_t), \alpha_t)|^2 |V_t|^2 d\lambda dt$$

$$\leq c\mathbb{E}\int_0^T \int_0^1 [(\lambda\epsilon)^2 |(V_t^\epsilon + V_t|^2 + |\beta_t|^2)|V_t|^2 d\lambda dt$$

$$\leq c\left[\left(\int_0^T \int_0^1 \mathbb{E}|\lambda\epsilon(V_t^\epsilon + V_t|^4 d\lambda dt\right)^{1/2} \left(\int_0^T \mathbb{E}|V_t|^4 dt\right)^{1/2}\right.$$
$$\left. + \left(\int_0^T \int_0^1 \mathbb{E}|\lambda\epsilon\beta_t|^4 d\lambda dt\right)^{1/2} \left(\int_0^T \mathbb{E}|V_t|^4 dt\right)^{1/2}\right],$$

which clearly converges to 0 as $\epsilon \searrow 0$ since the expectations are finite. A similar argument applies to $I_t^{\epsilon,2}$ and $I_t^{\epsilon,3}$. Next, we treat the diffusion part $V_t^{\epsilon,2}$ in the same way using Jensen's inequality and the Burkholder–Davis–Gundy inequality to control the quadratic variation of the stochastic integrals. Consequently, going back to (4.13), we see that

$$\mathbb{E}\sup_{0\leq t\leq T}|V_t^\epsilon|^2 dt \leq c\left(\int_0^T \mathbb{E}\sup_{0\leq s\leq t}|V_s^\epsilon|^2 dt + \int_0^T \sup_{0\leq s\leq t}\mathbb{E}|V_s^\epsilon|^2 dt\right) + \delta_\epsilon$$
$$\leq c\int_0^T \mathbb{E}\sup_{0\leq s\leq t}|V_s^\epsilon|^2 dt + \delta_\epsilon,$$

where, as usual, $c > 0$ is a generic constant which can change from line to line, and $\lim_{\epsilon \searrow 0} \delta_\epsilon = 0$. Finally, we get the desired result applying Gronwall's inequality. $\square$

Gâteaux Derivative

Lemma 4.21. *The function $\alpha \hookrightarrow J(\alpha)$ is Gâteaux differentiable and its derivative in the direction β is given by*

$$\frac{d}{d\epsilon}J(\alpha + \epsilon\beta)\Big|_{\epsilon=0} = \mathbb{E}\int_0^T [\partial_x f(t, X_t, \mathcal{L}(X_t), \alpha_t)V_t$$
$$+ \tilde{\mathbb{E}}\partial_\mu f(t, X_t, \mathcal{L}(X_t), \alpha_t)(\tilde{X}_t) \cdot \tilde{V}_t + \partial_\alpha f(t, X_t, \mathcal{L}(X_t), \alpha_t)\beta_t]dt$$
$$+ \mathbb{E}[\partial_x g(X_T, \mathcal{L}(X_t))V_T + \tilde{\mathbb{E}}\partial_\mu g(X_T, \mathcal{L}(X_t))(\tilde{X}_T) \cdot \tilde{V}_T]$$

Proof. We use freely the notation introduced in the proof of the previous lemma.

$$\frac{d}{d\epsilon}J(\alpha + \epsilon\beta)\Big|_{\epsilon=0} = \lim_{\epsilon \searrow 0}\frac{1}{\epsilon}\mathbb{E}\int_0^T [f(t, X_t^\epsilon, \mathcal{L}(X_t^\epsilon), \alpha_t^\epsilon) - f(t, X_t, \mathcal{L}(X_t), \alpha_t)]dt$$
$$+ \lim_{\epsilon \searrow 0}\frac{1}{\epsilon}\mathbb{E}[g(X_T^\epsilon, \mathcal{L}(X_T^\epsilon)) - g(X_T, \mathcal{L}(X_t))]$$

Computing the two limits separately we get

$$\lim_{\epsilon \searrow 0} \mathbb{E} \int_0^T [f(t, X_t^\epsilon, \mathcal{L}(X_t^\epsilon), \alpha_t^\epsilon) - f(t, X_t, \mathcal{L}(X_t), \alpha_t)] dt$$
$$= \lim_{\epsilon \searrow 0} \mathbb{E} \int_0^T \int_0^1 \frac{d}{d\lambda} f(t, X_t^{\lambda,\epsilon}, \mathcal{L}(X_t^{\lambda,\epsilon}), \alpha_t^{\lambda,\epsilon}) d\lambda dt$$
$$= \lim_{\epsilon \searrow 0} \mathbb{E} \int_0^T \int_0^1 [\partial_x f(t, X_t^{\lambda,\epsilon}, \mathcal{L}(X_t^{\lambda,\epsilon}), \alpha_t^{\lambda,\epsilon})(V_t^\epsilon + V_t)$$
$$+ \tilde{\mathbb{E}} \partial_\mu f(t, X_t^{\lambda,\epsilon}, \mathcal{L}(X_t^{\lambda,\epsilon}), \alpha_t^{\lambda,\epsilon})(\tilde{X}_t + \lambda \epsilon (\tilde{V}_t^\epsilon + \tilde{V}_t)) \cdot (\tilde{V}_t^\epsilon + \tilde{V}_t)$$
$$+ \partial_\alpha f(t, X_t^{\lambda,\epsilon}, \mathcal{L}(X_t^{\lambda,\epsilon}), \alpha_t^{\lambda,\epsilon}) \beta_t] d\lambda dt$$
$$= \mathbb{E} \int_0^T [\partial_x f(t, X_t, \mathcal{L}(X_t), \alpha_t) V_t$$
$$+ \tilde{\mathbb{E}} \partial_\mu f(t, X_t, \mathcal{L}(X_t), \alpha_t)(\tilde{X}_t) \cdot \tilde{V}_t + \partial_\alpha f(t, X_t, \alpha_t) \beta_t] dt,$$

using the hypothesis on the continuity and boundedness of the derivatives of f, the uniform convergence proven in the previous lemma, and Lebesgue's dominated convergence theorem. Similarly,

$$\lim_{\epsilon \searrow 0} \frac{1}{\epsilon} \mathbb{E}[g(X_T^\epsilon, \mathcal{L}(X_T^\epsilon)) - g(X_T, \mathcal{L}(X_t))]$$
$$= \lim_{\epsilon \searrow 0} \mathbb{E} \int_0^1 \frac{d}{d\lambda} g(X_T^{\lambda,\epsilon}, \mathcal{L}(X_T^{\lambda,\epsilon})) d\lambda$$
$$= \lim_{\epsilon \searrow 0} \mathbb{E} \int_0^1 [\partial_x g(X_T^{\lambda,\epsilon}, \mathcal{L}(X_T^{\lambda,\epsilon}))(V_T^\epsilon + V_T)$$
$$+ \tilde{\mathbb{E}} \partial_\mu g(X_T^{\lambda,\epsilon}, \mathcal{L}(X_T^{\lambda,\epsilon}))(\tilde{X}_T + \lambda \epsilon (\tilde{V}_T^\epsilon + \tilde{V}_T)) \cdot (\tilde{V}_T^\epsilon + \tilde{V}_T)] d\lambda$$
$$= \mathbb{E}[\partial_x g(X_T, \mathcal{L}(X_t)) V_T + \tilde{\mathbb{E}} \partial_\mu g(X_T, \mathcal{L}(X_t))(\tilde{X}_T) \cdot \tilde{V}_T],$$

which completes the proof. □

Duality

The duality relationship is given as follows.

Lemma 4.22. *It holds that*

$$\mathbb{E}[Y_T V_T] = \mathbb{E} \int_0^T [Y_t \partial_\alpha b(t, X_t, \mathcal{L}(X_t), \alpha_t) \beta_t - \partial_x f(t, X_t, \mathcal{L}(X_t), \alpha_t) V_t \quad (4.62)$$
$$- \tilde{\mathbb{E}} \partial_\mu f(t, X_t, \mathcal{L}(X_t), \alpha_t)(\tilde{X}_t) \cdot \tilde{V}_t + Z_t \partial_\alpha \sigma(t, X_t, \mathcal{L}(X_t), \alpha_t) \beta_t] dt.$$

Proof. Using the definitions (4.10) of the variation process **V** and (4.18) of the adjoint process **Y**, integration by parts gives

$$Y_T V_T = Y_0 V_0 + \int_0^T Y_t dV_t + \int_0^T dY_t V_t + \int_0^T d[\mathbf{Y}, \mathbf{V}]_t$$
$$= \int_0^T [Y_t \partial_x b(t, X_t, \mathcal{L}(X_t), \alpha_t) V_t + Y_t \tilde{\mathbb{E}} \partial_\mu b(t, X_t, \mathcal{L}(X_t), \alpha_t)(\tilde{X}_t) \cdot \tilde{V}_t$$
$$+ Y_t \partial_\alpha b(t, X_t, \mathcal{L}(X_t), \alpha_t) \beta_t$$
$$- \partial_x H(t, X_t, \mathcal{L}(X_t), Y_t, Z_t, \alpha_t) V_t - \tilde{\mathbb{E}} \partial_\mu H(t, \tilde{X}_t, \mathcal{L}(\tilde{X}_t), \tilde{Y}_t, \tilde{Z}_t, \tilde{\alpha}_t)(X_t) \cdot V_t$$
$$+ Z_t \partial_x \sigma(t, X_t, \mathcal{L}(X_t), \alpha_t) V_t + Z_t \tilde{\mathbb{E}} \partial_\mu \sigma(t, X_t, \mathcal{L}(X_t), \alpha_t)(\tilde{X}_t) \cdot \tilde{V}_t$$
$$+ Z_t \partial_\alpha \sigma(t, X_t, \mathcal{L}(X_t), \alpha_t) \beta_t] dt + M_T,$$

4.4. Optimal Control of McKean–Vlasov Dynamics

where $(M_t)_{0 \le t \le T}$ is a mean-zero square integrable martingale. By taking expectations on both sides, using the fact that $\mathbb{E} M_T = 0$ and (again because of Fubini's theorem)

$$\mathbb{E}\tilde{\mathbb{E}} \partial_\mu H(t, \tilde{X}_t, \mathcal{L}(\tilde{X}_t), \tilde{Y}_t, \tilde{Z}_t, \tilde{\alpha}_t)(X_t) \cdot V_t$$
$$= \mathbb{E}\tilde{\mathbb{E}} \partial_\mu H(t, X_t, \mathcal{L}(X_t), Y_t, Z_t, \alpha_t)(\tilde{X}_t) \cdot \tilde{V}_t$$
$$= \mathbb{E}\tilde{\mathbb{E}}[\partial_\mu b(t, X_t, \mathcal{L}(X_t), \alpha_t)(\tilde{X}_t) \cdot \tilde{V}_t Y_t + \partial_\mu \sigma(t, X_t, \mathcal{L}(X_t), \alpha_t)(\tilde{X}_t) \cdot \tilde{V}_t Z_t$$
$$+ \partial_\mu f(t, X_t, \mathcal{L}(X_t), \alpha_t)(\tilde{X}_t) \cdot \tilde{V}_t],$$

and using the fact that the inner products of matrices are given by computing traces, and that the traces are invariant under cyclic permutations of the matrices which are multiplied under the trace, cancellations occur and we get the desired equality (4.62). □

By putting together the duality relation (4.62) together with the formula for the Gâteaux derivative of J we get the following.

Corollary 4.23. *The Gâteaux derivative of J at α in the direction of β can be written as*

$$\frac{d}{d\epsilon} J(\alpha + \epsilon \beta)\big|_{\epsilon=0} = \mathbb{E}\left[\int_0^T [\partial_\alpha H(t, X_t, \mathcal{L}(X_t), Y_t, \alpha_t) \beta_t] \, dt\right]. \qquad (4.63)$$

Proof. Using Fubini's theorem, the second expectation appearing in the expression (4.62) of the Gâteaux derivative of J given in Lemma 4.8 can be rewritten as

$$\mathbb{E}[\partial_x g(X_T, \mathcal{L}(X_t)) V_T + \tilde{\mathbb{E}} \partial_\mu g(X_T, \mathcal{L}(X_t))(\tilde{X}_T) \cdot \tilde{V}_T]$$
$$= \mathbb{E}[\partial_x g(X_T, \mathcal{L}(X_t)) V_T] + \mathbb{E}\tilde{\mathbb{E}}[\partial_\mu g(\tilde{X}_T, \mathcal{L}(\tilde{X}_T))(X_T) \cdot V_T]$$
$$= \mathbb{E}[Y_T \cdot V_T],$$

and using the expression derived in Lemma 4.10 for $\mathbb{E}[Y_T \cdot V_T]$ in (4.62) gives the desired result. □

The necessary form of the Pontryagin stochastic maximum principle is contained in the following theorem.

Theorem 4.24. *Under the above assumptions, if we assume further that the coefficients b, σ, and f are twice continuously differentiable with respect to α, the admissible control $\alpha \in \mathbb{A}$ is optimal, $\mathbf{X}$ is the associated (optimally) controlled state, and $(\mathbf{Y}, \mathbf{Z})$ are the associated adjoint processes solving the adjoint equation (4.17), then for any other admissible control $\beta \in \mathbb{A}$, we have*

$$\forall \alpha \in A, \quad H(t, X_t, Y_t, Z_t, \alpha_t) \le H(t, X_t, Y_t, Z_t, \alpha) \quad \text{a.e. in } t \in [0, T], \; \mathbb{P}\text{-a.s.} \quad (4.64)$$

Proof. Since A is convex, given $\beta \in \mathbb{A}$ we can choose the perturbation $\alpha_t^\epsilon = \alpha_t + \epsilon(\beta_t - \alpha_t)$ which is still in $\mathbb{A}$ for $0 \le \epsilon \le 1$. Since α is optimal, we have the inequality

$$\frac{d}{d\epsilon} J(\alpha + \epsilon(\beta - \alpha))\big|_{\epsilon=0} = \mathbb{E}\int_0^T \partial_\alpha H(t, X_t, Y_t, Z_t, \alpha_t) \cdot (\beta_t - \alpha_t) \ge 0.$$

By convexity of the Hamiltonian with respect to the control variable $\alpha \in A$, we conclude that

$$\mathbb{E}\int_0^T [H(t, X_t, Y_t, Z_t, \beta_t) - H(t, X_t, Y_t, Z_t, \alpha_t)] dt \ge 0$$

for all β. Now, if for a given (deterministic) $\alpha \in A$ we choose β in the following way,

$$\beta_t = (\omega) \begin{cases} \alpha & \text{if } (t,\omega) \in C, \\ \alpha_t(\omega) & \text{otherwise} \end{cases}$$

for an arbitrary adapted set $C \subset [0.T] \times \Omega$, we see that

$$\mathbb{E}\int_0^T \mathbf{1}_C[H(t,X_t,Y_t,Z_t,\alpha) - H(t,X_t,Y_t,Z_t,\alpha_t)]dt \geq 0.$$

From this we conclude that

$$H(t,X_t,Y_t,Z_t,\alpha) - H(t,X_t,Y_t,Z_t,\alpha_t)] \geq 0, \qquad dt \otimes d\mathbb{P}\text{-a.e.},$$

which is the desired conclusion. □

4.4.5 ▪ A Sufficient Condition

The necessary condition for optimality identified in the previous subsection can be turned into a sufficient condition for optimality under the usual convexity assumptions.

Theorem 4.25. *Under the same assumptions of regularity on the coefficients as before, let* $\alpha \in \mathbb{A}$ *be an admissible control,* $\mathbf{X} = \mathbf{X}^\alpha$ *the corresponding controlled state process, and* $(\mathbf{Y},\mathbf{Z})$ *the adjoint processes. Let us also assume that for each* $t \in [0,T]$

1. $(x,\mu) \hookrightarrow g(x,\mu)$ *is convex;*

2. $(x,\mu,\alpha) \hookrightarrow H(t,x,\mu,Y_t,Z_t,\alpha)$ *is convex for* $dt \otimes d\mathbb{P}$-*a.e.*

Moreover, if

$$H(t,X_t,\mathcal{L}(X_t),Y_t,Z_t,\alpha_t) = \inf_{\alpha \in A} H(t,X_t,\mathcal{L}(X_t),Y_t,Z_t,\alpha), \quad dt \otimes d\mathbb{P}\text{-a.e.} \quad (4.65)$$

then α *is an optimal control, i.e.,*

$$J(\alpha) = \inf_{\alpha' \in \mathbb{A}} J(\alpha'). \qquad (4.66)$$

Proof. Let $\alpha' \in \mathbb{A}$ be a generic admissible control, and $\mathbf{X}' = \mathbf{X}^{\alpha'}$ the corresponding controlled state. By definition of the objective function of the control problem, we have

$$\begin{aligned}J(\alpha) - J(\alpha') &= \mathbb{E}[g(X_T,\mathcal{L}(X_t)) - g(X'_T,\mathcal{L}(X'_T))] \\ &\quad + \mathbb{E}\int_0^T [f(t,X_t,\mathcal{L}(X_t),\alpha_t) - f(t,X'_t,\mathcal{L}(X'_t),\alpha'_t)]dt \\ &= \mathbb{E}[g(X_T,\mathcal{L}(X_t)) - g(X'_T,\mathcal{L}(X'_T))] \\ &\quad + \mathbb{E}\int_0^T [H(t,X_t,\mathcal{L}(X_t),Y_t,Z_t,\alpha_t) - H(t,X'_t,\mathcal{L}(X'_t),Y_t,Z_t,\alpha'_t)]dt \\ &\quad - \mathbb{E}\int_0^T [b(t,X_t,\mathcal{L}(X_t),\alpha_t) - b(t,X'_t,\mathcal{L}(X'_t),\alpha'_t))] \cdot Y_t dt \\ &\quad - \mathbb{E}\int_0^T [\sigma(t,X_t,\mathcal{L}(X_t),\alpha_t) - \sigma(t,X'_t,\mathcal{L}(X'_t),\alpha'_t))] \cdot Z_t dt\end{aligned}$$

4.4. Optimal Control of McKean–Vlasov Dynamics

by definition of the Hamiltonian. g being convex, we have

$$g(x,\mu) - g(x',\mu') \leq (x - x') \cdot \partial_x g(x) + \tilde{\mathbb{E}}[\partial_\mu g(x,\mu)(\tilde{X}) \cdot (\tilde{X} - \tilde{X}')]$$

so that

$$\begin{aligned}
\mathbb{E}[g(X_T, &\mathcal{L}(X_t)) - g(X'_T, \mathcal{L}(X'_T))] \\
&\leq \mathbb{E}[(X_T - X'_T) \cdot \partial_x g(X_T, \mathcal{L}(X_t)) + \mathbb{E}\tilde{\mathbb{E}}[\partial_\mu g(X_T, \mathcal{L}(X_T))(\tilde{X}_T) \cdot (-\tilde{X} - \tilde{X}')] \\
&\leq \mathbb{E}[(X_T - X'_T) \cdot \partial_x g(X_T, \mathcal{L}(X_t)) + \tilde{\mathbb{E}}\mathbb{E}[\partial_\mu g(\tilde{X}_T, \mathcal{L}(X_T))(X_T) \cdot (-\tilde{X} - \tilde{X}')] \\
&\leq \mathbb{E}[(X_T - X'_T) \cdot Y_T],
\end{aligned}$$

where we used Fubini's theorem and the fact that the "tilde random variables" are independent copies of the "nontilde variables." Using the adjoint equation we get

$$\begin{aligned}
(X_T - X'_T) \cdot Y_T &= \int_0^T (X_t - X'_t) dY_t + \int_0^T Y_t d[X_t - X'_t] \\
&\quad + \int_0^T [\sigma(t, X_t, \mathcal{L}(X_t), \alpha_t) - \sigma(t, X'_t, \mathcal{L}(X'_t), \alpha'_t)] \cdot Z_t dt \\
&= -\int_0^T (X_t - X'_t) \partial_x H(t, X_t, \mathcal{L}(X_t), Y_t, Z_t, \alpha_t) dt + \int_0^T (X_t - X'_t) Z_t dW_t \\
&\quad - \int_0^T (X_t - X'_t) \tilde{\mathbb{E}} \partial_\mu H(t, \tilde{X}, \mathcal{L}(\tilde{X}_t), \tilde{Y}_t, \tilde{Z}_t, \tilde{\alpha}_t)(X_t) dt \\
&\quad + \int_0^T Y_t \cdot [b(t, X_t, \mathcal{L}(X_t), \alpha_t) - b(t, X'_t, \mathcal{L}(X'_t), \alpha'_t)] dt \\
&\quad + \int_0^T Y_t \cdot [\sigma(t, X_t, \mathcal{L}(X_t), \alpha_t) - \sigma(t, X'_t, \mathcal{L}(X'_t), \alpha'_t)] dW_t \\
&\quad + \int_0^T [\sigma(t, X_t, \mathcal{L}(X_t), \alpha_t) - \sigma(t, X'_t, \mathcal{L}(X'_t), \alpha'_t)] \cdot Z_t dt,
\end{aligned}$$

where we used integration by parts and the fact that Y_t solves the adjoint equation. Taking expectations, we get

$$\begin{aligned}
\mathbb{E}[(X_T - X'_T) \cdot Y_T] &= -\mathbb{E} \int_0^T (X_t - X'_t) \partial_x H(t, X_t, \mathcal{L}(X_t), Y_t, Z_t, \alpha_t) dt \\
&\quad - \mathbb{E} \int_0^T (X_t - X'_t) \tilde{\mathbb{E}} \partial_\mu H(t, \tilde{X}, \mathcal{L}(\tilde{X}_t), \tilde{Y}_t, \tilde{Z}_t, \tilde{\alpha}_t)(X_t) dt \\
&\quad + \mathbb{E} \int_0^T Y_t \cdot [b(t, X_t, \mathcal{L}(X_t), \alpha_t) - b(t, X'_t, \mathcal{L}(X'_T), \alpha'_t)] dt \\
&\quad + \mathbb{E} \int_0^T [\sigma(t, X_t, \mathcal{L}(X_t), \alpha_t) - \sigma(t, X'_t, \mathcal{L}(X'_T), \alpha'_t)] \cdot Z_t dt.
\end{aligned}$$

Consequently,

$$J(\alpha) - J(\alpha') \leq \mathbb{E} \int_0^T [H(t, X_t, \mathcal{L}(X_t), Y_t, Z_t, \alpha_t) - H(t, X'_t, \mathcal{L}(X'_T), Y_t, Z_t, \alpha'_t)]dt$$

$$- \mathbb{E} \int_0^T (X_t - X'_t) \partial_x H(t, X_t, \mathcal{L}(X_t), Y_t, Z_t, \alpha_t) dt$$

$$- \mathbb{E} \int_0^T (X_t - X'_t) \tilde{\mathbb{E}} \partial_\mu H(t, \tilde{X}, \mathcal{L}(\tilde{X}_t), \tilde{Y}_t, \tilde{Z}_t, \tilde{\alpha}_t)(X_t) dt$$

$$\leq 0,$$

because of the convexity assumption on the Hamiltonian. Indeed, using Fubini's theorem and the fact that $(\tilde{\alpha}_t, \tilde{X}_t, \tilde{Y}_t, \tilde{Z}_t)$ is an independent copy of $(\alpha_t, X_t, Y_t, Z_t)$, the third expectation in the right-hand side can be rewritten as

$$\int_0^T \mathbb{E}\tilde{\mathbb{E}}(X_t - X'_t) \tilde{\mathbb{E}} \partial_\mu H(t, \tilde{X}, \mathcal{L}(\tilde{X}_t), \tilde{Y}_t, \tilde{Z}_t, \tilde{\alpha}_t)(X_t) dt$$

$$= \int_0^T \mathbb{E}\tilde{\mathbb{E}} \partial_\mu \mathbf{H}(t, X_t, X_t, \mathcal{L}(X_t), Y_t, Z_t, \alpha_t)(\tilde{X}_t) \cdot (\tilde{X}_t - \tilde{X}'_t) dt,$$

and we conclude because of our assumptions and the definition (4.53) of convexity, since the first derivative with respect to α does not need to be present because of the criticality of the admissible control $\boldsymbol{\alpha} = (\alpha_t)_{0 \leq t \leq T}$. □

4.4.6 ▪ Examples

Using the formulas derived in the *Scalar Interactions* paragraph of Subsection 4.4.3, it is plain to check that the necessary and sufficient parts of the stochastic maximum principle proven in [7] are a particular case of the general principle proved above. In this subsection, we show how this principle can be used to solve explicit models of mean-field control problems.

Mean-Variance Portfolio Optimization

Again, we use the market model introduced in Subsection 2.1.4 of Chapter 2. For the sake of simplicity, we assume that the coefficients $\mathbf{r}$, $\boldsymbol{\mu}$, and $\boldsymbol{\sigma}$ are deterministic functions of time. As before, we denote by X_t the wealth of the investor at time t and by α_t^i the amount of wealth invested in the ith risky asset at time t. In this setup, the dynamics of the wealth of the investor are given by

$$dX_t = \left[X_t r_t + \alpha_t \cdot (\mu_t - r_t \mathbf{1})\right] dt + \alpha_t \cdot \sigma_t dW_t. \quad (4.67)$$

These dynamics are given by a standard affine SDE. However, the objective depends not only on the value of the terminal wealth, but also upon its distribution. Indeed, as explained earlier, for each fixed $\gamma \in \mathbb{R}$, we try to minimize the objective

$$J^\gamma(\boldsymbol{\alpha}) = -\mathbb{E}[X_T] + \gamma \text{var}[X_T] = \mathbb{E}g(X_T, \mathcal{L}(X_T)) \quad (4.68)$$

over all the admissible portfolios $\boldsymbol{\alpha}$. Here, the function g is given by

$$g(x, \mu) = \int_\mathbb{R} (\gamma x^2 - x) d\mu(x) - \gamma \left(\int_\mathbb{R} x d\mu(x)\right)^2.$$

4.4. Optimal Control of McKean–Vlasov Dynamics

Accordingly, the partial derivative of g with respect to the variable μ is given by

$$\partial_\mu g(x,\mu)(x') = 2\gamma x' - 1 - 2\gamma \int_\mathbb{R} x d\mu(x) = 2\gamma x' - 1 - 2\gamma\overline{\mu}, \qquad \mu\text{-a.e.} \quad x' \in \mathbb{R},$$

if we use the notation $\overline{\mu}$ for the mean of the probability measure μ. While the adjoint equation

$$dY_t = -r_t Y_t dt + Z_t dW_t$$

is the same as before, the terminal condition is now $Y_T = 2\gamma(X_T - \overline{x}_T) - 1$ (recall that $\partial_x g(x,\mu) = 0$) if we use the notation $\overline{x}_t = \mathbb{E}[X_t]$ for the mean of X_t. Taking expectations on both sides of the forward equation (4.67) of the wealth dynamics, we get

$$d\overline{x}_t = \left[r_t \overline{x}_t + \mathbb{E}[\alpha_t] \cdot (\mu_t - r_t \mathbf{1}) \right] dt. \tag{4.69}$$

For reasons already given in Subsection 4.3.4, the control problem is singular in the sense that the Hamiltonian, being linear in α, cannot be minimized with respect to the control variable α. But as before, the necessary part of the stochastic maximum principle implies

$$(\mu_t - r_t \mathbf{1})Y_t + Z\sigma_t = 0, \qquad 0 \leq t \leq T, \text{ a.s.} \tag{4.70}$$

and since we expect the decoupling field to be affine, given the specific form of the terminal condition of the adjoint equation, we look for a solution of the form $Y_t = \eta_t(X_t - \overline{x}_t) + \chi_t$ for two deterministic smooth functions η and χ of the time variable t. Computing the differential of Y_t from this ansatz and identifying term by term with the adjoint equation (4.44), we found that

$$Z_t = \eta_t \alpha_t \cdot \sigma_t,$$

which, together with (4.70), gives

$$\hat{\alpha}_t = -[\eta_t \sigma_t^\dagger \sigma_t]^{-1}(\mu_t - r_t \mathbf{1})(\eta_t[X_t - \overline{x}_t] + \chi_t), \tag{4.71}$$

from which we compute

$$\mathbb{E}[\hat{\alpha}_t] = -[\eta_t \sigma_t^\dagger \sigma_t]^{-1}(\mu_t - r_t \mathbf{1})\chi_t.$$

Identifying the bounded variation parts of dY_t, we get

$$(\dot{\eta}_t + \eta_t r_t)(X_t - \overline{x}_t) + \eta_t \alpha_t \cdot (\mu_t - r_t \mathbf{1}) + \dot{\chi}_t - \eta_t \mathbb{E}[\alpha_t] \cdot (\mu_t - r_t \mathbf{1}) = -r_t \eta_t(X_t - \overline{x}_t) - r_t \chi_t.$$

If we plug in this equality, the values we just found for the optimal control process, and its expectation, and if we identify the coefficients of $(X_t - \overline{x}_t)$ and the constant terms, we get

$$\begin{cases} \dot{\eta}_t + \eta_t(2r_t - \rho_t) = 0, \\ \dot{\chi}_t + r_t \chi_t = 0, \end{cases} \tag{4.72}$$

where ρ_t was defined in (4.49). We recall its definition for convenience:

$$\rho_t = (\mu_t - r_t \mathbf{1})^\dagger [\sigma_t^\dagger \sigma_t]^{-1}(\mu_t - r_t \mathbf{1}).$$

Recalling the terminal condition $\eta_T = 2\gamma$, the first of the two ODEs in (4.72) can be solved explicitly. We get

$$\eta_t = 2\gamma e^{\int_t^T (2r_s - \rho_s)ds}.$$

Using the terminal condition $\chi_T = -1$, the function χ_t is given by

$$\chi_t = -e^{\int_t^T r_s ds}.$$

Finally, we compute $\bar{x}_t$ by solving (4.69) in order to identify the final form of the adjoint process Y_t and the optimal control α_t. Using the expressions we just found for η_t and χ_t, (4.69) rewrites as

$$d\bar{x}_t = \left[r_t \bar{x}_t + \frac{\rho_t}{2\gamma} e^{-\int_t^T (r_s - \rho_s) ds} \right] dt,$$

whose solution is

$$\bar{x}_t = x_0 e^{\int_0^t r_s ds} + \frac{1}{2\gamma} e^{-\int_t^T (r_s - \rho_s) ds} \left(e^{\int_0^t \rho_s ds} - 1 \right),$$

which we can inject in (4.71) to obtain the optimal portfolio in closed loop feedback (Markovian) form.

Optimal Control of Linear-Quadratic (LQ) McKean–Vlasov Dynamics

We apply the Pontryagin stochastic maximum principle to analyze directly a class of LQ models of McKean–Vlasov type. Our goal is to illustrate the use of the maximum principle, so we do not aim at the most general LQ problems. In particular, we assume that the volatility is a positive constant σ, the drift is given by

$$b(t, x, \mu, \alpha) = b_1(t) x + b_2(t) \alpha + \bar{b}_1(t) \bar{\mu},$$

the running cost function by

$$f(t, x, \mu, \alpha) = \frac{1}{2} \left(x^\dagger q(t) x + \alpha^\dagger r(t) \alpha \right),$$

and the terminal cost by

$$g(x, \mu) = \frac{1}{2} \left(x^\dagger q x + (x - s\bar{\mu})^\dagger \bar{q} (x - s\bar{\mu}) \right).$$

We assume that b_1, b_2, and $\bar{b}_1$ are deterministic continuous functions on $[0, T]$ with values in $\mathbb{R}^{d \times d}$, $\mathbb{R}^{d \times k}$, and $\mathbb{R}^{d \times d}$, respectively. Similarly, we assume that q, r, and $\bar{q}$ are deterministic continuous functions on $[0, T]$ with values in $\mathbb{R}^{d \times d}$, $\mathbb{R}^{k \times k}$, $\mathbb{R}^{d \times k}$, and $\mathbb{R}^{d \times d}$, respectively, $q(t)$ being symmetric and nonnegative definite, and $r(t)$ being symmetric and positive definite (hence invertible). We also assume that the $d \times d$ matrices q and $\bar{q}$ are symmetric and nonnegative definite.

In the present setup, the stochastic control problem is to solve the optimization problem

$$\inf_{\alpha \in \mathbb{A}} \mathbb{E}\left[\frac{1}{2} \left(X_T^\dagger q X_T + (X_T - s\mathbb{E}[X_T])^\dagger \bar{q} (X_T - s\mathbb{E}[X_T]) \right) \right.$$
$$\left. + \frac{1}{2} \int_0^T \left(X_t^\dagger q(t) X_t + \alpha_t^\dagger r(t) \alpha_t \right) dt \right] \quad (4.73)$$

subject to $\quad dX_t = \left[b_1(t) X_t + b_2(t) \alpha_t + \bar{b}_1(t) \mathbb{E}[X_t] \right] dt + \sigma dW_t, \quad X_0 = x_0.$

This particular model is covered by the general result presented in the next subsection. However, we use the very special structure of the model to show how we can solve the

4.4. Optimal Control of McKean–Vlasov Dynamics

problem with *bare hands*. The Hamiltonian is given by

$$H(t, x, \mu, y, \alpha) = [b_1(t)x + b_2(t)\alpha + \overline{b}_1(t)\overline{\mu}] \cdot y + \frac{1}{2}\left(x^\dagger q(t)x + \alpha^\dagger r(t)\alpha\right).$$

This Hamiltonian is minimized for

$$\hat{\alpha} = \hat{\alpha}(t, x, \mu, y) = -r(t)^{-1}b_2(t)^\dagger y, \qquad (4.74)$$

which is independent of the measure argument μ. Its derivative with respect to μ is given by

$$\partial_\mu H(t, x, \mu, y)(v) = \overline{b}_1(t)y,$$

which is a constant function of the variable v. The McKean–Vlasov FBSDE derived from the Pontryagin stochastic maximum principle reads

$$\begin{cases} dX_t = \left(b_1(t)X_t - b_2(t)r(t)^{-1}b_2(t)^\dagger Y_t + \overline{b}_1(t)\mathbb{E}[X_t]\right)dt + \sigma dW_t, \\ dY_t = -\left(b_1(t)^\dagger Y_t + q(t)X_t + \overline{b}_1(t)\mathbb{E}[Y_t]\right)dt + Z_t dW_t \end{cases} \qquad (4.75)$$

with initial condition X_0 and terminal condition $Y_T = [q+\overline{q}]X_T + [s^\dagger \overline{q}s - (\overline{q}s + s^\dagger \overline{q})]\mathbb{E}[X_T]$. Taking expectations of both sides of (4.75) and using the notation $\overline{x}_t$ and $\overline{y}_t$ for the expectations $\mathbb{E}[X_t]$ and $\mathbb{E}[Y_t]$, respectively, we find that

$$\begin{cases} \dot{\overline{x}}_t = \left([b_1(t) + \overline{b}_1(t)]\overline{x}_t - b_2(t)r(t)^{-1}b_2(t)^\dagger \overline{y}_t\right)dt, & \overline{x}_0 = \mathbb{E}[X_0], \\ \dot{\overline{y}}_t = \left(-q(t)\overline{x}_t - [b_1(t) + \overline{b}_1(t)]\overline{y}_t\right)dt, & \overline{y}_T = [q + \overline{q} + s^\dagger \overline{q}s - (\overline{q}s + s^\dagger \overline{q})]\overline{x}_T. \end{cases} \qquad (4.76)$$

We rewrite this system in the form

$$\begin{cases} \dot{\overline{x}}_t = a_t \overline{x}_t + b_t \overline{y}_t, & \overline{x}_0 = \mathbb{E}[X_0], \\ \dot{\overline{y}}_t = c_t \overline{x}_t + d_t \overline{y}_t, & \overline{y}_T = e\overline{x}_T, \end{cases} \qquad (4.77)$$

which can be solved if we are able to solve the matrix Riccati equation

$$\dot{\overline{\eta}}_t + \overline{\eta}_t a_t - d_t \overline{\eta}_t + \overline{\eta}_t b_t \overline{\eta}_t - c_t = 0, \qquad \eta_T = e. \qquad (4.78)$$

The solution of (4.77) is obtained by solving

$$\dot{\overline{x}}_t = [a_t + b_t \eta_t]\overline{x}_t, \qquad \overline{x}_0 = \mathbb{E}[X_0] \qquad (4.79)$$

and setting $\overline{y}_t = \overline{\eta}_t \overline{x}_t$. Assuming momentarily that the matrix Riccati equation (4.78) has a unique solution η_t, and plugging $\mathbb{E}[X_t] = \overline{x}_t$ and $\mathbb{E}[Y_t] = \overline{y}_t$ into the McKean–Vlasov FBSDE (4.75), we reduce the problem to the solution of the affine FBSDE

$$\begin{cases} dX_t = [\mathfrak{a}_t X_t + \mathfrak{b}_t Y_t + \mathfrak{c}_t]dt + \sigma dW_t, & X_0 = x_0, \\ dY_t = [\mathfrak{m}_t X_t - \mathfrak{a}_t^\dagger Y_t + \mathfrak{d}_t]dt + Z_t dW_t, & Y_T = \mathfrak{q}X_T + \mathfrak{r}, \end{cases} \qquad (4.80)$$

with

$$\begin{cases} \mathfrak{a}_t = b_1(t), & \mathfrak{b}_t = -b_2(t)r(t)^{-1}b_2(t)^\dagger, & \mathfrak{c}_t = \overline{b}_1(t)\overline{x}_t, \\ \mathfrak{m}_t = -q(t), & \mathfrak{d}_t = -\overline{b}_1(t)\overline{y}_t, \\ \mathfrak{q} = q + \overline{q}, & \mathfrak{r} = [s^\dagger \overline{q}s - (\overline{q}s + s^\dagger \overline{q})]\overline{x}_T. \end{cases}$$

The affine structure of the FBSDE suggests that the decoupling field is affine as well, so we search for deterministic differentiable functions η_t and χ_t such that

$$Y_t = \eta_t X_t + \chi_t, \quad t \in [0,T]. \tag{4.81}$$

Computing dY_t from this ansatz, using the expression of dX_t given by the first equation of (4.80), and identifying term by term with the expression of dY_t given in (4.80), we get

$$\begin{cases} \dot{\eta}_t + \eta_t \mathfrak{b}_t \eta_t + \mathfrak{a}_t^\dagger \eta_t + \eta_t \mathfrak{a}_t - \mathfrak{m}_t = 0, & \eta_T = \mathfrak{q}, \\ \dot{\chi}_t + (\mathfrak{a}_t + \eta_t \mathfrak{b}_t)\chi_t - \mathfrak{d}_t + \eta_t \mathfrak{c}_t = 0, & \chi_T = \mathfrak{r}, \\ Z_t = \sigma \eta_t. \end{cases} \tag{4.82}$$

The first equation is a matrix Riccati equation. If and when it can be solved, the third equation is solved automatically, and the second equation becomes a first-order linear ODE, though not homogeneous this time, which can be solved by standard methods. Notice that the quadratic terms of the two Riccati equations are the same since $\mathfrak{b}_t = \mathfrak{b}_t = -b_2(t)r(t)^{-1}b_2(t)^\dagger$. However, the terminal conditions are different since the terminal condition in (4.82) is given by $\mathfrak{q} = q + \bar{q}$ while it was given by $e = q + \bar{q} + s^\dagger \bar{q} s - (\bar{q}s + s^\dagger \bar{q})$ in (4.78). Notice also that the first-order terms are different as well. As in the case of the linear-quadratic models, existence and uniqueness for a solution of this matrix Riccati equation is equivalent to the unique solvability of a deterministic control problem, proving that once more, we have existence and uniqueness of a solution to the LQ McKean–Vlasov control problem whenever the matrix coefficients are continuous, q, $\bar{q}$, $q(t)$, and $\bar{q}(t)$ are nonnegative definite, and $r(t)$ is strictly positive definite.

The optimally controlled state is Gaussian despite the nonlinearity due to the McKean–Vlasov nature of the dynamics. Moreover, because of the linearity of the ansatz, the adjoint process $Y = (Y_t)_{0 \le t \le T}$ is also Gaussian. Finally, using again the form of the ansatz, we see that the optimal control α_t, which was originally expected to be an open loop control, is in fact in closed loop feedback form $\hat{\alpha}_t = \varphi(t, X_t)$ since it can be rewritten as

$$\hat{\alpha}_t = -r(t)^{-1}b_2(t)^\dagger \eta_t X_t - r(t)^{-1}b_2(t)^\dagger \chi_t,$$

which incidentally shows that the optimal control is also a Gaussian process.

4.4.7 • Control of McKean–Vlasov Dynamics: General Results

Given the availability of the necessary and sufficient parts of the Pontryagin stochastic maximum principle, one can reformulate the strategy of the second probabilistic approach to the solution of optimal control problems in the case of McKean–Vlasov dynamics. Indeed, if we can find for each t, x, μ, y, and z a control $\hat{\alpha} = \hat{\alpha}(t, x, \mu, y, z) \in A$ minimizing the Hamiltonian H so that

$$\hat{\alpha}(t, x, \mu, y, z) \in \arg\min_{\alpha \in A} H(t, x, \mu, y, z, \alpha),$$

and if the function $(t, x, \mu, y, z) \hookrightarrow \hat{\alpha}(t, x, \mu, y, z)$ is smooth enough, then the stochastic maximum principle suggests that we will have solved the optimal control problem if we can solve the following forward/backward system of stochastic differential equations

$$\begin{cases} dX_t = b(t, X_t, \mathcal{L}(X_t), \hat{\alpha}(t, X_t, \mathcal{L}(X_t), Y_t, Z_t))dt \\ \qquad + \sigma(t, X_t, \mathcal{L}(X_t), \hat{\alpha}(t, X_t, \mathcal{L}(X_t), Y_t, Z_t))dW_t \\ dY_t = -\partial_x H(t, X_t, Y_t, Z_t, \hat{\alpha}(t, X_t, \mathcal{L}(X_t), Y_t, Z_t))dt + Z_t dW_t \\ \qquad - \tilde{\mathbb{E}}[\partial_\mu H(t, \tilde{X}_t, \mathcal{L}(X_t), \tilde{Y}_t, \tilde{Z}_t, \hat{\alpha}(t, \tilde{X}_t, \mathcal{L}(\tilde{X}_t), \tilde{Y}_t, \tilde{Z}_t))(X_t)]dt \end{cases} \tag{4.83}$$

4.4. Optimal Control of McKean–Vlasov Dynamics

with the appropriate initial condition $X_0 = x$ and terminal condition $Y_T = \partial_x g(X_T, \mathcal{L}(X_t))$ $+ \tilde{\mathbb{E}}[\partial_\mu g(\tilde{X}_T, \mathcal{L}(\tilde{X}_t))(X_T)]$, at least as long as the convexity assumptions of the sufficient part of the stochastic maximum principle are satisfied. The marginal distributions of (X_t, Y_t, Z_t) appearing in the coefficients, the above system appears as the epitome of an FBSDE of McKean–Vlasov type.

We gave two examples of models which could be explicitly solved using this approach. In this final subsection, we give explicit assumptions (see **(B.1)–(B.4)** below) under which the same approach leads to solutions. Despite the fact that these assumptions appear to be rather restrictive, the level of technicality of the proofs remains very high and we shall refrain from giving detailed proofs. We refer the interested reader to the Notes & Complements section at the end of the chapter for precise references.

As is most often the case in applications of the stochastic maximum principle, we choose $A = \mathbb{R}^k$ and we consider a *linear* model for the forward dynamics of the state.

(B.1) The drift b and the volatility σ are linear in μ, x, and α in the sense that

$$b(t, x, \mu, \alpha) = b_0(t) + b_1(t)\overline{\mu} + b_2(t)x + b_3(t)\alpha,$$
$$\sigma(t, x, \mu, \alpha) = \sigma_0(t) + \sigma_1(t)\overline{\mu} + \sigma_2(t)x + \sigma_3(t)\alpha$$

for some bounded measurable deterministic functions b_0, b_1, b_2, and b_3 with values in $\mathbb{R}^d$, $\mathbb{R}^{d \times d}$, $\mathbb{R}^{d \times d}$, and $\mathbb{R}^{d \times k}$, and σ_0, σ_1, σ_2, and σ_3 with values in $\mathbb{R}^{d \times m}$, $\mathbb{R}^{(d \times m) \times d}$, $\mathbb{R}^{(d \times m) \times d}$, and $\mathbb{R}^{(d \times m) \times k}$ (the parentheses around $d \times m$ indicating that $\sigma_i(t)u_i$ is seen as an element of $\mathbb{R}^{d \times m}$ whenever $u_i \in \mathbb{R}^d$ with $i = 1, 2$, or $u_i \in \mathbb{R}^k$ with $i = 3$), and where we use the notation $\overline{\mu} = \int x \, d\mu(x)$ for the mean of a measure μ.

(B.2) The cost functions f and g are continuously differentiable with respect to (x, μ, α) and (x, μ), respectively, and there exists a constant $\hat{L}$ such that

$$|f(t, x', \mu', \alpha') - f(t, x, \mu, \alpha)| + |g(x', \mu') - g(x, \mu)|$$
$$\leq \hat{L}\big[1 + |x'| + |x| + |\alpha'| + |\alpha| + \|\mu\|_2 + \|\mu'\|_2\big]\big[|(x', \alpha') - (x, \alpha)| + W_2(\mu', \mu)\big].$$

(B.3) There exists a constant $\hat{c} > 0$ such that the derivatives of f and g with respect to (x, α) and x, respectively, are $\hat{c}$-Lipschitz continuous with respect to (x, α, μ) and (x, μ), respectively, the Lipschitz property in the variable μ being understood in the sense of the 2-Wasserstein distance. Moreover, for any $t \in [0, T]$, any $x, x' \in \mathbb{R}^d$, any $\alpha, \alpha' \in \mathbb{R}^k$, any $\mu, \mu' \in \mathcal{P}_2(\mathbb{R}^d)$, and any $\mathbb{R}^d$-valued random variables X and X' having μ and μ' as distributions,

$$\mathbb{E}\big[|\partial_\mu f(t, x', \mu', \alpha')(X') - \partial_\mu f(t, x, \mu, \alpha)(X)|^2\big]$$
$$\leq \hat{c}\big(|(x', \alpha') - (x, \alpha)|^2 + \mathbb{E}\big[|X' - X|^2\big]\big),$$
$$\mathbb{E}\big[|\partial_\mu g(x', \mu')(X') - \partial_\mu g(x, \mu)(X)|^2\big] \leq \hat{c}\big(|x' - x|^2 + \mathbb{E}\big[|X' - X|^2\big]\big).$$

(B.4) The function f is strongly convex with respect to (x, μ, α) for t fixed in the sense that, for some $\lambda > 0$,

$$f(t, x', \mu', \alpha') - f(t, x, \mu, \alpha) - \partial_{(x,\alpha)} f(t, x, \mu, \alpha) \cdot (x' - x, \alpha' - \alpha)$$
$$- \tilde{\mathbb{E}}\big[\partial_\mu f(t, x, \mu, \alpha)(\tilde{X}) \cdot (\tilde{X}' - \tilde{X})\big] \geq \lambda |\alpha' - \alpha|^2,$$

whenever $\tilde{X}, \tilde{X}' \in L^2(\tilde{\Omega}, \tilde{\mathcal{A}}, \tilde{\mathbb{P}}; \mathbb{R}^d)$ with distributions μ and μ', respectively. The function g is also assumed to be convex in (x, μ), though not strongly in the sense that $\lambda = 0$.

The drift and the volatility being linear, the Hamiltonian takes the particular form

$$H(t, x, \mu, y, z, \alpha) = \big[b_0(t) + b_1(t)\overline{\mu} + b_2(t)x + b_3(t)\alpha\big] \cdot y$$
$$+ \big[\sigma_0(t) + \sigma_1(t)\overline{\mu} + \sigma_2(t)x + \sigma_3(t)\alpha\big] \cdot z + f(t, x, \mu, \alpha)$$

for $t \in [0,T]$, $x,y \in \mathbb{R}^d$, $z \in \mathbb{R}^{d\times m}$, $\mu \in \mathcal{P}_2(\mathbb{R}^d)$, and $\alpha \in \mathbb{R}^k$. Given $(t,x,\mu,y,z) \in [0,T] \times \mathbb{R}^d \times \mathcal{P}_2(\mathbb{R}^d) \times \mathbb{R}^d \times \mathbb{R}^{d\times m}$, the function $\mathbb{R}^k \ni \alpha \mapsto H(t,x,\mu,y,z,\alpha)$ is strictly convex so that there exists a unique minimizer $\hat{\alpha}(t,x,\mu,y,z)$:

$$\hat{\alpha}(t,x,\mu,y,z) = \mathrm{argmin}_{\alpha \in \mathbb{R}^k} H(t,x,\mu,y,z,\alpha). \tag{4.84}$$

Lemma 4.26. *The function $[0,T] \times \mathbb{R}^d \times \mathcal{P}_2(\mathbb{R}^d) \times \mathbb{R}^d \times \mathbb{R}^{d\times m} \ni (t,x,\mu,y,z) \mapsto \hat{\alpha}(t,x,\mu,y,z)$ is measurable, locally bounded and Lipschitz continuous with respect to (x,μ,y,z) uniformly in $t \in [0,T]$, the Lipschitz constant depending only upon λ, the supremum norms of b_3 and σ_3, and the Lipschitz constant of $\partial_\alpha f$ in (x,μ).*

For each admissible control $\alpha = (\alpha_t)_{0 \le t \le T}$, if we denote the corresponding solution of the state equation by $X = (X_t)_{0 \le t \le T}$, then the adjoint BSDE (4.17) introduced in Definition 4.19 reads

$$\begin{aligned}dY_t = &-\partial_x f\bigl(t,X_t,\mathcal{L}(X_t),\alpha_t\bigr)dt - b_2^\dagger(t)Y_t dt - \sigma_2^\dagger(t)Z_t dt + Z_t dW_t \\ &- \tilde{\mathbb{E}}\bigl[\partial_\mu f\bigl(t,\tilde{X}_t,\mathcal{L}(X_t),\tilde{\alpha}_t\bigr)(X_t)\bigr]dt - b_1^\dagger(t)\mathbb{E}[Y_t]dt - \sigma_1^\dagger(t)\mathbb{E}[Z_t]dt.\end{aligned} \tag{4.85}$$

Given the necessary and sufficient conditions proven in the previous section, our goal is to use the control $\hat{\boldsymbol{\alpha}} = (\hat{\alpha}_t)_{0\le t \le T}$ defined by $\hat{\alpha}_t = \hat{\alpha}(t,X_t,\mathcal{L}(X_t),Y_t,Z_t)$, where $\hat{\alpha}$ is the minimizer function constructed above and the process $(X_t,Y_t,Z_t)_{0 \le t \le T}$ is a solution of the FBSDE

$$\begin{cases} dX_t = \bigl[b_0(t) + b_1(t)\mathbb{E}[X_t] + b_2(t)X_t + b_3(t)\hat{\alpha}(t,X_t,\mathcal{L}(X_t),Y_t,Z_t)\bigr]dt \\ \qquad + \bigl[\sigma_0(t) + \sigma_1(t)\mathbb{E}[X_t] + \sigma_2(t)X_t + \sigma_3(t)\hat{\alpha}(t,X_t,\mathcal{L}(X_t),Y_t,Z_t)\bigr]dW_t, \\ dY_t = -\bigl[\partial_x f\bigl(t,X_t,\mathcal{L}(X_t),\hat{\alpha}(t,X_t,\mathcal{L}(X_t),Y_t,Z_t))\bigr) + b_2^\dagger(t)Y_t + \sigma_2^\dagger(t)Z_t\bigr]dt + Z_t dW_t \\ \qquad - \bigl\{\tilde{\mathbb{E}}\bigl[\partial_\mu f\bigl(t,\tilde{X}_t,\mathcal{L}(X_t),\hat{\alpha}(t,\tilde{X}_t,\mathcal{L}(X_t),\tilde{Y}_t,\tilde{Z}_t))\bigr)(X_t)\bigr] + b_1^\dagger(t)\mathbb{E}[Y_t] + \sigma_1^\dagger(t)\mathbb{E}[Z_t]\bigr\}dt, \end{cases} \tag{4.86}$$

with the initial condition $X_0 = x_0$, for a given deterministic point $x_0 \in \mathbb{R}^d$, and the terminal condition $Y_T = \partial_x g(X_T, \mathcal{L}(X_t)) + \tilde{\mathbb{E}}[\partial_\mu g(\tilde{X}_T, \mathcal{L}(X_t))(X_T)]$.

Here is the main existence and uniqueness result.

Theorem 4.27. *Under assumptions* **(B.1)** *through* **(B.4)**, *the forward-backward system (4.86) is uniquely solvable.*

4.5 • Notes & Complements

The probabilistic approach to optimal control based on a stochastic version of the Pontryagin maximum principle has its origins in the works of Kushner [76], Bensoussan [14], and Bismut [22]. The reader interested in a historical perspective on the evolution of the probabilistic approach to the Pontryagin maximum principle can also check Elliott [56] and Haussmann [67]. In [26], Cadenillas and Karatzas consider the linear state dynamics (4.36) constraining the state X_t to remain in a fixed convex subset V of $\mathbb{R}^d$. They assume A is convex, and define the set $\mathbb{A}$ of admissible controls as the set of those α satisfying the same integrability property as in the text, but for which $X_t \in V$ $\mathbb{P}$-almost surely for all t. This set is convex since for $\alpha \in \mathbb{A}$, $\alpha' \in \mathbb{A}$, and $\lambda \in [0,1]$, we have $X_t^{\lambda\alpha + (1-\lambda)\alpha'} = \lambda X_t^\alpha + (1-\lambda)X_t^{\alpha'}$ by linearity of the SDE, and V being convex, we necessarily have $X_t^{\lambda\alpha + (1-\lambda)\alpha'} \in V$ and $\lambda\alpha + (1-\lambda)\alpha' \in \mathbb{A}$. On the other hand, they assume that the terminal cost $g : \Omega \times V \ni (\omega, x) \hookrightarrow g(\omega, x) \in \mathbb{R}$ is $\mathcal{F}_T$-measurable for each $x \in V$ fixed, and convex and continuously differentiable on V for $\omega \in \Omega$ fixed. Similarly, they assume that the running cost $f : [0,T] \times \Omega \times V \times A \ni (t,\omega,x,\alpha) \hookrightarrow f(t,\omega,x,\alpha) \in \mathbb{R}$ is progressively measurable for each $(x,\alpha) \in V \times A$ fixed, and convex and continuously differentiable on $V \times A$ for $(t,\omega) \in [0,T] \times \Omega$

4.5. Notes & Complements

fixed. While these assumptions do not compare well to ours, the same proofs lead to the existence and uniqueness of adjoint processes, the same expression for the Gâteaux derivative of the cost functional $J(\alpha)$, and the same Pontryagin maximum principle (i.e., the necessary and sufficient conditions for optimality).

It is possible to derive a Pontryagin maximum principle without the convexity assumption on the control set A and the set $\mathbb{A}$ of admissible controls. While it is still possible to derive necessary conditions for optimality in terms of dual objects, the structure of the adjoint processes is not as straightforward. Indeed, the derivation requires the manipulation of spike perturbations of the controls, the introduction of two sets of dual processes and the solution of two different adjoint BSDEs. While similar to those of the previous subsection, the ideas behind the proofs are more technical and, as a result, we did not give them in the text. We refer the interested reader to Peng's original paper [104], Quenez' informal presentation [108], or Yong and Zhou's textbook [117].

Several attempts have been made at controlling stochastic dynamics of the McKean–Vlasov type. In [2], Ahmed and Ding derive a form of the HJB equation using a general form of the optimization problem expressed in terms of Nisio nonlinear operator semigroups. The probabilistic approach based on the Pontryagin maximum principle was considered by Andersson and Djehiche in [7] for convex control sets A and for scalar interactions, i.e., dependence upon the distribution limited to expectations of functions. For similar models, an approach based on Malliavin calculus is given in [94] by Meyer-Brandis, Øksendal, and Zhou. Our presentation is borrowed from the work [34] of Carmona and Delarue.

Dynamic versions of the one-period Markowitz portfolio optimization problem have been proposed by several authors, but it seems that the continuous time formulation presented in this chapter first appeared in the paper [119] of Zhou and Li. It can be found in book form in [106] or [100]. We followed the presentation of [106]. We detailed the solution of the auxiliary LQ control problem associated with the mean-variance portfolio optimization. A proof of the duality argument which we used without proof can be found in Section 6.6 of [106]. We revisited the mean-variance portfolio selection problem a final time by emphasizing the formulation of the problem as the control of McKean–Vlasov dynamics in the spirit of the results of this chapter. Our solution is based on the form of the Pontryagin stochastic maximum principle which we proved in the text. Our presentation was inspired by the paper [7] by Andersson and Djehiche.

The existence and uniqueness result for FBSDEs of McKean–Vlasov type described in the final subsection of the chapter is taken from the paper of Carmona and Delarue [34], or their subsequent book [35], where detailed proofs and complements can be found.

Part III

Stochastic Differential Games

Chapter 5
Stochastic Differential Games

This chapter is devoted to the analysis of stochastic differential games. Even though most of the notation and definitions are mere extensions of those introduced in our discussion of stochastic control problems, we state them in full in order to render this introduction to stochastic differential games as self-contained as possible. We focus on probabilistic approaches (as opposed to those based on PDEs) and we show how BSDEs and FBSDEs can be brought to bear in the search for Nash equilibriums. In particular, we use versions of the stochastic maximum principle to construct open loop as well as closed loop equilibriums. Motivated by applications to systemic risk and predatory trading, we implement our theoretical results on several instances of linear-quadratic (LQ) models treated as a testbeds.

5.1 • Introduction and First Definitions

The purpose of this chapter is to introduce and develop the mathematical analysis of stochastic differential games. We denote by N the number of players. We label them by the integers $1, \cdots, N$. As a stochastic control problem can be viewed as a stochastic game with $N = 1$ player, a significant part of the mathematical setup for this chapter parallels the notations and definitions already introduced in Chapter 3.

As in the case of stochastic control problems, we assume that at each time t, the players act on a system whose state X_t they influence through their actions. The dynamics of the state of the system are given by a stochastic differential equation of the Itô type, which we proceed to describe. See the Notes & Complements section at the end of the chapter for references to extensions to models, including jumps in the state dynamics. The Itô process giving the state dynamics is driven by an m-dimensional Wiener process $\boldsymbol{W} = (W_t)_{0 \leq t \leq T}$ defined on a probability space $(\Omega, \mathcal{F}, \mathbb{P})$, the filtration $\mathbb{F} = (\mathcal{F}_t)_{0 \leq t \leq T}$ being assumed to be most of the time its natural filtration. As before, $\mathcal{P}$ denotes the progressive σ-field.

The major difference with the setup of a stochastic control problem is the fact that the set(s) of admissible strategies and cost functionals are now of a different nature, so we proceed to introduce the notions specific to the stochastic game setup. We denote by $A^1, \cdots, A^N$ the sets of actions that players $1, \cdots, N$ can take at any point in time. As in the case of control problems, the set A^i is typically a compact metric space or a subset of a Euclidean space, say $A^i \subset \mathbb{R}^{k_i}$, and we denote by $\mathcal{A}^i$ its Borel σ-field. We use the notation $\mathbb{A}$ for the set of admissible strategy profiles. The elements $\boldsymbol{\alpha}$ of $\mathbb{A}$ are N-tuples $\boldsymbol{\alpha} = (\boldsymbol{\alpha}^1, \cdots, \boldsymbol{\alpha}^N)$ where each $\boldsymbol{\alpha}^i = (\alpha_t^i)_{0 \leq t \leq T}$ is an A^i-valued adapted process. Moreover, these individual strategies will have to satisfy extra conditions (e.g., measurability and integrability constraints) which change from one application to another. In most of the cases considered here, we shall assume that these constraints can be defined player by player, independently of each other. To be more specific, we shall often assume that $\mathbb{A} = \mathbb{A}^1 \times \cdots \times \mathbb{A}^N$, where for each $i \in \{1, \cdots, N\}$, $\mathbb{A}^i$ is the space of controls/strategies which are deemed admissible to player i, irrespective of what the other players do. Typically, $\mathbb{A}^i$ will be a space of A^i-valued $\mathcal{P}$-measurable processes $\boldsymbol{\alpha}^i = (\alpha^i_t)_{0 \leq t \leq T}$, being either bounded or satisfying an integrability condition such as $\mathbb{E} \int_0^T |\alpha^i_t|^2 dt < \infty$. Again, as in the case of control problems, we will also add measurability conditions specifying the kind of information each player can use. We will also use the notations $A = A^1 \times \cdots \times A^N$ for the set of actions $\alpha = (\alpha^1, \cdots, \alpha^N)$ available to players $1, \cdots, N$ at any given time. Also, in order to emphasize the multivariate nature of the game models (as opposed to the single-agent nature of control problems) we use the term "strategy profile".

A Convenient Notation. If $\alpha = (\alpha^1, \cdots, \alpha^N) \in A$, $i \in \{1, \cdots, N\}$, and $\beta^i \in A^i$, we will denote by (α^{-i}, β^i) the collective set of actions where all players except player i keep the same actions, while player i switches from action α^i to β^i. Similarly, if $\boldsymbol{\alpha} = (\boldsymbol{\alpha}^1, \cdots, \boldsymbol{\alpha}^N) \in \mathbb{A}$ is a set of admissible strategies for the N players, and $\boldsymbol{\beta}^i \in \mathbb{A}^i$ is an admissible strategy for player i, then we denote by $(\boldsymbol{\alpha}^{-i}, \boldsymbol{\beta}^i)$ the new set of strategies where at each time t all players $j \neq i$ keep the same action α^j_t while player i switches from action α^i_t to β^i_t. In other words, $(\boldsymbol{\alpha}^{-i}, \boldsymbol{\beta}^i)^j_t$ is equal to α^j_t if $j \neq i$ and β^i_t otherwise.

For each choice of strategy profile $\boldsymbol{\alpha} = (\alpha_t)_{0 \leq t \leq T} \in \mathbb{A}$, it is assumed that the time evolution of the state $\boldsymbol{X} = \boldsymbol{X}^{\boldsymbol{\alpha}}$ of the system satisfies

$$\begin{cases} dX_t = b(t, X_t, \alpha_t)dt + \sigma(t, X_t, \alpha_t)dW_t, & 0 \leq t \leq T, \\ X_0 = x. \end{cases} \quad (5.1)$$

where

$$(b, \sigma) : [0.T] \times \Omega \times \mathbb{R}^d \times A \hookrightarrow \mathbb{R}^d \times \mathbb{R}^{d \times m}$$

satisfies

(a) $\forall x \in \mathbb{R}^d, \forall \alpha \in A, (b(t, x, \alpha))_{0 \leq t \leq T}$ and $(\sigma(t, x, \alpha))_{0 \leq t \leq T}$ are progressively measurable processes with values in $\mathbb{R}^d$ and $\mathbb{R}^{d \times m}$, respectively;

(b) $\exists c > 0, \forall t \in [0, T], \forall \alpha \in A, \forall \omega \in \Omega, \forall x, x' \in \mathbb{R}^d$,
$|b(t, \omega, x, \alpha) - b(t, \omega, x', \alpha)| + |\sigma(t, \omega, x, \alpha) - \sigma(t, \omega, x', \alpha)| \leq c|x - x'|$.

As usual, we omit ω from the notation whenever possible. However, notice that in many cases, the coefficients b and σ will be independent of ω in the sense that they will be deterministic functions on $[0.T] \times \mathbb{R}^d \times A$.

As in the case of stochastic control problems, many applications can be studied with models in which only the drift is affected by the actions of the players. See Section 5.3 below. These models will give us the opportunity to discuss the differences between weak and strong formulations of a stochastic differential game.

Information Structures, Admissible Actions, and Strategies

The information structure of a stochastic game is usually very complex as each player may have his/her own filtration $\{\mathcal{F}^i_t\}_{0 \leq t \leq T}$, formalizing the information he/she has access to at time t in order to make the choice of an action $\alpha^i_t \in A^i$. In general, there is no reason why different players should have access to the same information. The search for solutions and the analytical tools which can be brought to bear in this search depend strongly upon the kind of information each player has access to at any given time, and which subset of this information is allowed to be used in order to take action.

While it is tempting to believe that a generalization of the definitions introduced in our discussion of stochastic control problems to the case of stochastic games should be straightforward, the subtleties introduced by the interaction and competition between the players need special attention. For the sake of definiteness, we recall the choices of information structures most often used in the case of stochastic control problems. They are open loop (OL), closed loop perfect state (CLPS), memoryless perfect state (MPS) and feedback perfect state (FPS).

We shall refine this terminology once we introduce the cost functions and the optimization problems. In the context of game theory, solving a game often amounts to identifying one (or sometimes all) possible equilibriums of the game, and for us, equilibrium will mean *Nash equilibrium* most of the time.

One of the most annoying features of the search for Nash equilibriums is the lack of uniqueness, and the likelihood that an infinite number of Nash equilibriums could exist. According to Basar "... *One way of avoiding the occurrence of infinitely many Nash equilibriums would be to disallow the players to receive dynamic information (such as state information) correlated with the actions of the players in the past; the open-loop (OL) information structure does that, but in many applications it may be too restrictive. ...*". We show below how to implement this suggestion.

5.1.1 ▪ A Frequently Encountered Special Case

It often happens that the state of the system is the aggregation of private states of individual players, say $X_t = (X_t^1, \cdots, X_t^N)$, where $X_t^i \in \mathbb{R}^{d_i}$ can be interpreted as the private state of player $i \in \{1, \cdots, N\}$. Here $d = d_1 + \cdots + d_N$ and consequently, $\mathbb{R}^d = \mathbb{R}^{d_1} \times \cdots \times \mathbb{R}^{d_N}$. The main feature of such a special case is that we usually assume that the dynamics of the private states are given by stochastic differential equations driven by separate Wiener processes $\boldsymbol{W}^i = (W_t^i)_{0 \le t \le T}$, which are most often assumed to be independent of each other. See nevertheless our discussion of the *common noise case* later on. So, typically, we assume that

$$dX_t^i = b^i(t, X_t, \alpha_t)dt + \sigma^i(t, X_t, \alpha_t)dW_t^i, \qquad 0 \le t \le T,\ i=1,\cdots,N, \quad (5.2)$$

where the $\boldsymbol{W}^i = (W_t^i)_{0 \le t \le T}$ are m_i-dimensional independent Wiener processes giving the components of $\boldsymbol{W} = (W_t)_{0 \le t \le T}$, and where the functions

$$(b^i, \sigma^i) : [0.T] \times \Omega \times \mathbb{R}^d \times A \hookrightarrow \mathbb{R}^{d_i} \times \mathbb{R}^{d_i \times m_i}$$

satisfy the same assumptions as before. It is important to notice that these N dynamical equations are coupled by the fact that all the private states and all the actions enter into the coefficients of these N equations. We can use vector/matrix notation, and rewrite the system of N stochastic dynamical equations in Euclidean spaces of dimensions $d_1, \cdots, d_N$, respectively, as a stochastic differential equation giving the dynamics of the d-dimensional state X_t. So if we set

$$X_t = \begin{bmatrix} X_t^1 \\ X_t^2 \\ \vdots \\ X_t^N \end{bmatrix}, \quad b(t,x,\alpha) = \begin{bmatrix} b^1(t,x,\alpha) \\ b^2(t,x,\alpha) \\ \vdots \\ b^N(t,x,\alpha) \end{bmatrix}, \quad \text{and} \quad W_t = \begin{bmatrix} W_t^1 \\ W_t^2 \\ \vdots \\ W_t^N \end{bmatrix} \quad (5.3)$$

with $m = m_1 + \cdots + m_N$ and

$$\sigma(t,x,\alpha) = \begin{bmatrix} \sigma^1(t,x,\alpha) & 0 & \cdots & 0 \\ 0 & \sigma^2(t,x,\alpha) & \cdots & 0 \\ \vdots & \ddots & \ddots & \vdots \\ 0 & \cdots & 0 & \sigma^N(t,x,\alpha) \end{bmatrix},$$

we recover the dynamics of the state of the system given by (5.1). However, it is important to emphasize the special block-diagonal structure of the volatility matrix σ, and most importantly, the fact that when the Wiener processes W^i are not independent, the components of the Wiener process W are not independent, which is an assumption we often make when we use dynamics of the form (5.1).

The popularity of this formulation is due to the ease with which we can define the information structures and admissible strategy profiles of some specific games of interest. For example, in a game where each player can only use the information of the state of the system at time t when making a strategic decision at that time, the admissible strategy profiles will be of the form $\alpha_t^i = \phi^i(t, X_t)$ for some deterministic function ϕ^i. These strategies are said to be closed loop in feedback form, or Markovian. Moreover, if the information which can be used by player i at time t can only depend upon his/her own private state at time t, then the admissible strategy profiles will be of the form $\alpha_t^i = \phi^i(t, X_t^i)$. Such strategies are usually called distributed.

5.1.2 ▪ Cost Functionals and Notions of Optimality

In full analogy with the case of stochastic control involving only one controller, we assume that each player (controller) faces instantaneous and running costs. So for each $i \in \{1, \cdots, N\}$, we assume that we have

- an $\mathcal{F}_T$-measurable square integrable random variable $\xi^i \in L^2(\Omega, \mathcal{F}_T, \mathbb{P})$, usually called the *terminal cost*. Most often, ξ^i will be of the form $\xi^i = g^i(X_T)$ for some $\mathcal{F}_T \times \mathcal{B}_{\mathbb{R}^d}$-measurable function $g^i : \Omega \times \mathbb{R}^d \hookrightarrow \mathbb{R}$, which is assumed to grow at most quadratically;

- a function $f^i : [0,T] \times \Omega \times \mathbb{R}^d \times A \hookrightarrow \mathbb{R}$, called the *running cost*, satisfying the same assumption as the drift b;

from which we define the overall expected cost to player i:

- (*cost functional*) if the N players use the strategy profile $\boldsymbol{\alpha} \in \mathbb{A}$, the expected total cost to player i is defined as

$$J^i(\boldsymbol{\alpha}) = \mathbb{E}\left[\int_0^T f^i(s, X_s, \alpha_s)ds + \xi^i\right], \qquad \boldsymbol{\alpha} = (\alpha^1, \cdots, \alpha^N) \in \mathbb{A},$$
(5.4)

where X is the state of the system whose dynamics are given by (5.1).

Notice that, in the general situation considered here, the cost to a given player depends upon the strategies used by the other players indirectly through the values of the state X_t over time, but also directly as the specific actions α_t^j taken by the other players may appear explicitly in the expression of the running cost f^i of player i.

Each player i attempts to minimize his/her total expected cost J^i. If we introduce the notation

$$J(\boldsymbol{\alpha}) = (J^1(\boldsymbol{\alpha}), \cdots, J^N(\boldsymbol{\alpha})), \qquad \boldsymbol{\alpha} \in \mathbb{A},$$

5.1. Introduction and First Definitions

heuristically speaking, finding a solution to the game amounts to searching for a solution to the stochastic optimization problem for the functional J over the set $\mathbb{A}$ of admissible strategy profiles. The major difficulty is that J is taking values in a multidimensional Euclidean space, which is not totally ordered in a natural fashion. We need to specify how we compare the costs of different strategy profiles in order to clearly define the notion of optimality.

Definition 5.1. *A set of admissible strategies* $\boldsymbol{\alpha}^* = (\alpha^{*1}, \cdots, \alpha^{*N}) \in \mathbb{A}$ *is said to be Pareto optimal or Pareto efficient if there is no* $\boldsymbol{\alpha} = (\alpha_1, \cdots, \alpha_N) \in \mathbb{A}$ *such that*

$$\forall i \in \{1, \cdots, N\}, \quad J^i(\boldsymbol{\alpha}) \leq J^i(\boldsymbol{\alpha}^*), \tag{5.5}$$

$$\exists i_0 \in \{1, \cdots, N\}, \quad J^{i_0}(\boldsymbol{\alpha}) < J^{i_0}(\boldsymbol{\alpha}^*). \tag{5.6}$$

In other words, there is no strategy which makes *every player* at least as well off and *at least one player* strictly better off. A strategy $\boldsymbol{\alpha}^* \in \mathbb{A}$ is sometimes said to be weakly Pareto efficient when there is no $\boldsymbol{\alpha} \in \mathbb{A}$ such that

$$\forall i \in \{1, \cdots, N\}, \quad J^i(\boldsymbol{\alpha}) < J^i(\boldsymbol{\alpha}^*).$$

The notion of Pareto optimality is natural in problems of optimal allocation of resources and, as a result, very popular in economics literature and in operations research applications. However, we shall use the notion of optimality associated to the concept of a Nash equilibrium.

Definition 5.2. *A set of admissible strategies* $\boldsymbol{\alpha}^* = (\alpha^{*1}, \cdots, \alpha^{*N}) \in \mathbb{A}$ *is said to be a Nash equilibrium for the game if*

$$\forall i \in \{1, \cdots, N\}, \; \forall \alpha^i \in \mathbb{A}^i, \quad J^i(\boldsymbol{\alpha}^*) \leq J^i(\boldsymbol{\alpha}^{*-i}, \alpha^i).$$

NB: Nash equilibriums are not Pareto efficient in general!

The existence and uniqueness (or lack thereof) of equilibriums, as well as the properties of the corresponding optimal strategy profiles, strongly depend upon the information structures available to the players, and the types of actions they are allowed to take. So, rather than referring to a single game with several information structures and admissible strategy profiles for the players, we choose to talk about models, e.g., the open loop model or the closed loop model, or even the Markovian model for the game. We give precise definitions below. The published literature on stochastic differential games, at least the part addressing terminology issues, is rather limited, and there is no clear consensus on the names to give to the many notions of admissibility for strategy profiles. We warn the reader that the definitions we use reflect our own personal biases and, this disclaimer being out of the way, the best we can do is to pledge consistency with our choices.

Definition 5.3. *If the strategy profile* $\boldsymbol{\alpha}^* = (\alpha^{*1}, \cdots, \alpha^{*N}) \in \mathbb{A}$ *satisfies the conditions of Definition 5.2 without further restriction on the strategies* α^{*i} *and* α^i, *we say that* $\boldsymbol{\alpha}^*$ *is an open loop Nash equilibrium (OLNE) for the game, or equivalently, a Nash equilibrium for the open loop game model.*

If the filtration $\mathbb{F}$ is generated by the Wiener process $\boldsymbol{W}$, except possibly for the presence of independent events in $\mathcal{F}_0$, the strategy profiles used in an open loop game model can be viewed as given by controls of the form

$$\alpha_t^i = \varphi^i(t, X_0, W_{[0,t]})$$

for some deterministic functions $\varphi^1, \cdots, \varphi^N$, where we used the notation $W_{[0,t]}$ for the path of the Wiener process between time $t = 0$ and t. The definition of an open loop equilibrium warrants some caution. This definition is very natural and very convenient from a mathematical point of view, and we shall see that powerful existence results can be proved for these game models. However, it is rather unrealistic from a practical point of view. Indeed, it is very difficult to imagine situations in which the whole trajectory $W_{[0,t]}$ of the random shocks can be observed. So, using functions of this trajectory as strategies does not seem very constructive as an approach to the search for an equilibrium!

Definition 5.4. *If the strategy profile* $\boldsymbol{\alpha}^* = (\boldsymbol{\alpha}^{*1}, \cdots, \boldsymbol{\alpha}^{*N}) \in \mathbb{A}$ *satisfies the conditions of Definition 5.2 with the restriction that the strategies* $\boldsymbol{\alpha}^{*i}$ *and* $\boldsymbol{\alpha}^i$ *are deterministic functions of time and the initial state, we say that* $\boldsymbol{\alpha}^*$ *is a deterministic Nash equilibrium (DNE) for the game.*

The strategy profiles used in the search for a deterministic equilibrium are given by controls of the form
$$\alpha_t^i = \varphi^i(t, X_0)$$
for some deterministic functions $\varphi^1, \cdots, \varphi^N$ of the time variable t and the state of the system x. Given the fact that the Wiener process $\boldsymbol{W}$ is not present in deterministic game models, the above definitions are consistent with the standard terminology used in the classical analysis of deterministic games.

Definition 5.5. *If the strategy profile* $\boldsymbol{\alpha}^* = (\boldsymbol{\alpha}^{*1}, \cdots, \boldsymbol{\alpha}^{*N}) \in \mathbb{A}$ *satisfies the conditions of Definition 5.2 with the restriction that the strategies* $\boldsymbol{\alpha}^{*i}$ *and* $\boldsymbol{\alpha}^i$ *are deterministic functions of time and the trajectory of the state between time* $t = 0$ *and* t, *we say that* $\boldsymbol{\alpha}^*$ *is a closed loop Nash equilibrium (CLNE) for the game, or equivalently, a Nash equilibrium for the closed loop game model.*

The strategy profiles used in the search for a closed loop equilibrium are given by controls of the form
$$\alpha_t^i = \varphi^i(t, X_{[0,t]})$$
for some deterministic functions $\varphi^1, \cdots, \varphi^N$. Finally, we have the following definition.

Definition 5.6. *If the strategy profile* $\boldsymbol{\alpha}^* = (\boldsymbol{\alpha}^{*1}, \cdots, \boldsymbol{\alpha}^{*N}) \in \mathbb{A}$ *satisfies the conditions of Definition 5.2 with the restriction that the strategies* $\boldsymbol{\alpha}^{*i}$ *and* $\boldsymbol{\alpha}^i$ *are deterministic functions of time, the initial state and the trajectory of the state at time* t, *we say that* $\boldsymbol{\alpha}^*$ *is a closed loop Nash equilibrium in feedback form (CLFFNE) for the game.*

The strategy profiles used in the search for a closed loop equilibrium in feedback form are given by controls of the form
$$\alpha_t^i = \varphi^i(t, X_0, X_t)$$
for some deterministic functions $\varphi^1, \cdots, \varphi^N$. The most important example of an application of this notion of equilibrium, concerns the case of deterministic drift and volatility coefficients b and σ, and cost functions f and g. This case will be discussed in Subsection 5.1.4 treating Markovian diffusions, and we shall strengthen the notion of a closed loop equilibrium in feedback form into the notion of a Markovian equilibrium.

Remark 5.7. *While we went out on a limb in our choice for the definition of open loop models, the following remarks are consistent with the terminology used in most write-ups*

5.1. Introduction and First Definitions 171

on deterministic games. Typically, in the open loop model, players cannot observe the play of their opponents, while in the closed loop model, at each time, all past play is common knowledge. From a mathematical standpoint, open loop equilibriums are more tractable than closed loop equilibriums because players need not consider how their opponents would react to deviations from the equilibrium path. With this in mind, one should expect that when the impact of players on their opponents' costs/rewards is small, open loop and closed loop equilibriums should be the same. We shall see instances of this intuition in the large N limit of the linear-quadratic (LQ) model of systemic risk analyzed in Section 5.5 and in our discussion of mean-field games (MFGs) in Chapter 6.

5.1.3 ▪ A First Look at the Notion of Nash Equilibriums

This subsection gathers a certain number of important facts relevant to the understanding of Nash equilibriums. While the discussion remains at an informal level, it motivates the introduction of important concepts (and the accompanying notations) which will be made rigorous on a case by case basis in the specific models which we are able to analyze.

Nash Equilibriums and Stochastic Control

For the purpose of the present discussion, unless stated otherwise, we assume that the set $\mathbb{A}$ of admissible strategies is the product of the sets $\mathbb{A}^i$ of admissible strategies for the individual players. For example, in the case of open loop equilibriums, most of the time we shall choose this set to be equal to $\mathbb{H}^{2,k^i}$ for player $i \in \{1, \cdots, N\}$, if adding extra constraints is not needed. We shall specify these sets of admissible strategy profiles on a case by case basis. Recall that if $\boldsymbol{\alpha}^* \in \mathbb{A}$ is a Nash equilibrium, then for any fixed $i \in \{1, \cdots, N\}$,

$$J^i(\boldsymbol{\alpha}^*) \leq J^i((\boldsymbol{\alpha}^{*,-i}, \boldsymbol{\alpha}^i)), \qquad \boldsymbol{\alpha}^i \in \mathbb{A}^i,$$

or, in other words,

$$\boldsymbol{\alpha}^{*,i} \in \arg\inf_{\boldsymbol{\alpha}^i \in \mathbb{A}^i} \mathbb{E}\left[\int_0^T f^i(t, X_t^{(\boldsymbol{\alpha}^{*,-i},\boldsymbol{\alpha}^i)}, (\alpha_t^{*,-i}, \alpha_t^i,))dt + g^i(X_T^{(\boldsymbol{\alpha}^{*,-i},\alpha_t^i,)})\right].$$

So, if we assume that the strategies $\boldsymbol{\alpha}^{*,-i}$ are held fixed, then $\boldsymbol{\alpha}^{*,i}$ solves the stochastic optimization problem

$$\inf_{\boldsymbol{\alpha}^i \in \mathbb{A}^i} \mathbb{E}\left[\int_0^T f^i(t, X_t, (\alpha_t^{*,-i}, \alpha_t^i))dt + g^i(X_T)\right]$$

under the constraint

$$\begin{cases} dX_t = b(t, X_t, (\alpha_t^{*,-i}, \alpha_t^i))dt + \sigma(t, X_t, (\alpha_t^{*,-i}, \alpha_t^i))dW_t, \\ X_0 = x, \end{cases}$$

which can be interpreted as a stochastic control problem whose solution can be tackled by the methods introduced in Chapter 3. The special case of Markovian equilibriums for which the admissible strategies are deterministic functions of the state X_t is discussed below in Subsection 5.1.4. In any case, we shall introduce the Hamiltonian for each of these stochastic control problems, write the adjoint BSDEs, and identify the corresponding adjoint processes. Also, writing down the sufficient conditions of the stochastic maximum principle for these stochastic control problems, one should get a generalized form of the stochastic maximum principle for the N-player game.

Nash Equilibriums as Fixed Points of the Best Response Maps

The following is another informal discussion highlighting how the search for a Nash equilibrium can be recast as the search for a fixed point. This point will be especially useful in our understanding of mean-field games (MFGs).

We construct a map Φ from $\mathbb{A}$ into itself in the following way. For each admissible strategy profile $\boldsymbol{\alpha} \in \mathbb{A}$, we define the strategy profile $\boldsymbol{\beta} = \Phi(\boldsymbol{\alpha})$ as the set $\boldsymbol{\beta} = (\boldsymbol{\beta}^1, \cdots, \boldsymbol{\beta}^N)$ such that for each $i \in \{1, \cdots, N\}$, $\boldsymbol{\beta}^i$ solves the stochastic optimization problem

$$\inf_{\alpha \in \mathbb{A}^i} \mathbb{E}\left[\int_0^T f^i(t, X_t, (\alpha_t^{-i}, \alpha_t))dt + g^i(X_T)\right]$$

under the constraint

$$\begin{cases} dX_t &= b(t, X_t, (\alpha_t^{-i}, \alpha_t))dt + \sigma(t, X_t, (\alpha_t^{-i}, \alpha_t))dW_t, \\ X_0 &= x. \end{cases}$$

So $\boldsymbol{\beta} = (\boldsymbol{\beta}^1, \cdots, \boldsymbol{\beta}^N)$ is constructed from $\boldsymbol{\alpha} = (\boldsymbol{\alpha}^1, \cdots, \boldsymbol{\alpha}^N)$ by solving N stochastic control problems. Clearly, this mapping Φ will be well defined only under some restrictive assumptions to guarantee that these stochastic control problems can be solved uniquely, or at least that we can always *select* an optimal strategy. This function Φ goes naturally under the name of "best response map". In any case, the purpose of this informal discussion is to remark that if Φ can be defined in this way, then $\boldsymbol{\alpha} \in \mathbb{A}$ is a Nash equilibrium if and only if it is a fixed point of the map Φ, i.e., $\Phi(\boldsymbol{\alpha}) = \boldsymbol{\alpha}$.

Players' Hamiltonians

The above discussion and its reliance on the solution of stochastic control problems suggest the introduction of the following: for each player $i \in \{1, \cdots, N\}$, we define his/her Hamiltonian as the function H^i

$$[0,T] \times \Omega \times \mathbb{R}^d \times \mathbb{R}^d \times \mathbb{R}^{d \times m} \times A \ni (t, x, y, z, \alpha) \hookrightarrow H^i(t, x, y, z, \alpha) \in \mathbb{R} \quad (5.7)$$

defined by

$$H^i(t, x, y, z, \alpha) = \underbrace{b(t, x, \alpha) \cdot y}_{\substack{\text{inner product of} \\ \text{state drift } b \text{ and} \\ \text{covariable } y}} + \underbrace{\text{trace}\,[\sigma(t, x, \alpha)^\dagger z]}_{\substack{\text{inner product of} \\ \text{state volatility } \sigma \\ \text{and covariable } z}} + \underbrace{f^i(t, x, \alpha)}_{\substack{\text{running cost} \\ \text{of player } i}}.$$

As usual, we talk about reduced Hamiltonians when the volatility is uncontrolled, and we ignore the second term in the above right-hand side.

Generalized Isaacs (Min-Max) Condition

The following definition is motivated by the generalization to stochastic differential games of the necessary part of the stochastic maximum principle, which we give below.

Definition 5.8. *We say that the generalized Isaacs (min-max) condition holds if there exists a function*

$$\hat{\alpha} : [0,T] \times \mathbb{R}^d \times (\mathbb{R}^d)^N \times (\mathbb{R}^{d \times m})^N \ni (t, x, y, z) \hookrightarrow \hat{\alpha}(t, x, y, z) \in A$$

5.1. Introduction and First Definitions

satisfying, for every $i \in \{1, \cdots, N\}$, $t \in [0,T]$, $x \in \mathbb{R}^d$, $y = (y^1, \cdots, y^N) \in (\mathbb{R}^d)^N$, *and* $z = (z^1, \cdots, z^N) \in (\mathbb{R}^{d \times m})^N$,

$$H^i(t,x,y^i,z^i,\hat{\alpha}(t,x,y,z)) \leq H^i(t,x,y^i,z^i,(\hat{\alpha}(t,x,y,z)^{-i},\alpha^i)) \quad \text{for all } \alpha^i \in A^i. \tag{5.8}$$

Notice that, in this definition, the function $\hat{\alpha}$ could depend upon the random scenario $\omega \in \Omega$ if the Hamiltonians H^i do. In words, this definition says that for each set of dual variables $y = (y^1, \cdots, y^N)$ and $z = (z^1, \cdots, z^N)$, for each time t and state x at time t, and possibly random scenario ω, one can find a set of actions $\hat{\alpha} = (\hat{\alpha}^1, \cdots, \hat{\alpha}^N)$ depending on these quantities, such that if we fix $N-1$ of these actions, say $\hat{\alpha}^{-i}$, then the remaining one $\hat{\alpha}^i$ minimizes the ith Hamiltonian in the sense that

$$\hat{\alpha}^i \in \arg\inf_{\alpha^i \in A^i} H^i(t,x,y^i,z^i,(\hat{\alpha}^{-i},\alpha^i)) \quad \text{for all } i \in \{1,\cdots,N\}. \tag{5.9}$$

The notation can be lightened slightly when the volatility is not controlled. Indeed, minimizing the Hamiltonian gives the same $\hat{\alpha}$ as minimizing the reduced Hamiltonian, and the argument $\hat{\alpha}$ of the minimization is independent of z. So, when the volatility is not controlled we say that the generalized Isaacs (min-max) condition holds if there exists a function

$$\hat{\alpha} : [0,T] \times \mathbb{R}^d \times (\mathbb{R}^d)^N \ni (t,x,y,z) \hookrightarrow \alpha^*(t,x,y) \in A$$

satisfying

$$\forall i \in \{1,\cdots,N\}, \forall t \in [0,T], \forall x \in \mathbb{R}^d, \forall y = (y^1,\cdots,y^N) \in (\mathbb{R}^d)^N,$$
$$\tilde{H}^i(t,x,y^i,\hat{\alpha}(t,x,y)) \leq H^i(t,x,y^i,(\hat{\alpha}(t,x,y)^{-i},\alpha^i)) \quad \text{for all } \alpha^i \in A^i. \tag{5.10}$$

Clearly, the fact that individual players' Hamiltonians should be minimized is suggested by the construction of the best response function as solutions of N stochastic control problems and the necessary part of the Pontryagin stochastic maximum principle, which we derived in Chapter 4 for stochastic control problems. Accordingly, the fact that the same function $\hat{\alpha}$ minimizes *all* the Hamiltonians simultaneously is dictated by the search for a fixed point of the best response function.

5.1.4 ▪ The Particular Case of Markovian/Diffusion Dynamics

In most applications for which actual numerical computations are possible, the coefficients of the state dynamics (5.1) depend only upon the *present value* X_t of the state instead of the entire past $X_{[0,t]}$ of the state of the system, or of the Wiener process driving the evolution of the state. In this case, the dynamics of the state are given by a diffusion-like equation

$$dX_t = b(t,X_t,\alpha_t)dt + \sigma(t,X_t,\alpha_t)dW_t, \qquad 0 \leq t \leq T, \tag{5.11}$$

with initial condition $X_0 = x$, and for deterministic drift and volatility functions

$$(b,\sigma) : [0,T] \times \mathbb{R}^d \times A \ni (t,x,\alpha) \hookrightarrow (b(t,x,\alpha), \sigma(t,x,\alpha)) \in \mathbb{R}^d \times \mathbb{R}^{d \times m}.$$

So, except for the fact that the strategy profile α may depend upon the past feature (that we are about to discard), the solution of (5.11) should be like a Markov diffusion. For this reason, we shall concentrate on strategy profiles which are deterministic functions of time and the current value of the state to indeed force the controlled state process to be a Markov diffusion. In fact, we shall also assume that the running cost functions f^i and the terminal cost random variables ξ^i are Markovian in the sense that, like b and σ, f^i does not depend upon the random scenario $\omega \in \Omega$, but only upon the current values of the state and the actions taken by the players, so that we can have $f^i : [0,T] \times \mathbb{R}^d \times A \ni (t,x,\alpha) \hookrightarrow f^i(t,x,\alpha) \in \mathbb{R}$, and ξ^i is of the form $\xi^i = g^i(X_T)$ for some measurable function $g^i : \mathbb{R}^d \ni x \hookrightarrow g^i(x) \in \mathbb{R}$ with (at most) quadratic growth. So in the case of Markovian/diffusion dynamics, the cost functional of player i is of the form

$$J^i(\boldsymbol{\alpha}) = \mathbb{E}\left\{\int_0^T f^i(t, X_t, \alpha_t)dt + g^i(X_T)\right\}, \qquad \boldsymbol{\alpha} \in \mathbb{A}, \qquad (5.12)$$

and we tailor the notion of equilibrium to this situation by considering closed loop strategy profiles in feedback form, which *simultaneously* provide Nash equilibriums for all the games starting at times $t \in [0,T]$ (i.e., over the time periods $[t,T]$) and all the possible initial conditions $X_t = x$, as long as they share the same drift and volatility coefficients b and σ, and cost functions f^i and g^i.

Markovian Nash Equilibriums

Inspired by the notion of *subgame perfect* equilibriums, we introduce the strongest notion yet of Nash equilibriums.

Definition 5.9. *A set $\phi = (\varphi^1, \cdots, \varphi^N)$ of N deterministic functions $\varphi^i : [0,T] \times \mathbb{R}^d \hookrightarrow \mathbb{R}^k$ for $i = 1, \cdots, N$ is said to be a Markovian Nash equilibrium (MNE), or a Nash equilibrium for the Markovian game model, if for each $(t,x) \in [0,T] \times \mathbb{R}^d$, the strategy profile $\boldsymbol{\alpha}^* = (\boldsymbol{\alpha}^{*1}, \cdots, \boldsymbol{\alpha}^{*N}) \in \mathbb{A}$, defined for $s \in [t,T]$ by $\alpha_s^{*i} = \varphi^i(s, X_s^{t,x})$, where $\boldsymbol{X}^{t,x}$ is the unique solution of the stochastic differential equation*

$$dX_s = b(s, X_s, \phi(s, X_s))ds + \sigma(s, X_s, \phi(s, X_s))dW_s, \qquad t \leq s \leq T,$$

with initial condition $X_t = x$, satisfies the conditions of Definition 5.2 with the restriction that the strategy $\boldsymbol{\alpha}^i$ is also given by a deterministic function φ on $[t,T] \times \mathbb{R}^d$.

It goes without saying that regularity assumptions on the functions φ^i and φ are needed for the stochastic differential equations giving the dynamics of the controlled state to have a unique strong solution. We did not spell out these assumptions in order to avoid further cluttering the statement of the definition. Typically, in the Markovian case, we assume that

the coefficients b and σ are Lipschitz in (x, α) uniformly in $t \in [0,T]$.

The strategy profiles used in the above definition are called Markovian strategy profiles. Obviously, they are closed loop in feedback form.

A priori, the Hamiltonian of player i should be the same as the real-valued function H^i defined earlier on $[0,T] \times \mathbb{R}^d \times \mathbb{R}^d \times \mathbb{R}^{d \times m} \times A$ by the formula

$$H^i(t, x, y, z, \alpha) = y \cdot b(t, x, \alpha) + z \cdot \sigma(t, x, \alpha) + f^i(t, x, \alpha) \in \mathbb{R} \qquad (5.13)$$

for $\alpha = (\alpha^1, \cdots, \alpha^N) \in A = A^1 \times \cdots \times A^N$. However, if we want to have a way of checking that a set of Markov strategies $\phi = (\varphi^1, \cdots, \varphi^N)$ forms a Markovian Nash equilibrium, for each player i, the function φ^i should provide a closed loop control $\alpha_t^i = \varphi^i(t, X_t^\phi)$ in feedback form solving the stochastic control problem

$$\inf_{\boldsymbol{\alpha} \in \mathbb{A}^i} \mathbb{E}\left[\int_0^T f^i(t, X_t, (\varphi(t, X_t)^{-i}, \alpha_t))dt + g^i(X_T)\right] \qquad (5.14)$$

5.1. Introduction and First Definitions

under the constraint

$$dX_t = b(t, X_t, (\varphi(t, X_t)^{-i}, \alpha_t))dt + \sigma(t, X_t, (\varphi(t, X_t)^{-i}, \alpha_t))dW_t \quad (5.15)$$

with initial condition $X_0 = x$. With this in mind, it becomes natural to introduce and use the Hamiltonian H^{-i} defined as

$$H^{-i}(t, x, y, z, \alpha) = y \cdot b(t, x, (\varphi(t, x)^{-i}, \alpha)) + z \cdot \sigma(t, x, (\varphi(t, x)^{-i}, \alpha)) \\ + f^i(t, x, (\varphi(t, x)^{-i}, \alpha))$$

for $\alpha \in A^i$, instead of the Hamiltonian H^i defined above in (5.13).

Remark 5.10. *Even though we try to avoid such abuse of notation, for the sake of convenience, we shall sometimes use the same notation α for the Markovian strategy profile and for the deterministic function ϕ from which it is derived. So we often write $\alpha^i_t(\omega) = \alpha^i(t, X^\alpha_t(\omega))$ using the same symbol α^i for the deterministic functions $\alpha^i : [0.T] \times \mathbb{R}^d \hookrightarrow A_i$, which were denoted φ^i in the above definition.*

The PDE Approach

Being in a Markovian setup, it is natural to consider the PDE approach to the solution of the optimal control problems appearing in the construction of the best response function. This approach is based on the solution of the HJB equations, which we now introduce. With the very same rationale used to introduce the Hamiltonian H^{-i}, we introduce the operator symbol

$$L^{-i,\alpha}(t, x, y, z) = \frac{1}{2} \sum_{hk=1}^{d} a_{hk}^{-i,\alpha}(t, x) z_{h,k} + \sum_{h=1}^{d} b_h^{-i,\alpha}(t, x) y_h$$

and the corresponding second-order (linear) partial differential operator

$$\mathcal{L}_t^{-i,\alpha} = \frac{1}{2} \sum_{hk=1}^{d} a_{hk}^{-i,\alpha}(t, x) \frac{\partial^2}{\partial x_h \partial x_k} + \sum_{h=1}^{d} b_h^{-i,\alpha}(t, x) \frac{\partial}{\partial x_h},$$

whose coefficients are defined as

$$b_h^{-i,\alpha}(t, x) = b_h(t, x, (\varphi(t, x)^{-i}, \alpha)), \qquad h = 1, \cdots, d,$$

and

$$a_{hk}^{-i,\alpha}(t, x) = \sum_{\ell=1}^{m} \sigma_{h\ell}(t, x, (\varphi(t, x)^{-i}, \alpha)) \sigma_{k\ell}(t, x, (\varphi(t, x)^{-i}, \alpha)), \qquad h, k = 1, \cdots, d.$$

For each $i \in \{1, \cdots, N\}$, we denote by V^i the value function of player i, namely the function

$$(t, x) \hookrightarrow V^i(t, x) = \inf_{\alpha^i \in \mathbb{A}^i} \mathbb{E}\left[\int_t^T f^i(s, X_s, (\varphi^{*-i}(s, X_s), \alpha^i_s))ds + g^i(X_T) \Big| X_t = x \right].$$

It is important to keep in mind the fact that this function depends upon the feedback functions φ^{-i} of the other players. It is expected to satisfy the HJB equation

$$\partial_t V^i + L^{-i,*}(t, x, \partial_x V^i(t, x), \partial^2_{xx} V^i(t, x)) = 0 \quad (5.16)$$

with terminal condition $V^i(T,x) = g^i(x)$, whether in a classical sense or as a viscosity solution to this equation. Here, the operator symbol $L^{-i,*}$ is defined by

$$L^{-i,*}(t,x,y,z) = \inf_{\alpha \in A^i} L^{-i,\alpha}(t,x,y,z,\alpha).$$

The best response of player i will be completely determined if we can find a function $(t,x,y,z) \hookrightarrow \alpha^{i,*}(t,x,y,z) \in A^i$ satisfying

$$L^{-i,*}(t,x,y,z) = L^{-i,\alpha}(t,x,y,z,\alpha^{i,*}(t,x,y,z)) \tag{5.17}$$

for all (t,x,y,z). Notice the similarity with the Isaacs condition. Indeed, in that case, the feedback function

$$\varphi^i(t,x) = \alpha^{-i,*}(t,x,\partial_x V^i(t,x), \partial_{xx}^2 V^i(t,x)) \tag{5.18}$$

provides an optimal control. The N HJB equations (5.16) are highly coupled if we want the N feedback functions φ^i defined by (5.18) to form a Nash equilibrium. Indeed, the fact that a Nash equilibrium is a fixed point of the best response map creates internal dependencies between the optimal controls of the individual players, creating strong couplings between these HJB equations. And, even if we can compute the minimizers $\alpha^{i,*}(t,x,y,z)$ (we provide several examples below), solving the system of highly nonlinear coupled HJB equations is typically very difficult. Nevertheless, see the examples discussed later in the chapter.

The Stochastic Maximum Approach

Despite the strong appeal of the PDE approach recasting the search for Nash equilibriums as the solution of a system of coupled HJB partial differential equations, one may want to approach the individual optimal control problems entering the construction of the best response function via the Pontryagin stochastic maximum principle. If we choose to do so, once the Markovian feedback functions φ^{-i} are *frozen*, we should use the Hamiltonian function H^{-i} defined above, and the driver for the BSDE, whose solution gives the adjoint processes $(\boldsymbol{Y}^i, \boldsymbol{Z}^i)$ of player i, should be given by the negative of its (partial) derivative with respect to x, namely,

$$\begin{aligned}
\partial_x H^{-i}(t,x,y,z,\alpha) = &\ \partial_x b(t,x,(\alpha^{*-i}(t,x),\alpha))y + \partial_x \sigma(t,x,(\alpha^{*-i}(t,x),\alpha))z \\
&+ \partial_x f^i(t,x,(\alpha^{*-i}(t,x),\alpha)) \\
&+ \sum_{j=1, j\neq i}^{N} \Big[\partial_{\alpha^j} b(t,x,(\alpha^{*-i}(t,x),\alpha))y + \partial_{\alpha^j}\sigma(t,x,(\alpha^{*-i}(t,x),\alpha))z \\
&+ \partial_{\alpha^j} f^i(t,x,(\alpha^{*-i}(t,x),\alpha))\Big] \partial_x \alpha^j(t,x). \tag{5.19}
\end{aligned}$$

Obviously, for this driver to make sense, the feedback functions used to define the strategy profile should be differentiable. Moreover, and we believe that it is extremely important to emphasize this fact, the contributions of the partial derivatives appearing in the above summation are not present in the stochastic maximum approach to the search for open loop Nash equilibriums. We shall come back to this point and give more detailed explanations in Subsection 5.4.2 below.

5.2 • Specific Examples

We introduce several models which we use to test the theoretical tools of stochastic analysis developed for the purpose of studying stochastic differential games.

5.2.1 • Zero-Sum Games

The class of zero-sum games has been very popular among game theorists for two very specific reasons: (1) first and foremost, its simplicity and intuitive rationale; (2) its mathematical analysis can very often be pursued all the way to a full solution of the game. In the framework of this chapter, a zero-sum game is a game with $N = 2$ players for which $J^1 + J^2 = 0$, in which case one studies only the objective function $J = J^1 = -J^2$ of the first player. In other words, player 1 wants to minimize the cost $J(\alpha^1, \alpha^2)$, while player 2 wants to maximize the same quantity.

For a zero-sum game, a strategy profile $\alpha \in \mathbb{A}$ is a Nash equilibrium if

$$\inf_{\alpha^1 \in \mathbb{A}^1} \sup_{\alpha^2 \in \mathbb{A}^2} J(\alpha^1, \alpha^2) = \sup_{\alpha^2 \in \mathbb{A}^2} \inf_{\alpha^1 \in \mathbb{A}^1} J(\alpha^1, \alpha^2). \tag{5.20}$$

The study of zero-sum games is performed through the analysis of the lower and upper values of the game defined as

$$\underline{V}_0 = \sup_{\alpha^2 \in \mathbb{A}^2} \inf_{\alpha^1 \in \mathbb{A}^1} J(\alpha^1, \alpha^2) \quad \text{and} \quad \overline{V}_0 = \inf_{\alpha^1 \in \mathbb{A}^1} \sup_{\alpha^2 \in \mathbb{A}^2} J(\alpha^1, \alpha^2), \tag{5.21}$$

respectively. It is clear that $\underline{V}_0 \leq \overline{V}_0$, hence their names. The analysis of a zero-sum game usually focuses on the following two questions.

(i) When does the game value exist? By this we mean that $\underline{V}_0 = \overline{V}_0$, in which case we denote by V_0 the common value.

(ii) Given the existence of the game value, can we find optimal controls forming a saddle point? In other words, can we find admissible controls $\alpha^{1,*} \in \mathbb{A}^1$ and $\alpha^{2,*} \in \mathbb{A}^2$ such that

$$V_0 = J(\alpha^{1,*}, \alpha^{2,*}) = \sup_{\alpha^2 \in \mathbb{A}^2} J(\alpha^{1,*}, \alpha^2) = \inf_{\alpha^1 \in \mathbb{A}^1} J(\alpha^1, \alpha^{2,*})? \tag{5.22}$$

In the context of zero-sum games, the so-called Isaacs condition takes a slightly different form. It reads

$$\inf_{\alpha^1 \in A^1} \sup_{\alpha^2 \in A^2} \left[b(t, x, \alpha^1, \alpha^2) \cdot y + \frac{1}{2} a(t, x, \alpha^1, \alpha^2) \cdot z \right] \tag{5.23}$$

$$= \sup_{\alpha^2 \in A^2} \inf_{\alpha^1 \in A^1} \left[b(t, x, \alpha^1, \alpha^2) \cdot y + \frac{1}{2} a(t, x, \alpha^1, \alpha^2) \cdot z \right].$$

As usual, we use the notation a for the diffusion matrix $\sigma^\dagger \sigma$. If we define the operator symbol L^+ by

$$L^+(t, x, y, z) = \inf_{\alpha^1 \in A^1} \sup_{\alpha^2 \in A^2} \left[b(t, x, \alpha^2, \alpha^2) \cdot y + \frac{1}{2} a(t, x, \alpha^1, \alpha^2) \cdot z \right],$$

then the upper value function $\overline{V}$ of the game is a viscosity subsolution of the HJB–Isaacs PDE

$$\partial_t V(t, x) + L^+(t, x, \partial V(t, x), \partial^2_{x,x} V(t, x)) = 0$$

with terminal condition $V(T, x) = g(x)$. A similar statement holds true for the lower value function $\underline{V}$, which is a viscosity supersolution of the same PDE with the operator L^+ replaced by L^-, and whose symbol is given by the right-hand side of (5.23). When the Isaacs condition (5.23) holds, the two PDEs are the same and $V = \overline{V} = \underline{V}$ is a viscosity solution of the common PDE.

Buckdahn's Counterexample

The following discussion is based on Appendix E of [107]. We assume that $k_1 = k_2 = 2$ and that for $i = 1$ and $i = 2$, A^i is the ball centered at the origin of $\mathbb{R}^2$ with radius i. Next, we assume that $d = m = 2$, $X_0 = 0$, and $\sigma(t, x, \alpha^1, \alpha^2) = \sigma$ for a nonnegative constant σ, and that $b^1(t, x, \alpha^1, \alpha^2) = \alpha^1$ and $b^2(t, x, \alpha^1, \alpha^2) = \alpha^2$. In other words

$$X_t^1 = \int_0^t \alpha_s^1 \, ds + \sigma W_t^1 \quad \text{and} \quad X_t^2 = \int_0^t \alpha_s^2 \, ds + \sigma W_t^2.$$

In the open loop model, for $i = 1$ and $i = 2$, we assume that $\mathbb{A}^i$ is the set of all the progressively measurable processes with values in A^i. The objective function is given by

$$J(\boldsymbol{\alpha}^1, \boldsymbol{\alpha}^2) = \mathbb{E}\left[|a + X_T^1 - X_T^2|\right].$$

The main insight derived from this model is contained in the following statement.

Proposition 5.11. *Under the conditions above, if $0 \leq \sigma < \sqrt{T/2}$ and $|a| \leq T$, the lower and upper values $\underline{V}_0$ and $\overline{V}_0$ of the game as defined by (5.21) are different, so the value of the game does not exist for the open loop form of the game.*

Proof. For the purpose of this counterexample, player 1 tries to maximize and player 2 tries to minimize, which is different from the set-up we chose originally. As a consequence, the lower and upper value functions of the game are defined as:

$$\underline{V}_0 = \sup_{\boldsymbol{\alpha}^1 \in \mathbb{A}^1} \inf_{\boldsymbol{\alpha}^2 \in \mathbb{A}^2} J(\boldsymbol{\alpha}^1, \boldsymbol{\alpha}^2), \quad \overline{V}_0 = \inf_{\boldsymbol{\alpha}^2 \in \mathbb{A}^2} \sup_{\boldsymbol{\alpha}^1 \in \mathbb{A}^1} J(\boldsymbol{\alpha}^1, \boldsymbol{\alpha}^2).$$

For any $\boldsymbol{\alpha}^1 \in \mathbb{A}^1$, defining $\boldsymbol{\alpha}^2$ by $\alpha_t^2 = \alpha_t^1 + a/T$, we have that $\boldsymbol{\alpha}^2 \in \mathbb{A}^2$ and

$$a + X_T^1 - X_T^2 = a + \int_0^T \alpha_s^1 \, ds + \sigma W_T^1 - \int_0^T \left[\alpha_s^1 + \frac{a}{T}\right] ds - \sigma W_T^2 = \sigma[W_T^1 - W_T^2]$$

so that

$$J(\boldsymbol{\alpha}^1, \boldsymbol{\alpha}^2) = \sigma \mathbb{E}\left[|W_T^1 - W_T^2|\right] \leq \sigma\sqrt{2T},$$

which implies $\inf_{\boldsymbol{\alpha}^2 \in \mathbb{A}^2} J(\boldsymbol{\alpha}^1, \boldsymbol{\alpha}^2) \leq \sigma\sqrt{2T}$, which in turn implies that $\underline{V}_0 \leq \sigma\sqrt{2T}$ since $\boldsymbol{\alpha}^1$ was arbitrary. On the other hand, for any $\boldsymbol{\alpha}^2 \in \mathbb{A}^2$, if we define the constant process $\boldsymbol{\alpha}^1$ by

$$\alpha^1{}_t = \frac{a - \mathbb{E}[X_T^2]}{|a - \mathbb{E}[X_T^2]|} \mathbf{1}_{\{a - \mathbb{E}[X_T^2] \neq 0\}} + \mathbf{1}_{\{a - \mathbb{E}[X_T^2] = 0\}},$$

then it is easy to see that

$$\boldsymbol{\alpha}^1 \in \mathbb{A}^1, \quad |\alpha_t^1| = 1, \quad \text{and} \quad a - \mathbb{E}[X_T^2] = \alpha_0^1 |a - \mathbb{E}[X_T^2]|,$$

from which we get

$$\mathbb{E}\left[a + X_T^1 - X_T^2\right] = a + \alpha_0^1 T - \mathbb{E}[X_T^2] = \alpha_0^1 \left[T + |a - \mathbb{E}[X_T^2]|\right],$$

and consequently,

$$J(\alpha^1,\alpha^2) \geq \left|\mathbb{E}\left[a + X_T^1 - X_T^2\right]\right| = |\alpha_0^1|\left[T + |a - \mathbb{E}[X_T^2]|\right] = T + |a - \mathbb{E}[X_T^2]| \geq T.$$

Hence $\sup_{\alpha^1 \in \mathbb{A}^1} J(\alpha^1, \alpha^2) \geq T$, and since α^2 was arbitrary, this implies that $\overline{V}_0 \geq T$. Finally, this implies that $\underline{V}_0 < \overline{V}_0$ whenever $0 \leq \sigma < \sqrt{T/2}$. □

The reason why this model is called a counterexample is as follows. If one now considers the Markov version of the game, in the sense that the admissible controls α^1 and α^2 are now required to be of the form $\alpha_t^1 = \varphi^1(t, X_t^1, X_t^2)$ and $\alpha_t^2 = \varphi^2(t, X_t^1, X_t^2)$ for some deterministic functions $\varphi^i : [0,T] \times \mathbb{R}^2 \to A^i$ for $i = 1, 2$, and if we define the lower and upper values of the game accordingly, then it is possible to prove that these two values are equal! See the Notes & Complements section at the end of the chapter for references.

5.2.2 ▪ Linear-Quadratic (LQ) Stochastic Games

As in the case of stochastic control problems, linear-quadratic (LQ) models provide a major source of examples and applications of the mathematical theory of stochastic differential games. They offer a convenient testbed for which Nash equilibriums can be shown to exist and be studied. We introduce them here in as general a stochastic framework as our stochastic analysis tools can handle. Our discussion of the open loop equilibriums follows [65] and [105]. We shall discuss the case of Markov equilibriums separately.

In an LQ model, the dynamics of the state of the system are linear in the sense that the coefficients of the SDE (5.1) are affine functions:

$$dX_t = (A_t X_t + B_t \alpha_t + \beta_t)dt + (C_t X_t + D_t \alpha_t + \gamma_t)dW_t.$$

So, with the notation used so far, the drift and the volatility of the dynamics of the state are given by the affine functions

$$b(t, x, \alpha) = A_t x + B_t \alpha + \beta_t = A_t x + \sum_{i=1}^{N} B_t^i \alpha^i + \beta_t$$

and

$$\sigma(t, x, \alpha) = C_t x + D_t \alpha + \sigma_t = C_t x + \sum_{i=1}^{N} D_t^i \alpha^i + \gamma_t,$$

where

(a) $\mathbf{A} = (A_t)_{t \in [0,T]}$, $\mathbf{B}^i = (B_t^i)_{t \in [0,T]}$, and $\boldsymbol{\beta} = (\beta_t)_{t \in [0,T]}$ are bounded $\mathcal{F}_t$-adapted stochastic processes with values in $\mathbb{R}^{d \times d}$, $\mathbb{R}^{d \times k_i}$, and $\mathbb{R}^d$, respectively;

(b) $\mathbf{C} = (C_t)_{t \in [0,T]}$, $\mathbf{D}^i = (D_t^i)_{t \in [0,T]}$, and $\boldsymbol{\gamma} = (\gamma_t)_{t \in [0,T]}$ are bounded $\mathcal{F}_t$-adapted stochastic processes with values in $\mathbb{R}^{(d \times m) \times d}$, $\mathbb{R}^{(d \times m) \times k_i}$, and $\mathbb{R}^{d \times m}$, respectively.

The set $\mathbb{A}$ of admissible strategies will be the product of the sets $\mathbb{A}^i$ of strategies which are admissible to the individual players. Since we first study open loop games, we choose the individual $\mathbb{A}^i$ to be the sets of all square integrable progressively measurable processes with values in A^i. In the particular case when A^i is equal to the whole space $\mathbb{R}^{k_i}$, then $\mathbb{A}^i = \mathcal{H}^{2,k_i}$. Recall that this means that the processes $\boldsymbol{\alpha}^i = (\alpha_t^i)_{t \in [0,T]}$ are progressively measurable and satisfy $\mathbb{E} \int_0^T |\alpha_t^i|^2 dt < \infty$.

In LQ models, the running and terminal costs are quadratic. For the sake of simplicity of exposition, we choose not to include cross terms and pick them of the form

$$f^i(t, x, \alpha) = \frac{1}{2}[x^\dagger M_t^i x + \alpha^{i\dagger} N_t^i \alpha^i]$$

and

$$\xi^i = g^i(X_T) \quad \text{with} \quad g^i(x) = \frac{1}{2} x Q^i x,$$

where for each $i \in \{1, \cdots, N\}$,

(a) $\mathbf{M}^i = (M_t^i)_{t \in [0,T]}$ is a bounded $\mathcal{P}$-measurable stochastic processes with values in $\mathbb{R}^{d \times d}$;

(b) $\mathbf{N}^i = (N_t^i)_{t \in [0,T]}$ is a bounded $\mathcal{P}$-measurable stochastic processes with values in $\mathbb{R}^{k^i \times k^i}$;

(c) Q^i is a bounded $\mathcal{F}_T$-measurable random variable in $\mathbb{R}^{d \times d}$;

satisfying

(d) $M_t^i(\omega)$, $N_t^i(\omega)$, and $Q^i(\omega)$ are nonnegative symmetric matrices;

(e) $\delta \leq N_t^i(\omega)$ for some $\delta > 0$ independent of $i \in \{1, \cdots, N\}, t \in [0, T]$, and $\omega \in \Omega$;

(f) $\delta \leq \sum_{i=1}^N M_t^i(\omega)$ and $\delta \leq \sum_{i=1}^N Q^i(\omega)$.

Remark 5.12. *The boundedness assumptions may seem to be restrictive. They are made here for convenience. They can be dispensed with in many applications, but we want to focus the technicalities of the analysis on the search for equilibriums.*

Generalized Min-Max Isaacs Condition

For each player $i \in \{1, \cdots, N\}$, the Hamiltonian writes as

$$H^i(t, x, y, z, \alpha) = \left[A_t x + \sum_{j=1}^N B_t^j \alpha^j + \beta_t \right] \cdot y + \left[C_t x + \sum_{j=1}^N D_t^j \alpha^j + \gamma_t \right] \cdot z$$
$$+ \frac{1}{2} x^\dagger M_t^i x + \frac{1}{2} \alpha^{i\dagger} N_t^i \alpha^i, \tag{5.24}$$

from which we easily compute

$$\partial_{\alpha^i} H^i(t, x, y, z, \alpha) = y^\dagger B_t^i + z^\dagger D_t^i + \alpha^{i\dagger} N_t^i. \tag{5.25}$$

Setting this quantity to 0 gives N equations (one for each player) which are completely decoupled and can be solved separately, the solution α^i being only a function of the dual variables $y = y^i$ and $z = z^i$ of player i. Consequently, the function $\hat{\alpha} = (\hat{\alpha}^1, \cdots, \hat{\alpha}^N)$, defined by

$$\hat{\alpha}^i = \hat{\alpha}^i(t, x, (y^1, \cdots, y^N), (z^1, \cdot, z^N)) = -[N_t^i]^{-1}[B_t^{i\dagger} y^i + D_t^{i\dagger} z^i], \quad i = 1, \cdots, N, \tag{5.26}$$

can be used to prove that the generalized Isaacs condition holds. When the volatility σ does not depend upon α, the optimal strategies satisfying the min-max Isaacs condition are

$$\hat{\alpha}^i = \hat{\alpha}^i(t, x, (y^1, \cdots, y^N), (z^1, \cdot, z^N)) = -[N_t^i]^{-1} B_t^{i\dagger} y^i, \quad i = 1, \cdots, N. \tag{5.27}$$

5.2.3 ■ Semilinear Quadratic Games

Our next example is a variation on the LQ game model discussed above. We call this model the *semilinear* LQ game. An important feature of this model is the fact that the volatility matrix

$$\sigma(t, X, \alpha) = I_d$$

is deterministic and constant and equal to the identity matrix in d dimensions, and that the drift is of the form

$$b(t, X, \alpha) = \bar{b}(t, X_t) + \sum_{j=1}^{N} \alpha^j \qquad (5.28)$$

for some bounded measurable function $\bar{b} : [0, T] \times \Omega \times \mathbb{R}^d \hookrightarrow \mathbb{R}^d$ such that the $\mathbb{R}^d$-valued process $\{\bar{b}(t, x)\}_{t \in [0,T]}$ is $\mathcal{P}$-measurable for each fixed $x \in \mathbb{R}^d$. Anticipating the discussion of the weak formulation approach provided in the next section below, we set $X_t^0 = x + W_t$. We also assume $k_1 = k_2 = \cdots = k_N = d$ and $A^i = \mathbb{R}^d$ (so that $k = Nd$), and for a set of strategies $\alpha = (\alpha^1, \cdots, \alpha^N)$ to be admissible, in addition to the usual $\mathcal{P}$-measurability, we require that each of the processes $\alpha^i = (\alpha_t^i)_{t \in [0,T]}$ be bounded. So, for each admissible set of strategies $\alpha \in \mathbb{A}$, the probability measure $\mathbb{P}^\alpha$ is well-defined by its Radon–Nikodym derivative, and under the measure $\mathbb{P}^\alpha$, the process $X^0 = (X_t^0)_{t \in [0,T]}$ satisfies

$$dX_t^0 = \left[\bar{b}(t, X_t^0) + \sum_{j=1}^{N} \alpha_t^j\right] dt + dW_t^\alpha. \qquad (5.29)$$

We now define the costs to the players. They are defined as

$$f^i(t, x, \alpha) = \bar{f}^i(t, x) + \frac{\gamma}{2}|\alpha^i|^2 + \theta \alpha^i \sum_{j \ne i} \alpha^j \qquad (5.30)$$

for some real constants γ and $\theta \ne 0$, and for bounded measurable random functions $\bar{f}^i : [0, T] \times \Omega \times \mathbb{R}^d \hookrightarrow \mathbb{R}$ such that the processes $\{\bar{f}^i(t, x)\}_{t \in [0,T]}$ are $\mathcal{P}$-measurable for each fixed $x \in \mathbb{R}^d$ and $i \in \{1, \cdots, N\}$. Similarly, we assume that the terminal costs are given by measurable bounded random functions $g^i : \Omega \times \mathbb{R}^d \hookrightarrow \mathbb{R}$ such that the random variables $g^i(x)$ are $\mathcal{F}_T$-measurable for each fixed $x \in \mathbb{R}^d$ and $i \in \{1, \cdots, N\}$. So the final form of the total cost to player i reads

$$J^i(\alpha) = \mathbb{E}^{\mathbb{P}^\alpha}\left[g^i(X_T^0) + \int_0^T [\bar{f}^i(t, X_t^0) + \frac{\gamma}{2}|\alpha_t^i|^2 + \theta \alpha_t^i \sum_{j \ne i} \alpha_t^j]dt\right].$$

In order to check if the min-max Isaacs condition is satisfied, we note that the reduced Hamiltonian of player i is given by

$$\tilde{H}^i(t, x, y, \alpha) = y \cdot \bar{b}(t, x) + y \cdot \sum_{j=1}^{N} \alpha^j + \bar{f}^i(t, x) + \frac{\gamma}{2}|\alpha^i|^2 + \theta \alpha^i \sum_{j \ne i} \alpha^j. \qquad (5.31)$$

For fixed α^{-i}, the search for α^i minimizing the reduced Hamiltonian amounts to minimizing the function

$$\mathbb{R}^k \ni \alpha \hookrightarrow y \cdot \alpha + \frac{\gamma}{2}|\alpha|^2 + \theta \alpha \cdot \sum_{j \ne i} \alpha^j.$$

The first-order condition reads

$$y + \gamma\alpha + \theta \sum_{j \neq i} \alpha^j = 0,$$

which is quite different from the simple LQ model studied earlier. Indeed, the N equations obtained by writing the first-order conditions are coupled, and the solution α^i for player i does involve the adjoint variables y^j of other players. Finding $\alpha = (\alpha^1, \cdots, \alpha^N)$ satisfying the Isaacs condition turns out to be equivalent to solving the system of N equations

$$y^i + \gamma\alpha^i + \theta \sum_{j \neq i} \alpha^j = 0, \qquad i = 1, \cdots, N, \tag{5.32}$$

whose solution is easily found to be

$$\hat{\alpha}^i = \frac{1}{(N-1)\theta - \delta\gamma}\left[\delta y^i - \sum_{j \neq i} y^j\right], \qquad i = 1, \cdots, N, \tag{5.33}$$

with $\delta = (N-2) + \gamma/\theta$ whenever $(N-1)\theta^2 - (N-2)\theta - \gamma \neq 0$. So, in this case, substituting this value for δ, we see that the function $\hat{\alpha}$ defined by

$$\hat{\alpha}(y^1, \cdots, y^N) = \left(\frac{1}{(N-1)\theta^2 + (N-2)\gamma\theta + \gamma}[((N-2)\theta + \gamma)y^i - \sum_{j \neq i} y^j]\right)_{i=1,\cdots,N} \tag{5.34}$$

satisfies the generalized min-max Isaacs condition. Notice that $\hat{\alpha}$ does not depend upon the state variable x. Simple (though tedious) computations show that the minimal values of the reduced Hamiltonians are given by the formulas

$$\tilde{H}^i(t, x, y^i, \hat{\alpha}(y)) = y^i \cdot \overline{b}(t, x) + \overline{f}^i(t, x) - \frac{1}{\theta}|y^i|^2 - \frac{\gamma}{\theta} y^i \hat{\alpha}^i - \frac{\gamma}{2}|\hat{\alpha}^i|^2 \tag{5.35}$$

from which we clearly see that whenever $\gamma = 0$, the value of the reduced Hamiltonian $\tilde{H}^i(t, x, y^i, \hat{\alpha}(y^1, \cdots, y^N))$ depends only upon y^i and not on y^j for $j \neq i$, and is quadratic in y^i. We will take full advantage of this remark when we solve this game model later on.

5.3 • Weak Formulation and the Case of Uncontrolled Volatility

As we already explained in our discussion of stochastic control problems, in many models, the volatility of the state of the system is not affected by the actions of the controllers or players. If, moreover, the volatility is invertible, then using the so-called weak formulation approach becomes a viable solution strategy.

5.3.1 • Setup of the Weak Formulation

In this section, we assume that the volatility is given by a function $\sigma : [0, T] \times \Omega \times \mathbb{R}^d \hookrightarrow \mathbb{R}^{d \times m}$, which does not depend upon the actions $\alpha \in A$ chosen by the players. Moreover, as before, we assume that the dimension of the noise is the same as the dimension of the state,

5.3. Weak Formulation and the Case of Uncontrolled Volatility

in other words, $m = d$, that the volatility matrix σ is progressively measurable, Lipschitz in x uniformly in $t \in [0, T]$ and $\omega \in \Omega$, invertible, and satisfies

$$\|\sigma(t,x)\| \leq c \quad \text{and} \quad \|\sigma(t,x)^{-1}\| \leq c, \qquad (t,x) \in [0,T] \times \mathbb{R}^d, \tag{5.36}$$

for some constant $c > 0$ independent of t and x. As usual, we drop the dependence upon $\omega \in \Omega$ from our notation whenever possible. Under the above assumptions, it is convenient to first solve a driftless stochastic differential equation and use Girsanov-type changes of measures to shift the dependence upon the actions α from the actual sample scenarios of the state of the system to the likelihoods of these scenarios occurring. We make this statement fully rigorous by repeating the discussion of Subsection 3.1.7. So, for each $\alpha \in \mathbb{A}$, we assume that the function $(t,x,\alpha) \hookrightarrow \sigma(t,x)^{-1}b(t,x,\alpha)$ is bounded, and for each $\boldsymbol{\alpha} \in \mathbb{A}$, we define the probability measure $\mathbb{P}^\alpha$ by

$$\frac{d\mathbb{P}^\alpha}{d\mathbb{P}} = \exp\left[\int_0^T \sigma(t, X_t^0)^{-1} b(t, X_t^0, \alpha_t) dW_t - \frac{1}{2} \int_0^T |\sigma(t, X_t^0)^{-1} b(t, X_t^0, \alpha_t)|^2 dt \right],$$

where $\boldsymbol{X}^0 = (X_t^0)_{0 \leq t \leq T}$ is the unique strong solution of the SDE

$$dX_t = \sigma(t, X_t) dW_t, \qquad X_0 = x.$$

For each $\boldsymbol{\alpha} \in \mathbb{A}$, the probability measure $\mathbb{P}^\alpha$ is equivalent to $\mathbb{P}$ and on the probability space $(\Omega, \mathcal{F}, \mathbb{P}^\alpha)$, the process $\boldsymbol{W}^\alpha = (W_t^\alpha)_{0 \leq t \leq T}$ defined by

$$W_t^\alpha = W_t - \int_0^t \sigma(s, X_s^0)^{-1} b(s, X_s^0, \alpha_s) ds$$

is an $\mathbb{F}$-Brownian motion and $\boldsymbol{X}^0$ satisfies

$$dX_t^0 = b(t, X_t^0, \alpha_t) dt + \sigma(t, X_t^0) dW_t^\alpha, \qquad t \in [0, T].$$

Consequently, we have

$$\begin{aligned} J^i(\boldsymbol{\alpha}) &= \mathbb{E}^{\mathbb{P}} \left\{ \int_0^T f^i(s, X_s^\alpha, \alpha_s) ds + g^i(X_T^\alpha) \right\} \\ &= \mathbb{E}^{\mathbb{P}^\alpha} \left\{ \int_0^T f^i(s, X_s^0, \alpha_s) ds + g^i(X_T^0) \right\}, \end{aligned}$$

so that when the players use the admissible strategy profile $\boldsymbol{\alpha} = (\boldsymbol{\alpha}^1, \cdots, \boldsymbol{\alpha}^N) \in \mathbb{A}$, the cost to any player can be given either by an expectation over the original probability $\mathbb{P}$, the total cost being computed as a function of the path X^α taken by the state under the influence of the strategies $\boldsymbol{\alpha}$, or alternatively, by an expectation over the equivalent probability $\mathbb{P}^\alpha$ depending upon the strategies used by the players, the total cost being computed as a function of the path X^0, which is independent of the choices of the strategies $\boldsymbol{\alpha} = (\boldsymbol{\alpha}^1, \cdots, \boldsymbol{\alpha}^N) \in \mathbb{A}$.

Since the volatility is of the form $\sigma(t, x, \alpha) \equiv \sigma(t, x)$, as we already pointed out, the dual variable z entering the Hamiltonians through the inner product with the volatility is not needed, and for all practical purposes, we can limit ourselves to the use and minimizations of the reduced Hamiltonians

$$\tilde{H}^i(t, x, y, \alpha) = b(t, x, \alpha) \cdot y + f^i(t, x, \alpha). \tag{5.37}$$

The search for a function $\hat{\alpha}$ satisfying the min-max Isaacs condition can be reduced to the search for a function

$$\hat{\alpha} : [0, T] \times \mathbb{R}^d \times (\mathbb{R}^d)^N \ni (t, x, y) \hookrightarrow \hat{\alpha}(t, x, y) \in A$$

such that for each set of dual variables $y = (y^1, \cdots, y^N)$ for each time t, state x, and possibly each random scenario ω, $\hat{\alpha}(t, \omega, x, y)$ gives a set of actions $\hat{\alpha} = (\hat{\alpha}^1, \cdots, \hat{\alpha}^N)$ such that if we fix $N-1$ of them, say $\hat{\alpha}^{-i}$, then the remaining one $\hat{\alpha}^i$ minimizes the reduced ith Hamiltonian in the sense that

$$\hat{\alpha}^i(t, x, (y^1, \cdots, y^N)) \in \arg\inf_{\alpha^i \in A^i} \tilde{H}^i(t, x, y^i, (\hat{\alpha}(t, x, (y^1, \cdots, y^N))^{-i}, \alpha^i)) \tag{5.38}$$

for all $i \in \{1, \cdots, N\}$.

5.3.2 ▪ Open Loop Nash Equilibriums in the Weak Formulation

We now review the first of the two BSDE approaches followed earlier in the case of stochastic control problems, and adapt it to the present game framework.

Players' Value Functions as Solutions of BSDEs

We generalize to the game setup, the connection between solutions of BSDEs, and players' value functions. The game version is given in the following simple lemma. In this subsection, the assumptions spelled out in Subsection 5.3.1 above are in force, and for each $i \in \{1, \cdots, N\}$, the set $\mathbb{A}^i$ of admissible controls is $\mathbb{H}^{2,k^i}$.

Lemma 5.13. *Under the assumptions of Subsection 5.3.1 above, if $\boldsymbol{\alpha} = (\alpha^1, \cdots, \alpha^N) \in \mathbb{A}$ is an admissible strategy profile and if there exist $2N$ $\mathcal{P}$-measurable processes $(\boldsymbol{Y}^{\boldsymbol{\alpha}}, \boldsymbol{Z}^{\boldsymbol{\alpha}})$ $= ((Y^{1,\boldsymbol{\alpha}}, \cdots, Y^{N,\boldsymbol{\alpha}}), (Z^{1,\boldsymbol{\alpha}}, \cdots, Z^{N,\boldsymbol{\alpha}}))$ with values in $\mathbb{R}^N \times \mathbb{R}^{Nd}$ such that for each $i \in \{1 \cdots, N\}$, $Y^{i,\boldsymbol{\alpha}} \in \mathbb{H}^{2,1}$, $Z^{i,\boldsymbol{\alpha}} \in \mathbb{H}^{2,d}$, and*

$$\begin{cases} dY_t^{i,\boldsymbol{\alpha}} = -\tilde{H}^i(t, X_t^{(0)}, Z_t^{i,\boldsymbol{\alpha}}\sigma(t, X_t^{(0)})^{-1}, \alpha_t)dt + Z_t^{i,\boldsymbol{\alpha}}dW_t, \\ Y_T^{i,\boldsymbol{\alpha}} = g^i(X_T^{(0)}), \end{cases} \quad i = 1, \cdots, N, \tag{5.39}$$

holds, then

$$J^i(\boldsymbol{\alpha}) = Y_0^{i,\boldsymbol{\alpha}}, \qquad i = 1, \cdots, N.$$

Proof. Let us fix $i \in \{1, \cdots, N\}$. Then, assumption (5.39) implies

$$Y_t^{i,\boldsymbol{\alpha}} = g^i(X_T^{(0)}) + \int_t^T \tilde{H}^i(s, X_s^{(0)}, Z_s^{i,\boldsymbol{\alpha}}\sigma(s, X_s^{(0)})^{-1}, \alpha_s)ds - \int_t^T Z_s^{i,\boldsymbol{\alpha}}dW_s$$

$$= g^i(X_T^{(0)}) + \int_t^T f^i(s, X_s^{(0)}, \alpha_s)ds$$

$$\quad + \int_t^T Z_s^{i,\boldsymbol{\alpha}}[\sigma(s, X_s^{(0)})^{-1}b(s, X_s^{(0)}, \alpha_s)ds - dW_s]$$

$$= g^i(X_T^{(0)}) + \int_t^T f^i(s, X_s^{(0)}, \alpha_s)ds + \int_t^T Z_s^{i,\boldsymbol{\alpha}}dW_s^{\boldsymbol{\alpha}}.$$

5.3. Weak Formulation and the Case of Uncontrolled Volatility

Taking $\mathbb{P}^\alpha$ conditional expectations with respect to $\mathcal{F}_t$ on both sides we get

$$Y_t^{i,\alpha} = \mathbb{E}^{\mathbb{P}^\alpha}\left[g^i(X_T^{(0)}) + \int_t^T f^i(s, X_s^{(0)}, \alpha_s)ds \Big| \mathcal{F}_t\right]$$

because W^α is a $\mathbb{P}^\alpha$-Wiener process and $Z^{i,\alpha}$ is square integrable for $\mathbb{P}$ and $\mathbb{P}^\alpha$. So, for $t = 0$, $\mathcal{F}_0$ being trivial, $Y_0^{i,\alpha}$ is deterministic, and

$$Y_0^{i,\alpha} = \mathbb{E}^{\mathbb{P}^\alpha}\left[g^i(X_T^{(0)}) + \int_0^T f^i(s, X_s^{(0)}, \alpha_s)ds\right] = J^i(\boldsymbol{\alpha}),$$

which is the value function of player i if all the players use the strategies $\boldsymbol{\alpha} = (\alpha^1, \cdots, \alpha^N)$. □

Remark 5.14. *Notice that, because the volatility is not controlled, we used the players' reduced Hamiltonians in the BSDEs (5.39).*

The main result of this subsection is as follows.

Theorem 5.15. *Let us assume that the volatility σ is as above, that the deterministic function $\hat{\alpha}$ satisfies the min-max Isaacs condition (5.10) for the players' reduced Hamiltonians, and that there exist $2N$ $\mathcal{P}$-measurable processes $((\boldsymbol{Y}^1, \cdots, \boldsymbol{Y}^N), (\boldsymbol{Z}^1, \cdots, \boldsymbol{Z}^N))$ with values in $\mathbb{R}^N \times \mathbb{R}^{Nd}$ such that for each $i \in \{1 \cdots, N\}$, $\boldsymbol{Y}^i \in \mathcal{B}$, $\boldsymbol{Z}^i \in \mathbb{H}^{2,d}$ and*

$$dY_t^i = -\tilde{H}^i(t, X_t^{(0)}, \sigma(s, X_s^{(0)})^{-1}Z_t^i, \hat{\alpha}(t, X_t^{(0)}, (Z_t^1\sigma(t, X_t^{(0)})^{-1}, \cdots, Z_t^N\sigma(t, X_t^{(0)})^{-1}))dt + Z_t^i dW_t \quad (5.40)$$

holds. Then the strategy profile $\hat{\boldsymbol{\alpha}}$ defined by

$$\hat{\alpha}_t = \hat{\alpha}\big(t, X_t^{(0)}, (Z_t^1\sigma(t, X_t^{(0)})^{-1}, \cdots, Z_t^N\sigma(t, X_t^{(0)})^{-1})\big)$$

is an open loop Nash equilibrium, and the optimal value of the individual players in equilibrium are given by

$$J^i(\boldsymbol{\alpha}) = Y_0^i, \qquad i = 1, \cdots, N.$$

Proof. For each fixed $i \in \{1, \cdots, N\}$, we need to prove that for any $\beta^i \in \mathbb{A}^i$, we have $J^i(\boldsymbol{\alpha}) \leq J^i((\boldsymbol{\alpha}^{-i}, \beta^i))$, or equivalently (using Lemma 5.13 above), $Y_0^{i,\hat{\boldsymbol{\alpha}}} \leq Y_0^{i,(\boldsymbol{\alpha}^{-i},\beta^i)}$ if we use the notation $(Y^{i,\alpha}, Z^{i,\alpha})_{i=1,\cdots,N}$ for the solutions of the BSDEs (5.39). Notice that both BSDEs share the same terminal condition $\xi_T = g^i(X_T^{(0)})$, so we prove the desired inequality by comparing the drivers of these BSDEs and using Theorem 2.8, comparing solutions of BSDEs with the same terminal condition ξ_T and comparable drivers. Indeed, the driver of the first equation is

$$\Psi^{(1)}(t, \omega, y, z) = \tilde{H}^i(t, X_t^{(0)}(\omega), z\sigma(t, X_t^{(0)}(\omega))^{-1}, \hat{\alpha}_t(\omega)),$$

while the driver of the second equation is

$$\Psi^{(2)}(t, \omega, y, z) = \tilde{H}^i(t, X_t^{(0)}(\omega), z\sigma(t, X_t^{(0)}(\omega))^{-1}, (\hat{\alpha}_t^{-i}(\omega), \beta_t^i(\omega))).$$

But the fact that $\hat{\alpha}$ satisfies the generalized min-max Isaacs condition (5.10) implies that
$$\Psi^{(1)}(t,y,z) \leq \Psi^{(2)}(t,y,z), \quad \text{a.s.}$$
which in turn implies $Y_0^{(1)} \leq Y_0^{(2)}$ a.s. by the comparison in Theorem 2.4. The proof is complete. □

Remark 5.16. *It is important to remark that even though we wrote them separately for each player, the BSDEs (5.40) form a system of coupled BSDEs because of the coupling through the Z_t^i's. In other words, if we want to use a BSDE existence result to apply Theorem 5.15, we can only use the existence result of the Lipschitz case given in Theorem 2.2, and not the quadratic growth result in Theorem 2.7, which is limited to the one-dimensional case and does not apply to multivariate systems in general. This should shed some light on the restrictive conditions we need to have in the application we shall discuss later on.*

Remark 5.17. *Even though we obtained a rather explicit expression for a function satisfying the min-max Isaacs condition, the result above cannot be used in the case of general LQ games unless the volatility matrix is constant (i.e., $C_t = D_t = 0$).*

Application to Semilinear Quadratic Games

We use Theorem 5.15, our first result on the connection between stochastic games and BSDEs, to express the value functions of individual players as solutions of specific BSDEs, and find a Nash equilibrium using the function $\hat{\alpha}$ satisfying the min-max Isaacs condition which we identified in Subsection 5.2.3. As pointed out in Remark 5.16 above, straightforward applications of this theorem are rather rare, and difficult to come by.

Assuming that $\gamma = 0$, we check that the $\hat{\alpha}$ provides a Nash equilibrium. Indeed, in this case, the special form (5.34) of the functions satisfying the Isaacs condition implies that the system of BSDEs to be checked for existence reduces to a set of N completely decoupled scalar BSDEs. Hence, one can use Kobylanski's result to obtain the existence of solutions to these individual equations despite their quadratic growths. More precisely, for $\gamma = 0$ according to (5.35), the system considered in Theorem 5.15 rewrites as

$$dY_t^i = -Z_t^i \cdot \overline{b}(t, X_t) - \overline{f}^i(t, X_t) + \frac{1}{\theta}|Z_t^i|^2 + Z_t^i dW_t, \quad i = 1, \cdots, N,$$

which is a set of N decoupled BSDEs which can be solved separately, since the ith equation depends only upon Z_t^i and not on the Z_t^j for $j \neq i$. These equations are scalar, and have a solution by the Kobylanski theorem (Theorem 2.7). But before we can conclude that the strategies

$$\alpha_t^{*i} = \alpha^{*i}(t, X, (Z_t^1, \cdots, Z_t^N)) = \frac{1}{(N-1)\theta}\left[(N-2)Z_t^i - \sum_{j \neq i} Z_t^j\right], \quad i = 1, \cdots, N,$$

form an adapted (*open loop*) Nash equilibrium for the semilinear stochastic game, we need to prove that they are admissible according to the definition of the set $\mathbb{A}$ which we chose. In order to do that, we need to prove that the above α_t^* are bounded. This is proven in [83] using Malliavin calculus. Such a proof is beyond the scope of this book.

5.4 ▪ Game Versions of the Stochastic Maximum Principle

Proving generalizations of the Pontryagin maximum principle to stochastic games is not as straightforward as one would like. While many versions are possible, we limit ourselves to

5.4. Game Versions of the Stochastic Maximum Principle

open loop and Markovian equilibriums for the sake of simplicity. We treat them separately to emphasize the differences, especially because we do not know of any published treatment of the Markov case based on the stochastic maximum principle. Throughout this section, we assume that the drift and volatility functions, as well as the running and terminal cost functions, are deterministic functions which are differentiable with respect to the variable x, and that the partial derivatives $\partial_x b$, $\partial_x \sigma$, $\partial_x f^i$, and $\partial_x g^i$ for $i = 1, \cdots, N$ are uniformly bounded. Notice that since b takes values in $\mathbb{R}^d$ and $x \in \mathbb{R}^d$, $\partial_x b$ is an element of $\mathbb{R}^{d \times d}$, in other words, a $d \times d$ matrix whose entries are the partial derivatives of the components b^j of b with respect to the components x^i of x. Analog statements can be made concerning $\partial_x \sigma$ which has the interpretation of a tensor.

5.4.1 ▪ Open Loop Equilibriums

The generalization of the stochastic Pontryagin maximum principle to open loop stochastic games can be approached in a very natural way, and forms of the open loop sufficient condition for the existence and identification of a Nash equilibrium have been used in the case of linear-quadratic models. See the Notes & Complements section at the end of the chapter for references.

Definition 5.18. *Given an open loop admissible strategy profile $\boldsymbol{\alpha} \in \mathbb{A}$ and the corresponding evolution $\boldsymbol{X} = \boldsymbol{X}^{\boldsymbol{\alpha}}$ of the state of the system, a set of N couples $(\boldsymbol{Y}^{i,\boldsymbol{\alpha}}, \boldsymbol{Z}^{i,\boldsymbol{\alpha}}) = (Y_t^{i,\boldsymbol{\alpha}}, Z_t^{i,\boldsymbol{\alpha}})_{t \in [0,T]}$ of processes in $\mathbb{S}^{2,d}$ and $\mathbb{H}^{2,d \times m}$, respectively, for $i = 1, \cdots, N$, is said to be a set of adjoint processes associated with $\boldsymbol{\alpha} \in \mathbb{A}$ if for each player $i \in \{1, \cdots, N\}$ they satisfy the BSDEs*

$$\begin{cases} dY_t^{i,\boldsymbol{\alpha}} = -\partial_x H^i(t, X_t, Y_t^{i,\boldsymbol{\alpha}}, Z_t^{i,\boldsymbol{\alpha}}, \alpha_t) dt + Z_t^{i,\boldsymbol{\alpha}} dW_t, \\ Y_T^{i,\boldsymbol{\alpha}} = -\partial_x g^i(X_T). \end{cases} \quad (5.41)$$

The existence and uniqueness of the adjoint processes is not an issue under the present hypotheses. Indeed, given $\boldsymbol{\alpha} \in \mathbb{A}$ and the corresponding state evolution $\boldsymbol{X} = \boldsymbol{X}^{\boldsymbol{\alpha}}$, equation (5.41) can be viewed as a BSDE with random coefficients, terminal condition in L^2, and driver

$$\psi(t, \omega, y, z)$$
$$= -\partial_x b(t, X_t(\omega), \alpha_t(\omega)) \cdot y - \partial_x \sigma(t, X_t(\omega), \alpha_t(\omega)) \cdot z - \partial_x f^i(t, X_t(\omega), \alpha_t(\omega)),$$

which is an affine function of y and z with uniformly bounded random coefficients and an L^2 intercept. So, for each $i \in \{1, \cdots, N\}$, the existence and uniqueness of a solution $(Y_t^{i,\boldsymbol{\alpha}}, Z_t^{i,\boldsymbol{\alpha}})_{0 \le t \le T}$ follows from Theorem 2.2.

The following result is the open loop *game analog* of the necessary part of the Pontryagin maximum principle, which we proved in the case of stochastic control problems with convex sets of admissible controls. Its proof can be conducted along the lines of the derivation given in Chapter 3. We do not give it as we only use this result as a rationale for the search for a function satisfying the min-max Isaacs condition.

Theorem 5.19. *Under the above conditions, if $\boldsymbol{\alpha}^* \in \mathbb{A}$ is a Nash equilibrium for the open loop game, and if we denote by $\boldsymbol{X}^* = (X_t^*)_{0 \le t \le T}$ the corresponding controlled state of the* system, and by $(\boldsymbol{Y}^*, \boldsymbol{Z}^*) = ((Y^{*1}, \cdots, Y^{*N}), (Z^{*1}, \cdots, \hat{Z}^{*N}))$ the adjoint processes, then the generalized min-max Isaacs conditions hold along the optimal paths in the sense that, for each $i \in \{1, \cdots, N\}$,

$$H^i(t, X_t^*, Y_t^{*i}, Z_t^{*i}, \alpha_t^*) = \inf_{\alpha^i \in A^i} H^i(t, X_t^*, Y_t^{*i}, Z_t^{*i}, (\alpha^{*-i}, \alpha^i)), \quad dt \otimes d\mathbb{P}\text{-a.s.} \quad (5.42)$$

We now state and prove the sufficient condition which we will use in many applications.

Theorem 5.20. *Assuming that the coefficients of the game are twice continuously differentiable in $(x, \alpha) \in \mathbb{R}^d \times A$ with bounded partial derivatives, that $\hat{\boldsymbol{\alpha}} \in \mathbb{A}$ is an admissible adapted (open loop) strategy profile, $\hat{\boldsymbol{X}} = (\hat{X}_t)_{0 \leq t \leq T}$ is the corresponding controlled state, and $(\hat{\boldsymbol{Y}}, \hat{\boldsymbol{Z}}) = ((\hat{\boldsymbol{Y}}^1, \cdots, \hat{\boldsymbol{Y}}^N), (\hat{\boldsymbol{Z}}^1, \cdots, \hat{\boldsymbol{Z}}^N))$ is the corresponding adjoint processes, if we also assume that, for each $i \in \{1, \cdots, N\}$,*

1. *$(x, \alpha) \hookrightarrow H^i(t, x, \hat{Y}_t^i, \hat{Z}_t^i, \alpha)$ is a convex function, $dt \otimes d\mathbb{P}$-a.s.;*

2. *g^i is convex $\mathbb{P}$-a.s.*

and if, moreover, for every $i \in \{1, \cdots, N\}$ we have

$$H^i(t, \hat{X}_t, \hat{Y}_t^i, \hat{Z}_t^i, \hat{\alpha}_t) = \inf_{\alpha^i \in A^i} H^i(t, \hat{X}_t, \hat{Y}_t^i, \hat{Z}_t^i, (\hat{\alpha}^{-i}, \alpha^i)), \quad dt \otimes d\mathbb{P}\text{-a.s.}, \quad (5.43)$$

then $\hat{\boldsymbol{\alpha}}$ is a Nash equilibrium for the open loop game.

Proof. We fix $i \in \{1, \cdots, N\}$, a generic (adapted) $\alpha^i \in \mathbb{A}^i$, and for the sake of simplicity, we denote by $\boldsymbol{X}$ the state $\boldsymbol{X}^{(\hat{\alpha}^{-i}, \alpha^i)}$ controlled by the strategies $(\hat{\alpha}^{-i}, \alpha^i)$. The function g^i being convex almost surely, we have

$$g^i(\hat{X}_T) - g^i(X_T)$$
$$\leq (\hat{X}_T - X_T) \partial_x g^i(\hat{X}_T)$$
$$= (\hat{X}_T - X_T) \hat{Y}_T^i$$
$$= \int_0^T (\hat{X}_t - X_t) d\hat{Y}_t^i + \int_0^T \hat{Y}_t^i d(\hat{X}_t - X_t) + \int_0^T [\sigma(t, \hat{X}_t, \hat{\alpha}_t) - \sigma(t, X_t, (\hat{\alpha}^{-i}, \alpha^i))] \cdot \hat{Z}_t^i dt$$
$$= -\int_0^T (\hat{X}_t - X_t) \partial_x H^i(t, \hat{X}_t, \hat{Y}_t^i, \hat{Z}_t^i, \hat{\alpha}_t) dt + \int_0^T \hat{Y}_t^i [b(t, \hat{X}_t, \hat{\alpha}_t) - b(t, X_t, (\hat{\alpha}^{-i}, \alpha^i))] dt$$
$$+ \int_0^T [\sigma(t, \hat{X}_t, \hat{\alpha}_t) - \sigma(t, X_t, (\hat{\alpha}^{-i}, \alpha^i))] \cdot \hat{Z}_t^i dt + \text{martingale} \quad (5.44)$$

So that, taking expectations of both sides and plugging the result into

$$J^i(\hat{\boldsymbol{\alpha}}) - J^i((\hat{\boldsymbol{\alpha}}^{-i}, \boldsymbol{\alpha}^i))$$
$$= \mathbb{E}\left\{\int_0^T [f^i(t, \hat{X}_t, \hat{\alpha}_t) - f^i(t, X_t, (\hat{\alpha}^{-i}, \alpha^i))] dt\right\} + \mathbb{E}\{g^i(\hat{X}_T) - g^i(X_T)\}$$

5.4. Game Versions of the Stochastic Maximum Principle 189

we get

$$J^i(\hat{\boldsymbol{\alpha}}) - J^i((\hat{\boldsymbol{\alpha}}^{-i}, \alpha^i))$$
$$= \mathbb{E}\left\{\int_0^T [H^i(t, \hat{X}_t, \hat{Y}_t^i, \hat{Z}_t^i, \hat{\alpha}_t) - H^i(t, X_t, \hat{Y}_t^i, \hat{Z}_t^i, (\hat{\alpha}^{-i}, \alpha^i))]dt\right\}$$
$$- \mathbb{E}\left\{\int_0^T \hat{Y}_t^i[b(t, \hat{X}_t, \hat{\alpha}_t) - b(t, X_t, (\hat{\alpha}^{-i}, \alpha^i))]\, dt\right\}$$
$$- \mathbb{E}\left\{\int_0^T [\sigma(t, \hat{X}_t, \hat{\alpha}_t) - \sigma(t, X_t, (\hat{\alpha}^{-i}, \alpha^i))]\cdot \hat{Z}_t^i\, dt\right\}$$
$$+ \mathbb{E}\{g^i(\hat{X}_T) - g^i(X_T)\}$$
$$\leq \mathbb{E}\bigg\{\int_0^T [H^i(t, \hat{X}_t, \hat{Y}_t^i, \hat{Z}_t^i, \hat{\alpha}_t) - H^i(t, X_t, \hat{Y}_t^i, \hat{Z}_t^i, (\hat{\alpha}^{-i}, \alpha^i))$$
$$- (\hat{X}_t - X_t)\partial_x H^i(t, \hat{X}_t, \hat{Y}_t^i, \hat{Z}_t^i, \hat{\alpha}_t)]\, dt\bigg\}$$
$$\leq 0, \qquad (5.45)$$

because the above integrand is nonpositive for $dt \otimes d\mathbb{P}$ almost all $(t, \omega) \in [0, T] \times \Omega$. Indeed, this is easily seen by a second-order Taylor expansion as a function of (x, α), using the convexity assumption and the fact that $\hat{\alpha}$ is a critical point (where the first-order derivative vanishes) because it satisfies the generalized Isaacs condition by assumption. □

Implementation Strategy

We shall try to use this sufficient condition in the following manner. When the coefficients of the model are differentiable with respect to the state variable x with bounded derivatives, if the convexity assumptions (1) and (2) of the above theorem are satisfied, we shall search for a deterministic function $\hat{\alpha}$

$$(t, x, (y^1, \cdots, y^N), (z^1, \cdots, z^N)) \hookrightarrow \hat{\alpha}(t, x, (y^1, \cdots, y^N), (z^1, \cdots, z^N)) \in A$$

on $[0, T] \times \mathbb{R}^d \times \mathbb{R}^{dN} \times \mathbb{R}^{dmN}$ which satisfies the Isaacs conditions. Next, we replace the *adapted* controls $\boldsymbol{\alpha}$ in the forward dynamics of the state as well as in the adjoint BSDEs by

$$\hat{\alpha}(t, X_t, (Y_t^1, \cdots, Y_t^N), (Z_t^1, \cdots, Z_t^N)).$$

This creates a large FBSDE comprising a forward equation in dimension d and N backward equations in dimension d. The couplings between these equations may be highly nonlinear, and this system may be very difficult to solve. However, if we find processes $\boldsymbol{X}$, $(\boldsymbol{Y}^1, \cdots, \boldsymbol{Y}^N)$, and $(\boldsymbol{Z}^1, \cdots, \boldsymbol{Z}^N)$ solving this FBSDE

$$\begin{cases} dX_t = b(t, X_t, \hat{\alpha}(t, X_t, (Y_t^1, \cdots, Y_t^N), (Z_t^1, \cdots, Z_t^N)))dt \\ \qquad\quad + \sigma(t, X_t, \hat{\alpha}(t, X_t, (Y_t^1, \cdots, Y_t^N), (Z_t^1, \cdots, Z_t^N)))dW_t, \\ dY_t^1 = -\partial_x H^1(t, X_t, Y_t^1, Z_t^1, \hat{\alpha}(t, X_t, (Y_t^1, \cdots, Y_t^N), (Z_t^1, \cdots, Z_t^N)))dt + Z_t^1 dW_t, \\ \cdots \;\, = \;\, \cdots \\ dY_t^N = -\partial_x H^N(t, X_t, Y_t^N, Z_t^N, \hat{\alpha}(t, X_t, (Y_t^1, \cdots, Y_t^N), (Z_t^1, \cdots, Z_t^N)))dt + Z_t^N dW_t, \end{cases}$$
$$(5.46)$$

with initial condition $X_0 = x$ for the forward equation, and for each $i \in \{1, \cdots, N\}$, terminal condition $Y_T^i = \partial_x g^i(X_T)$ for the backward equation, the above sufficient condition says that the strategy profile $\hat{\alpha}$ defined by $\hat{\alpha}_t = \hat{\alpha}(t, X_t, (Y_t^1, \cdots, Y_t^N), (Z_t^1, \cdots, Z_t^N))$ is an open loop Nash equilibrium.

5.4.2 ▪ Markovian Nash Equilibriums

We now consider the case of Markovian equilibriums. As we are about the see, the sufficient condition of the stochastic maximum principle takes a different form. The first order of business is to introduce an appropriate notion of the adjoint process.

Let us assume that $\phi = (\varphi^1, \cdots, \varphi^N)$ is a jointly measurable function from $[0, T] \times \mathbb{R}^d$ into $A = A^1 \times \cdots \times A^N$, which is differentiable in $x \in \mathbb{R}^d$ for $t \in [0, T]$ fixed, with derivatives uniformly bounded in $(t, x) \in [0, T] \times \mathbb{R}^d$. Assuming that the drift and volatility functions b and σ are Lipschitz in (x, α) uniformly in $t \in [0, T]$, we denote by $\boldsymbol{X}^\phi$ the unique strong solution of the state equation

$$dX_t = b(t, X_t, \phi(t, X_t))dt + \sigma(t, X_t, \phi(t, X_t))dW_t \tag{5.47}$$

with initial condition $X_0 = x$. We consider only the case of the interval $[0, T]$ for the sake of simplicity.

Definition 5.21. *A set of N couples $(\boldsymbol{Y}^{\phi,i}, \boldsymbol{Z}^{\phi,i}) = (Y_t^{\phi,i}, Z_t^{\phi,i})_{t \in [0,T]}$ of processes in $\mathbb{S}^{2,d}$ and $\mathbb{H}^{2,d \times m}$, respectively, for $i = 1, \cdots, N$, is said to be a set of adjoint processes associated with ϕ if for each $i \in \{1, \cdots, N\}$ they satisfy the BSDEs*

$$\begin{cases} dY_t^{\phi,i} = -[\partial_x H^i(t, X_t^\phi, Y_t^{\phi,i}, Z_t^{\phi,i}, \phi(t, X_t)) \\ \qquad + \sum_{j=1, j \neq i}^N \partial_{\alpha^j} H^i(t, X_t^\phi, Y_t^{\phi,i}, Z_t^{\phi,i}, \phi(t, X_t)) \partial_x \varphi^j(t, X_t^\phi)]dt + Z_t^{\phi,i} dW_t \\ Y_T^{\phi,i} = -\partial_x g^i(X_T^\phi). \end{cases} \tag{5.48}$$

We drop the superscript ϕ when no confusion is possible. Given the current assumptions on the coefficients of the model and the functions φ^i, the existence and uniqueness of the adjoint processes follow the same argument as in the open loop case.

We now state and prove the sufficient condition which we shall use in applications. We give an overly detailed proof because this form of the stochastic maximum principle is not standard, and despite systematic searches, we could not find it in the existing literature on stochastic differential games.

Theorem 5.22. *We assume that the coefficients of the game are twice continuously differentiable in $(x, \alpha) \in \mathbb{R}^d \times A$ with bounded partial derivatives, that the function $\phi = (\varphi^1, \cdots, \varphi^N)$ is continuously differentiable in $x \in \mathbb{R}^d$ for $t \in]0, T]$ fixed with bounded partial derivatives, and we denote by $(\boldsymbol{Y}^\phi, \boldsymbol{Z}^\phi) = ((\boldsymbol{Y}^{\phi,1}, \cdots, \boldsymbol{Y}^{\phi,N}), (\boldsymbol{Z}^{\phi,1}, \cdots, \boldsymbol{Z}^{\phi,N}))$ the adjoint processes associated with ϕ. We also assume that, for each $i \in \{1, \cdots, N\}$,*

1. *the function $\mathbb{R}^d \times A^i \ni (x, \alpha^i) \hookrightarrow h^i(x, \alpha^i) = H^i(t, x, Y_t^{\phi,i}, Z_t^{\phi,i}, (\phi(t, x)^{-i}, \alpha^i))$ is convex, $dt \otimes d\mathbb{P}$-a.s.,*

2. *g^i is convex,*

and that $dt \otimes d\mathbb{P}$-a.s.:

$$H^i(t, X_t^\phi, Y_t^{\phi,i}, Z_t^{\phi,i}, \phi(t, X_t^\phi)) = \inf_{\alpha^i \in A^i} H^i(t, X_t^\phi, Y_t^{\phi,i}, Z_t^{\phi,i}, (\phi(t, X_t^\phi)^{-i}, \alpha^i)). \tag{5.49}$$

Then ϕ is a Markovian Nash equilibrium for the game.

5.4. Game Versions of the Stochastic Maximum Principle

Proof. The proof is along the same lines as in the case of open loop equilibriums. The differences will become clear below. As before, we fix $i \in \{1, \cdots, N\}$, a generic feedback function $(t,x) \hookrightarrow \psi(t,x) \in A^i$, and for the sake of simplicity, we denote by $\boldsymbol{X}$ the solution $\boldsymbol{X}^{(\phi^{-i}, \psi)}$ of the state equation (5.47) with the feedback function (ϕ^{-i}, ψ) in lieu of ϕ. Starting from

$$J^i(\phi) - J^i((\phi^{-i}, \psi)) = \mathbb{E}\left[\int_0^T [f^i(t, X_t^\phi, \phi(t, X_t^\phi)) - f^i(t, X_t, (\phi(t, X_t)^{-i}, \psi(t, X_t)))]dt\right]$$
$$+ \mathbb{E}[g^i(X_T^\phi) - g^i(X_T)]$$

we use the definition of the Hamiltonian H^i to replace f^i. We get

$$J^i(\phi) - J^i((\phi^{-i}, \psi))$$
$$= \mathbb{E}\left[\int_0^T [H^i(t, X_t^\phi, Y_t^{\phi,i}, Z_t^{\phi,i}, \phi(t, X_t^\phi))\right.$$
$$\left. - H^i(t, X_t, Y_t^{\phi,i}, Z_t^{\phi,i}, (\phi(t, X_t)^{-i}, \psi(t, X_t)))]dt\right]$$
$$- \mathbb{E}\left[\int_0^T [b(t, X_t^\phi, \phi(t, X_t^\phi)) - b(t, X_t, (\phi(t, X_t)^{-i}, \psi(t, X_t)))] \cdot Y_t^{\phi,i} \, dt\right]$$
$$- \mathbb{E}\left[\int_0^T [\sigma(t, X_t^\phi, \phi(t, X_t^\phi)) - \sigma(t, X_t, (\phi(t, X_t)^{-i}, \psi(t, X_t)))] \cdot Z_t^{\phi,i} \, dt\right]$$
$$+ \mathbb{E}[g^i(X_T^\phi) - g^i(X_T)]. \tag{5.50}$$

We bound this last expectation using the convexity of g^i:

$$g^i(X_T^\phi) - g^i(X_T)$$
$$\leq (X_T^\phi - X_T)\partial_x g^i(X_T^\phi)$$
$$= (X_T^\phi - X_T) Y_T^{\phi,i}$$
$$= \int_0^T (X_t^\phi - X_t) \, dY_t^{\phi,i} + \int_0^T Y_t^{\phi,i} \, d(X_t^\phi - X_t)$$
$$+ \int_0^T [\sigma(t, X_t^\phi, \phi(t, X_t^\phi)) - \sigma(t, X_t, (\phi(t, X^t)^{-i}, \psi(t, X_t)))] \cdot Z_t^{\phi,i} \, dt$$
$$= -\int_0^T (X_t^\phi - X_t)[\partial_x H^i(t, X_t^\phi, Y_t^{\phi,i}, Z_t^{\phi,i}, \phi(t, X_t^\phi)) \tag{5.51}$$
$$- \sum_{j=1, j\neq i}^N \partial_{\alpha^j} H^i(t, X_t^\phi, Y_t^{\phi,i}, Z_t^{\phi,i}, \phi(t, X_t^\phi)) \partial_x \phi^j(t, X_t^\phi)] \, dt$$
$$+ \int_0^T Y_t^{\phi,i} \cdot [b(t, X_t^\phi, \phi(t, X_t^\phi)) - b(t, X_t, (\phi(t, X_t)^{-i}, \psi(t, X_t)))] \, dt$$
$$+ \int_0^T [\sigma(t, X_t^\phi, \phi(t, X_t^\phi)) - \sigma(t, X_t, (\phi(t, X_t)^{-i}, \psi(t, X_t)))] \cdot Z_t^{\phi,i} \, dt + \text{martingale},$$

where we used the special form (5.48) of the adjoint BSDE. Putting together (5.50) and

(5.51), we get

$$J^i(\phi) - J^i((\phi^{-i}, \psi))$$
$$\leq \mathbb{E}\Bigg[\int_0^T \Big(H^i(t, X_t^\phi, Y_t^{\phi,i}, Z_t^{\phi,i}, \phi(t, X_t^\phi))$$
$$- H^i(t, X_t, Y_t^{\phi,i}, Z_t^{\phi,i}, (\phi(t, X_t)^{-i}, \psi(t, X_t)))$$
$$- (X_t^\phi - X_t)[\partial_x H^i(t, X_t^\phi, Y_t^{\phi,i}, Z_t^{\phi,i}, \phi(t, X_t^\phi))$$
$$- \sum_{j=1, j \neq i}^N \partial_{\alpha^j} H^i(t, X_t^\phi, Y_t^{\phi,i}, Z_t^{\phi,i}, \phi(t, X_t^\phi))\partial_x \phi^j(t, X_t^\phi)]\Big)dt\Bigg] \quad (5.52)$$

and we conclude using the convexity assumption on the function $(x, \alpha^i) \hookrightarrow h^i(x, \alpha^i)$ for (t, ω) fixed. We use this convexity assumption in the form

$$h^i(x, \alpha^i) - h^i(x', \alpha^{i'}) - (x - x')\partial_x h^i(x, \alpha^i) - (\alpha^i - \alpha^{i'})\partial_{\alpha^i} h^i(x, \alpha^i) \leq 0,$$

which we apply to $x = X_t^\phi$, $x' = X_t$, $\alpha^i = \varphi^i(t, X_t^\phi)$, and $\alpha^{i'} = \psi(t, X_t)$. Notice that the assumption that the minimum of the Hamiltonian is attained along the (candidate for the) optimal path gives

$$\frac{\partial h^i}{\partial \alpha^i}(X_t^\phi, \varphi^i(t, X_t^\phi)) = 0,$$

and altogether this gives

$$H^i(t, X_t^\phi, Y_t^{\phi,i}, Z_t^{\phi,i}, (\phi(t, X_t^\phi)^{-i}, \varphi^i(t, X_t^\phi)))$$
$$- H^i(t, X_t, Y_t^{\phi,i}, Z_t^{\phi,i}, (\phi(t, X_t^\phi)^{-i}, \psi^i(t, X_t)))$$
$$- (X_t^\phi - X_t)\Big[\partial_x H^i(t, X_t^\phi, Y_t^{\phi,i}, Z_t^{\phi,i}, (\phi(t, X_t^\phi)^{-i}, \varphi^i(t, X_t^\phi)))$$
$$+ \sum_{j=1, j \neq i}^N \partial_{\alpha^j} H^i(t, X_t^\phi, Y_t^{\phi,i}, Z_t^{\phi,i}, (\phi(t, X_t^\phi)^{-i}, \varphi^i(t, X_t^\phi)))\partial_x \phi^j(t, X_t^\phi)\Big] \leq 0,$$

which is what we needed to conclude that

$$J^i(\phi) - J^i((\phi^{-i}, \psi)) \leq 0,$$

which completes the proof of the fact that ϕ is a Markovian Nash equilibrium. □

Implementation Strategy

As one can expect, the systematic use of the above sufficient condition to construct equilibriums is much more delicate than in the open loop case. Still, we may try to use it in the following circumstances. When the coefficients of the model are differentiable with respect to the couple (x, α) with bounded derivatives, if the convexity assumptions (1) and (2) of the above theorem are satisfied, we should, as in the open loop case, search for a deterministic function $\hat{\alpha}$

$$(t, x, (y^1, \cdots, y^N), (z^1, \cdots, z^N)) \hookrightarrow \hat{\alpha}(t, x, (y^1, \cdots, y^N), (z^1, \cdots, z^N)) \in A$$

on $[0, T] \times \mathbb{R}^d \times \mathbb{R}^{dN} \times \mathbb{R}^{dmN}$ which satisfies the Isaacs conditions. If such a function is found, and if for example, it does not depend upon $z = (z^1, \cdots, z^N)$, we replace the instances of the controls in the forward dynamics of the state as well as in the adjoint BSDEs

5.4. Game Versions of the Stochastic Maximum Principle

by $\hat{\alpha}(t, X_t, (Y_t^1, \cdots, Y_t^N))$, looking for an FBSDE which could be solved. Unfortunately, while this idea was reasonable in the open loop case, it cannot be implemented in a straightforward manner for Markov games, because the adjoint equations require the derivatives of the controls. A possible workaround would be to use the fact that, if such a Markovian FBSDE can be derived and solved, at least in the regular cases, the solution Y_t of the backward equation should be of the form $Y_t = u(t, X_t)$ for some deterministic function u, called the FBSDE value function or the decoupling field. Itô's formula implies that, at least when u is smooth, $Z_t = \partial_x u(t, X_t)$. So, if we want to use $\phi(t, x) = \hat{\alpha}(t, x, y)$, we can, in the BSDE giving the adjoint processes, use the quantity $\partial_y \hat{\alpha}(t, x, u(t, x)) \partial_x u(t, x)$ instead of the term $\partial_x \phi(t, x)$, and when we replace x by X_t, we might as well replace the instances of $u(t, x)$ by Y_t and those of $\partial_x u(t, x)$ by Z_t. As in the case of the open loop models, this creates a large FBSDE which we need to solve in order to obtain a Markovian Nash equilibrium.

As an illustration of the above strategy, we shall use the Pontryagin stochastic maximum principle to construct Markovian Nash equilibriums for the toy model of systemic risk studied in Section 5.5.

Important Disclaimer. The approach to the construction of Markovian Nash equilibriums based on the use of the Pontryagin stochastic maximum principle is in the face of the common wisdom which says that "*... when searching for Nash equilibriums for a stochastic differential game, use the stochastic maximum principle if you are searching for open loop Nash equilibriums, and the HJB equations if you are searching for closed loop equilibriums*". Moreover, while not stated explicitly, it is tacitly implied that these two alternatives are exclusive. The conclusion of the above discussion is that this adage is grossly misleading, if not false! Indeed, we proved that one can search, at least under some reasonable regularity conditions, for Markovian Nash equilibriums using the stochastic maximum principle. Even though the regularity conditions may look very restrictive, the reader should keep in mind that constructing Markovian Nash equilibriums is usually very difficult, and most often requires, restrictive assumptions, whichever approach one chooses.

5.4.3 ▪ Application to LQ Stochastic Differential Games

Earlier in this section, we said that the existence of the adjoint processes was a simple consequence of the existence and uniqueness of solutions of BSDEs with affine drivers. However, as we already pointed out, things are much more delicate when we want the control strategies to be given by a function satisfying the generalized Isaacs condition. Indeed, such a function typically depends upon the adjoint variables y^i and z^i, and injecting such a function into the dynamics of the controlled state creates couplings between the forward and backward components. In other words, the existence issue now concerns an FBSDE instead of a plain BSDE, and we know that this is a much more difficult question.

Existence of a Nash Equilibrium

Proof. Given our derivation of the min-max Isaacs condition, we introduce the following notation for the sake of convenience:

$$\tilde{B}_t^i = B_t^i [N_t^i]^{-1} B_t^{i\dagger} \qquad \text{and} \qquad \tilde{D}_t^i = D_t^i [N_t^i]^{-1} D_t^{i\dagger}. \qquad (5.53)$$

We also add to the assumptions (a)–(f) articulated in Subsection 5.2.2 the following extra assumption:

(g) $\tilde{B}_t^i(\omega) = \tilde{B}_t(\omega)$ is independent of $i \in \{1, \cdots, N\}$.

With the specific form of $\hat{\alpha}$ found above in (5.26), the forward state dynamics become

$$dX_t = \left[A_t X_t - \sum_{j=1}^N (\tilde{B}_t^j Y_t^j + B_t^j [N_t^j]^{-1} D_t^j Z_t^j) + \beta_t \right] dt$$

$$+ \left[C_t X_t - \sum_{j=1}^N (D_t^j [N_t^j]^{-1} B_t^{j\dagger} Y_t^j + \tilde{D}_t^j Z_t^j) + \gamma_t \right] dW_t.$$

Since the notation is quickly becoming cumbersome, we restrict ourselves to the simpler case where the volatility is not controlled by the players, in other words, to the case $D_t^i \equiv 0$. In this case, the controlled dynamics of the state become

$$dX_t = \left[A_t X_t - \sum_{j=1}^N \tilde{B}_t^j Y_t^j + \beta_t \right] dt + [C_t X_t + \gamma_t] dW_t. \tag{5.54}$$

We now determine the special form of the backward components. Each player $1 \leq i \leq N$ has a couple of adjoint processes $(\boldsymbol{Y}^i, \boldsymbol{Z}^i) = (Y_t^i, Z_t^i)_{0 \leq t \leq T}$ with values in $\mathbb{R}^d \times \mathbb{R}^{d \times m}$. We first derive the BSDE satisfied by each $(\boldsymbol{Y}^i, \boldsymbol{Z}^i)$, and we put them together into a single BSDE for $(\boldsymbol{Y}, \boldsymbol{Z})$, where the vector processes $\boldsymbol{Y} = (Y_t)_{0 \leq t \leq T}$ and $\boldsymbol{Z} = (Z_t)_{0 \leq t \leq T}$ are with values in $\mathbb{R}^{Nd}$ and $\mathbb{R}^{Ndm}$, respectively. Notice that because of their *dual* nature, the adjoint processes Y^i should be viewed as row vectors (as opposed to X_t, which is a column vector in matrix notation).

The partial derivatives of the Hamiltonian of player i are given by

$$\partial_x H^i(t, x, y, z, \alpha) = y^\dagger A_t + z^\dagger C_t + x^\dagger M_t^i,$$

which does not depend upon the players' strategies $\boldsymbol{\alpha}$, so the BSDE defining the adjoint processes $(\boldsymbol{Y}^i, \boldsymbol{Z}^i)$ reads

$$dY_t^i = -\partial_x H^i(t, X_t, Y_t^i, Z_t^i, \hat{\alpha}_t) dt + Z_t^i dW_t$$

$$= -[Y_t^{i\dagger} A_t + Z_t^{i\dagger} C_t + X_t^\dagger M_t^i] dt + \sum_{j=1}^m Z_t^{ij} dW_t^j. \tag{5.55}$$

Notice the special structure of (5.55). While the forward dynamics (5.54) involve *all* the adjoint processes Y^j for $j \in \{1, \cdots, N\}$, except for the presence of X_t, the backward equations are decoupled. Motivated by the result of the Pontryagin stochastic maximum principle in the form of a single BSDE for an Nd-dimensional column vector $\boldsymbol{Y}$ obtained by piling up the transposes of the row vectors Y^i, we get

$$\begin{cases} dY_t = -[\tilde{A}_t Y_t + \tilde{C}_t X_t + \tilde{M}_t Z_t] dt + Z_t dW_t, \\ Y_T = g_{PW}(X_T), \end{cases} \tag{5.56}$$

where we denote by $\tilde{A}_t$ (resp., $\tilde{C}_t$, $\tilde{M}$) the $(Nd) \times (Nd)$ block-diagonal matrix with N copies of $A_t^\dagger$ (resp., the matrices $C_t^\dagger$, M_t^i) on the diagonal, and by g_{PW} the function defined on $\mathbb{R}^d$ by

$$g_{PW}(x) = [\partial_x g^1(x), \cdots, \partial_x g^N(x)]^\dagger = [Q^1 x, \cdots, Q^N x]^\dagger = \tilde{Q} x, \tag{5.57}$$

where we denote by $\tilde{Q}$ the $(Nd) \times d$ matrix obtained by piling up the N matrices Q^i on top of each other.

5.4. Game Versions of the Stochastic Maximum Principle

We now prove the existence and uniqueness of a solution for the FBSDEs (5.54)–(5.56) by checking the hypotheses of Theorem 2.20. We use an $(Nd) \times d$ matrix G obtained by piling up N copies of the identity matrix I_d in d dimensions. Assumption **(H1)** is clearly satisfied since the forward and backward equations as well as the terminal condition are linear with bounded coefficients. The second condition of assumption **(H3)** is easy to check. Indeed

$$\langle g_{PW}(x) - g_{PW}(\overline{x}), G(x - \overline{x})\rangle = \langle \tilde{Q}((x - \overline{x}), G(x - \overline{x})\rangle$$
$$= \sum_{i=1}^{N} \langle Q^i((x - \overline{x}), x - \overline{x}\rangle$$
$$\geq \delta \|x - \overline{x}\|^2$$

by assumption on the matrices Q^i. Checking the first condition is slightly more involved:

$$\langle A(t, u) - A(t, \overline{u}), u - \overline{u}\rangle$$
$$= \langle -G^{\dagger}F(t,u) + G^{\dagger}F(t,\overline{u}), x - \overline{x}\rangle$$
$$+ \langle Gb(t,u) + Gb(t,\overline{u}), y - \overline{y}\rangle$$
$$+ \langle G\sigma(t,u)G\sigma(t,\overline{u}), z - \overline{z}\rangle$$
$$= -\langle \tilde{A}_t(y - \overline{y}) + \tilde{C}_t(x - \overline{x}) + \tilde{M}_t(z - \overline{z}), G(x - \overline{x})\rangle$$
$$+ \langle GA_t(x - \overline{x}), y - \overline{y}\rangle - \sum_{i=1}^{N}\langle G\tilde{B}_t^i(y^i - \overline{y}^i), y - \overline{y}\rangle$$
$$+ \langle GC_t(x - \overline{x}), z - \overline{z}\rangle$$
$$= -\sum_{i=1}^{N}\langle A_t^{\dagger}(y^i - \overline{y}^i), x - \overline{x}\rangle - \sum_{i=1}^{N}\langle C_t^{\dagger}(z^i - \overline{z}^i), x - \overline{x}\rangle - \sum_{i=1}^{N}\langle M_t^i(x - \overline{x}), (x - \overline{x})\rangle$$
$$+ \sum_{i=1}^{N}\langle A_t(x - \overline{x}), y^i - \overline{y}^i\rangle + \sum_{i,j=1}^{N} \langle \tilde{B}_t^i(y^i - \overline{y}^i), y^j - \overline{y}^j\rangle$$
$$+ \langle GC_t(x - \overline{x}), z - \overline{z}\rangle$$
$$= -\sum_{i=1}^{N}\langle M_t^i(x - \overline{x}), x - \overline{x}\rangle - \left\langle \sum_{j=1}^{N}\tilde{B}_t^j(y^j - \overline{y}^j), \sum_{i=1}^{N}(y^i - \overline{y}^i)\right\rangle$$
$$\leq -\delta\|x - \overline{x}\|^2 - \left\langle \sum_{j=1}^{N}\tilde{B}_t^j(y^j - \overline{y}^j), \sum_{i=1}^{N}(y^i - \overline{y}^i)\right\rangle.$$

Now, under the extra assumption that $\tilde{B}_t^i(\omega)$ as defined in (5.53) is independent of $i \in \{1, \cdots, N\}$, we have that

$$\left\langle \sum_{j=1}^{N}\tilde{B}_t^j(y^j - \overline{y}^j), \sum_{i=1}^{N}(y^i - \overline{y}^i)\right\rangle = \left\langle \tilde{B}_t \sum_{j=1}^{N}(y^j - \overline{y}^j), \sum_{i=1}^{N}(y^i - \overline{y}^i)\right\rangle \geq 0$$

from which we conclude the proof of the second condition of assumption **(H3)** by just taking $\mu_1 = \beta_1 = \delta > 0$ and $\beta_2 = 0$. □

Optimal Strategies in Closed Loop Feedback Form

The derivation above proved the existence of an open loop Nash equilibrium given by the optimal strategies

$$\hat{\alpha}_t^i = -[N_t^i]^{-1} B_t^{i\dagger} Y_t^i, \qquad i = 1, \cdots, N,$$

where the Y_t^i are the adjoint processes whose existence was guaranteed by our analysis of the Pontryagin FBSDE. However, as the backward components of the solutions of FBSDEs are most often given by functions (the so-called decoupling fields) of the forward component, we expect the above open loop Nash equilibrium to in fact be expressed in closed loop feedback form. We propose searching for such a closed loop representation by conjecturing that for each $i \in \{1, \cdot, N\}$, the adjoint process Y^i is of the form

$$Y_t^i = P_t^i X_t + Q_t^i, \qquad 0 \leq t \leq T, \tag{5.58}$$

for some bounded variation adapted processes $P^i = (P_t^i)_{0 \leq t \leq T}$ and $Q^i = (Q_t^i)_{0 \leq t \leq T}$ with values in $\mathbb{R}^{d \times d}$ and $\mathbb{R}^d$, respectively. Applying Itô's formula to (5.58), we get

$$\begin{aligned}
dY_t^i &= \dot{P}_t^i X_t dt + P_t^i dX_t + \dot{Q}_t^i dt \\
&= \dot{P}_t^i X_t dt + P_t^i A_t X_t dt - P_t^i \left(\sum_{j=1}^N \tilde{B}_t^j Y_t^j \right) dt \\
&\quad + P_t^i \beta_t dt + P_t^i [C_t X_t + \gamma_t] dW_t + \dot{Q}_t^i dt \\
&= \dot{P}_t^i X_t dt + P_t^i A_t X_t dt - P_t^i \sum_{j=1}^N \tilde{B}_t^j (P_t^j X_t + Q_t^j) dt \\
&\quad + P_t^i \beta_t dt + P_t^i [C_t X_t + \gamma_t] dW_t + \dot{Q}_t^i dt \\
&= \left[\left(\dot{P}_t^i + P_t^i A_t - P_t^i \sum_{j=1}^N \tilde{B}_t^j P_t^j \right) X_t - P_t^i \left(\sum_{j=1}^N \tilde{B}_t^j Q_t^j + \beta_t \right) + \dot{Q}_t^i \right] dt \\
&\quad + P_t^i [C_t X_t + \gamma_t] dW_t.
\end{aligned}$$

Identifying term by term with the corresponding backward component of the FBSDE

$$\begin{aligned}
dY_t^i &= -[A_t^\dagger Y_t^i + C^\dagger Z_t^i + M_t^i X_t] dt + Z_t^i dW_t \\
&= -[(A_t^\dagger P_t^i + M_t^i) X_t + C^\dagger Z_t^i + A_t^\dagger Q_t^i] dt + Z_t^i dW_t
\end{aligned}$$

we get

$$Z_t^i = P_t^i [C_t X_t + \gamma_t]. \tag{5.59}$$

Similarly

$$-P_t^i \left(\sum_{j=1}^N \tilde{B}_t^j Q_t^j + \beta_t \right) + \dot{Q}_t^i = -C^\dagger Z_t^i - A_t^\dagger Q_t^i,$$

which gives the ordinary differential equation

$$\dot{Q}_t^i = A_t^\dagger Q_t^i + P_t^i \sum_{j=1}^N \tilde{B}_t^j Q_t^j + P_t^i \beta_t - C^\dagger P_t^i [C_t X_t + \gamma t], \qquad Q_T^i = 0. \tag{5.60}$$

Finally,

$$\dot{P}_t^i + P_t^i A_t - P_t^i \sum_{j=1}^N \tilde{B}_t^j P_t^j = -A_t^\dagger P_t^i - M_t^i,$$

5.4. Game Versions of the Stochastic Maximum Principle

which gives the ordinary differential equation

$$\dot{P}_t^i + P_t^i A_t + A_t^\dagger P_t^i - P_t^i \sum_{j=1}^N \tilde{B}_t^j P_t^j + M_t^i = 0, \qquad P_T^i = Q^i, \qquad (5.61)$$

where Q^i is the matrix defining the terminal cost of the LQ game. The strategy is now clear.

1. For each $i \in \{1, \cdots, N\}$ solve the matrix Riccati equation (5.61).
2. Plug the solutions into (5.59).
3. Insert the N solutions P_t^i and the Z_t^i from (5.59) into the system of N coupled ordinary differential equations (5.60) and solve for the N functions Q_t^i.

Remark 5.23. *Note that, when the coefficients are random, these steps are taken for ω fixed, so the randomness of the original coefficients of the LQ model is not an issue.*

5.4.4 ▪ Application to the Case of Semilinear Quadratic Games

We use the function satisfying the Isaacs min-max condition identified earlier and the strategy outlined above in an attempt to conclude the analysis of the game model, which we called the semilinear quadratic model. As we are about to see, this attempt will not be fully successful given the technology at hand. The forward dynamics of the state read

$$dX_t = \left[\bar{b}(t, X_t) + \sum_{i=1}^N \hat{\alpha}^i(Y_t)\right] dt + dW_t,$$

where the $\hat{\alpha}^i$ are the functions (5.34) found to satisfy the Isaacs condition. Summing (5.33) over i, we find that

$$\sum_{i=1}^N \alpha^i = \frac{1}{(N-1)\theta - \delta\gamma} \left(\frac{\gamma}{\theta} - 1\right) \sum_{i=1}^N y^i,$$

so the controlled state dynamics (i.e., the forward component of the FBSDE) becomes

$$dX_t = \left[\bar{b}(t, X_t) + \frac{1}{(N-1)\theta - \delta\gamma} \left(\frac{\gamma}{\theta} - 1\right) \sum_{i=1}^N Y_t^i\right] dt + dW_t. \qquad (5.62)$$

Since

$$\partial_x H^i(t, x, y, z, \alpha) = \partial_x \bar{b}(t, x) \cdot y + \partial_x \bar{f}^i(t, x)$$

is independent of z and α, we define $F(t, u)$ as the Nd-dimensional column vector obtained by piling up the d-dimensional vectors $\partial_x \bar{b}(t, x) \cdot y^i + \partial_x \bar{f}^i(t, x)$ (recall that we use the notation $u = (x, y, z)$), and the backward component of the FBSDE given by the Pontryagin stochastic maximum principle reads

$$dY_t = -F(t, X_t, Y_t)dt + Z_t dW_t \qquad (5.63)$$

with the terminal condition

$$Y_T = [\partial_x g^1(X_T), \cdots, \partial_x g^N(X_T)]^\dagger.$$

Proving the existence of a solution for the FBSDE (5.62)–(5.63) would allow us to use Theorem 5.20 and conclude that the functions $\hat{\alpha}$ provide a Nash equilibrium. However, despite its deceptive simplicity, more assumptions are required to obtain a solution for this FBSDE by means of the tools we have developed so far.

5.5 • A Simple Model for Systemic Risk

We consider a network of N banks and for each $i \in \{1, \cdot, N\}$, we denote by $X_t^{(i)}$ the logarithm of the cash reserve of bank $i \in \{1, \cdots, N\}$ at time t. The following simple model for borrowing and lending between banks through the drifts of their log-cash reserves, while unrealistic, will serve our pedagogical objectives perfectly. For independent Wiener processes W^i, $i = 0, 1, \ldots, N$, and a positive constant $\sigma > 0$ we assume that

$$dX_t^i = \frac{a}{N} \sum_{j=1}^{N} (X_t^j - X_t^i) \, dt + \sigma dB_t^i$$
$$= a(\overline{X}_t - X_t^i) \, dt + \sigma dB_t^i, \quad i = 1, \ldots, N, \quad (5.64)$$

where
$$dB_t^i = \sqrt{1 - \rho^2} dW_t^i + \rho dW_t^0$$

for some $\rho \in [-1, 1]$. In other words, we assume that the log-cash reserves are Ornstein–Uhlenbeck processes reverting to their sample mean $\overline{X}_t$ at a rate $a > 0$. This sample mean represents the interaction between the various banks. We also consider a negative constant $D < 0$ which represents a critical liability threshold under which a bank is considered in a state of default.

The following are easy consequences of the above assumptions. Summing up equations (5.64) shows that the sample mean $\overline{X}_t$ is a Wiener process with volatility $\sigma/\sqrt{N}$. Simple Monte Carlo simulations show that stability of the system is easily achieved by increasing the rate a of borrowing and lending. Moreover, it is plain to compute analytically the loss distribution, i.e., the distribution of the number of firms whose log-cash reserves cross the level D, and large deviation estimates (which are mere Gaussian tail probability estimates in the present situation) show that increasing a increases systemic risk understood as the simultaneous default of a large number of banks.

While attractive, these conclusions depend strongly on the choice of the model and its specificities. We now consider a modification of the model which will hopefully lead to an equilibrium in which we hope that similar conclusions could hold. We consider the new dynamics

$$dX_t^i = \left[a(\overline{X}_t - X_t^i) + \alpha_t^i \right] dt + \sigma dB_t^i, \quad i = 1, \cdots, N,$$

where α^i is understood as the control of bank i, say the amount of lending and borrowing outside of the N bank network (e.g., issuing debt, borrowing at the Fed window, etc). In this modified model, firm i tries to *minimize*

$$J^i(\alpha^1, \cdots, \alpha^N) = \mathbb{E}\left[\int_0^T \left[\frac{1}{2}|\alpha_t^i|^2 - q\alpha_t^i(\overline{X}_t - X_t^i) + \frac{\epsilon}{2}(\overline{X}_t - X_t^i)^2 \right] dt \right.$$
$$\left. + \frac{c}{2}(\overline{X}_T - X_T^i)^2 \right]$$

for some positive constants ϵ and c, which balance the individual costs of borrowing and lending with the average behavior of the other banks in the network. The parameter $q > 0$ weights the contributions of the relative sizes of these components, imposing the choice of sign of α_t^i and the decision of whether to borrow or lend. The choice of q is likely to be the regulator's prerogative. Notice that

- if X_t^i is small (relative to the empirical mean $\overline{X}_t$), then bank i will want to borrow and choose $\alpha_t^i > 0$;
- if X_t^i is large, then bank i will want to lend and set $\alpha_t^i < 0$.

5.5. A Simple Model for Systemic Risk

We assume that
$$q^2 \leq \epsilon$$
to guarantee the convexity of the running cost functions. In this way, the problem is an instance of an LQ game, and we proceed to solve several forms of the model explicitly.

Solving for an Open Loop Nash Equilibrium by the Stochastic Maximum Approach

For each player, the set of admissible strategies is the space $\mathbb{H}^2$ of square integrable adapted processes. In the present situation, we use reduced Hamiltonians since the volatility depends neither upon the state X_t, nor the controls α_t. For each $i \in \{1, \cdots, N\}$, the reduced Hamiltonian of player i reads

$$\tilde{H}^i(x, y, \alpha) = \sum_{j=1}^{N}[a(\overline{x} - x^j) + \alpha^j]y^j + \frac{1}{2}(\alpha^i)^2 - q\alpha^i(\overline{x} - x^i) + \frac{\epsilon}{2}(\overline{x} - x^i)^2$$

and the value of α^i minimizing this reduced Hamiltonian with respect to α^i, when all the other variables, including α^j for $j \neq i$, are fixed, is given by

$$\hat{\alpha}^i = \hat{\alpha}^i(x, y) = -y^i + q(\overline{x} - x^i). \tag{5.65}$$

Now, given an admissible strategy profile $\boldsymbol{\alpha} = (\alpha^1, \cdots, \alpha^N)$ and the corresponding controlled state $X_t = X_t^{\alpha}$, the adjoint processes associated to $\boldsymbol{\alpha}$ are the processes $(\boldsymbol{Y}, \boldsymbol{Z}) = ((Y^1, \cdots, Y^N), (Z^1, \cdots, Z^N))$ solving the system of BSDEs

$$dY_t^{i,j} = -\partial_{x^j}\tilde{H}^i(X_t, Y_t^i, \alpha_t)dt + \sum_{k=0}^{N} Z_t^{i,j,k}dW_t^k, \tag{5.66}$$

$$= -\left[\sum_{k=1}^{N} a\left(\frac{1}{N} - \delta_{k,j}\right)Y_t^{i,k} - q\alpha_t^i\left(\frac{1}{N} - \delta_{i,j}\right) + \epsilon(\overline{X}_t - X_t^i)\left(\frac{1}{N} - \delta_{i,j}\right)\right]dt$$

$$+ \sum_{k=0}^{N} Z_t^{i,j,k}dW_t^k$$

for $i, j = 1, \cdots, N$ with terminal conditions $Y_T^i = c(\overline{X}_T - X_T^i)(\frac{1}{N} - 1)$. According to the strategy outlined earlier, we replace all the occurrences of the controls α_t^i in the forward equations, giving the dynamics of the state, and the backward adjoint equations by $\hat{\alpha}^i(X_t, Y_t^i) = -Y_t^{i,i} + q(\overline{X}_t - X_t^i)$, and we try to solve the resulting system of forward-backward equations; if we do, the strategy profile $\boldsymbol{\alpha} = (\alpha^1, \cdots, \alpha^N)$ defined by

$$\alpha_t^i = \hat{\alpha}^i(X_t, Y_t^i) = -Y_t^{i,i} + q(\overline{X}_t - X_t^i) \tag{5.67}$$

will provide an open loop Nash equilibrium. In the present situation, the FBSDEs read

$$\begin{cases} dX_t^i = [(a+q)(\overline{X}_t - X_t^i) - Y_t^{i,i}]dt + \sigma\rho dW_t^0 + \sigma\sqrt{1-\rho^2}dW_t^i, & i = 1, \cdots, N, \\ dY_t^{i,j} = -\left[a\sum_{k=1}^{N}(\frac{1}{N} - \delta_{k,j})Y_t^{i,k} + q[Y_t^{i,i} - q(\overline{X}_t - X_t^i)](\frac{1}{N} - \delta_{i,j}) \right. \\ \qquad\qquad \left. + \epsilon(\overline{X}_t - X_t^i)(\frac{1}{N} - \delta_{i,j})\right] + \sum_{k=0}^{N} Z_t^{i,j,k}dW_t^k, \\ Y_T^{i,j} = c(\overline{X}_T - X_T^i)(\frac{1}{N} - \delta_{i,j}), \qquad i, j = 1, \ldots, N. \end{cases} \tag{5.68}$$

This is an affine FBSDE, so as explained in Chapter 2, we expect the decoupling field to be affine as well, so we look for a solution Y_t given by an affine function of X_t. However, since the couplings between all these equations depend only upon quantities of the form $\overline{X}_t - X_t^i$, we search for a solution of the form

$$Y_t^{i,j} = \eta_t(\overline{X}_t - X_t^i)\left(\frac{1}{N} - \delta_{i,j}\right) \tag{5.69}$$

for some smooth deterministic function $t \hookrightarrow \eta_t$ to be determined. Computing the differential $dY_t^{i,j}$ from (5.69), we get

$$dY_t^{i,j} = \left(\frac{1}{N} - \delta_{i,j}\right)(\overline{X}_t - X_t^i)\left[\dot{\eta}_t - \eta_t\left(a + q + \left(1 - \frac{1}{N}\right)\eta_t\right)\right] \tag{5.70}$$
$$+ \sigma\sqrt{1-\rho^2}\,\eta_t\left(\frac{1}{N} - \delta_{i,j}\right)\left(\frac{1}{N}\sum_{k=1}^{N} dW_t^k - dW_t^i\right).$$

Evaluating the right-hand side of the BSDE part of (5.68) using the ansatz (5.69), we get

$$dY_t^{i,j} = -\left[a\sum_{k=1}^{N}\left(\frac{1}{N} - \delta_{k,j}\right)\eta_t(\overline{X}_t - X_t^i)\left(\frac{1}{N} - \delta_{i,k}\right)\right.$$
$$+ q[\eta_t(\overline{X}_t - X_t^i)\left(\frac{1}{N} - 1\right) - q(\overline{X}_t - X_t^i)]\left(\frac{1}{N} - \delta_{i,j}\right)$$
$$\left. + \epsilon(\overline{X}_t - X_t^i)\left(\frac{1}{N} - \delta_{i,j}\right)\right]dt + \sum_{k=0}^{N} Z_t^{i,j,k} dW_t^k \tag{5.71}$$
$$= \left(\frac{1}{N} - \delta_{i,j}\right)(\overline{X}_t - X_t^i)\left[(a+q)\eta_t - \frac{1}{N}q\eta_t + q^2 - \epsilon\right]dt + \sum_{k=0}^{N} Z_t^{i,j,k} dW_t^k.$$

Identifying the two Itô decompositions of $Y_t^{i,j}$ given in (5.70) and (5.71), we get

$$Z_t^{i,j,0} = 0, \quad Z_t^{i,j,k} = \sigma\sqrt{1-\rho^2}\,\eta_t\left(\frac{1}{N} - \delta_{i,j}\right)\left(\frac{1}{N} - \delta_{i,k}\right), \quad k = 1,\cdots,N,$$

and

$$\dot{\eta}_t - \eta_t\left(a + q + \left(1 - \frac{1}{N}\right)\eta_t\right) = (a+q)\eta_t - \frac{1}{N}q\eta_t + q^2 - \epsilon,$$

which we rewrite as a standard scalar Riccati equation

$$\dot{\eta}_t = \left[2(a+q) - \frac{1}{N}q\right]\eta_t + \left(1 - \frac{1}{N}\right)\eta_t^2 + q^2 - \epsilon \tag{5.72}$$

with terminal condition $\eta_T = c$. Under the condition $\epsilon \geq q^2$ (which guarantees the convexity of the running cost function f^i), this Riccati equation admits a unique solution.

Now that the existence of such a function η_t has been demonstrated, one can conclude that the strategy profile

$$\alpha_t^i = \left[q + \left(1 - \frac{1}{N}\right)\eta_t\right](\overline{X}_t - X_t^i), \tag{5.73}$$

obtained by plugging the value (5.69) of $Y_t^{i,j}$ into (5.67), is an open loop Nash equilibrium. Notice that the controls (5.73) entering the *optimal* strategy profile are in feedback form,

5.5. A Simple Model for Systemic Risk

since they only depend upon the current value of the state X_t at time t. Note also that in equilibrium, the dynamics of the state X_t are given by the stochastic differential equations

$$dX_t^i = \left[a+q+\left(1-\frac{1}{N}\right)\eta_t\right](\overline{X}_t-X_t^i)dt+\sigma\rho dW_t^0+\sigma\sqrt{1-\rho^2}dW_t^i, \qquad i=1,\cdots,N, \tag{5.74}$$

which are exactly the equations we started the subsection with in order to describe the model, except for the fact that the mean reversion coefficient a is replaced by the time-dependent mean reversion rate $a + q + (1 - \frac{1}{N})\eta_t$.

Important Remark. The strategy profile given by (5.73) happens to be in closed loop form since each α_t^i is a deterministic function of time and the state X_t of the system at time t. It was constructed to satisfy the sufficient condition for an open loop Nash equilibrium given by the stochastic maximum principle, and as such, it is indeed an open loop Nash equilibrium. However, even though it is in closed loop form, or even Markovian form, there is a priori no reason, except possibly wishful thinking, to believe that it could also be a closed loop or a Markovian Nash equilibrium, simply because of the definition we chose for the Markovian equilibriums.

Solving for a Markovian Nash Equilibrium by the Stochastic Maximum Approach

We search for a set $\phi = (\varphi^1, \cdots, \varphi^N)$ of feedback functions φ^i forming a Nash equilibrium for the Markov model of the game. For each player $i \in \{1, \cdots, N\}$, the reduced Hamiltonian (recall that the volatility depends neither on the state x nor on the controls) reads

$$\tilde{H}^{-i}(x,y,\alpha) = \sum_{k=1, k\neq i}^{N} [a(\overline{x} - x^k) + \varphi^k(t,x)]y^k + [a(\overline{x} - x^i) + \alpha]y^i$$
$$+ \frac{1}{2}\alpha^2 - q\alpha(\overline{x} - x^i) + \frac{\epsilon}{2}(\overline{x} - x^i)^2,$$

and the value of α minimizing this Hamiltonian (when all the other variables are fixed) is the same $\hat{\alpha}$ as given by (5.65). Using this formula and the fact that the adjoint equations will lead to an affine FBSDE where the couplings depend upon quantities of the form $\overline{X}_t - X_t^i$, as before, we suspect that Markov feedback functions could be based on the same rationale as above, leading to the forms (5.65) and (5.69). For these reasons, we define the feedback functions φ^i by

$$\varphi^i(t,x) = \left[q - \eta_t\left(\frac{1}{N} - 1\right)\right](\overline{x} - x^i), \qquad (t,x) \in [0,T] \times \mathbb{R}^d, \; i=1,\cdots,N, \tag{5.75}$$

for some deterministic function $t \hookrightarrow \eta_t$, and we try to determine such a function in order for these feedback functions φ^i to form a Markovian Nash equilibrium.

Using formula (5.19) for the partial derivative of the Hamiltonian, we learn that solving the Markovian model by the Pontryagin approach is done by solving the FBSDE

$$\begin{cases} dX_t^i = [(a+q)(\overline{X}_t - X_t^i) - Y_t^{i,i}]dt + \sigma\rho dW_t^0 + \sigma\sqrt{1-\rho^2}dW_t^i, \qquad i=1,\cdots,N, \\ dY_t^{i,j} = -\Big[a\sum_{k=1}^{N}(\frac{1}{N} - \delta_{k,j})Y_t^{i,k} + a\sum_{k=1,k\neq i}^{N}\partial_{x^j}\varphi^k(t,X_t)Y_t^{i,k} \\ \qquad\qquad - q[Y_t^{i,i} - q(\overline{X}_t - X_t^i)](\frac{1}{N} - \delta_{i,j}) + \epsilon(\overline{X}_t - X_t^i)(\frac{1}{N} - \delta_{i,j})\Big]dt + \sum_{k=0}^{N}Z_t^{i,j,k}dW_t^k, \\ Y_T^{i,j} = c(\overline{X}_T - X_T^i)(\frac{1}{N} - \delta_{i,j}), \qquad i,j=1,\ldots,N. \end{cases} \tag{5.76}$$

For the particular choice (5.75) of feedback functions, we have

$$\partial_{x^j}\varphi^k(t,x) = \left(\frac{1}{N} - \delta_{j,k}\right)\left[q - \eta_t\left(\frac{1}{N} - 1\right)\right], \tag{5.77}$$

and the backward component of the BSDE rewrites

$$dY_t^{i,j} = -\left[a\sum_{k=1}^N \left(\frac{1}{N} - \delta_{k,j}\right)Y_t^{i,k} + a\sum_{k=1,k\neq i}^N \left(\frac{1}{N} - \delta_{j,k}\right)\left[q - \eta_t\left(\frac{1}{N} - 1\right)\right]Y_t^{i,k}\right.$$
$$\left. - q[Y_t^{i,i} - q(\overline{X}_t - X_t^i)]\left(\frac{1}{N} - \delta_{i,j}\right) + \epsilon(\overline{X}_t - X_t^i)\left(\frac{1}{N} - \delta_{i,j}\right)\right]dt + \sum_{k=0}^N Z_t^{i,j,k}dW_t^k. \tag{5.78}$$

For the same reasons as in the open loop case (couplings depending only upon $\overline{X}_t - X_t^i$), we make the same ansatz on the form of $Y_t^{i,j}$, which provides consistency between (5.67), (5.69), (5.73), and the choice (5.75). So we search for a solution of the FBSDE (5.76) in the form (5.69) with the same function $t \hookrightarrow \eta_t$ as in (5.75). Evaluating the right-hand side of the BSDE part of (5.78) using the ansatz (5.69), we get

$$dY_t^{i,j} = -\left[a\sum_{k=1}^N \left(\frac{1}{N} - \delta_{k,j}\right)\eta_t(\overline{X}_t - X_t^i)\left(\frac{1}{N} - \delta_{k,j}\right)\right.$$
$$+ a\sum_{k=1,k\neq i}^N \left(\frac{1}{N} - \delta_{j,k}\right)\left[q - \eta_t\left(\frac{1}{N} - 1\right)\right]\eta_t(\overline{X}_t - X_t^i)\left(\frac{1}{N} - \delta_{k,j}\right)$$
$$- q\left[\eta_t(\overline{X}_t - X_t^i)\left(\frac{1}{N} - \delta_{i,j}\right) - q(\overline{X}_t - X_t^i)\right]\left(\frac{1}{N} - \delta_{i,j}\right)$$
$$\left. + \epsilon(\overline{X}_t - X_t^i)\left(\frac{1}{N} - \delta_{i,j}\right)\right] + \sum_{k=0}^N Z_t^{i,j,k}dW_t^k.$$
$$= \left(\frac{1}{N} - \delta_{i,j}\right)(\overline{X}_t - X_t^i)\left[\left(a + q - \frac{q}{N}\right)\eta_t + q^2 - \epsilon\right]dt + \sum_{k=0}^N Z_t^{i,j,k}dW_t^k. \tag{5.79}$$

Equating with the differential $dY_t^{i,j}$ obtained in (5.70) from the ansatz, we get the same identification for the $Z_t^{i,j,k}$ and the following Riccati equation for η_t:

$$\dot{\eta}_t = 2(a+q)\eta_t + \left(1 - \frac{1}{N^2}\right)\eta_t^2 + q^2 - \epsilon \tag{5.80}$$

with the same terminal condition $\eta_T = c$. We solve this equation under the same condition $\epsilon \geq q^2$. Figure 5.1 gives the plots of this solution for a few values of the parameters.

Clearly, the function η_t obtained in our search for Markovian Nash equilibriums is different from the function giving the open loop Nash equilibrium found earlier.

The Large Game Limit

While clear differences exist for the finite player games, these differences seem to disappear in the limit $N \to \infty$. Indeed, both solutions of the Riccati equations (5.72) and (5.80) converge toward the solution of

$$\dot{\eta}_t = 2(a+q)\eta_t + \eta_t^2 + q^2 - \epsilon. \tag{5.81}$$

5.5. A Simple Model for Systemic Risk

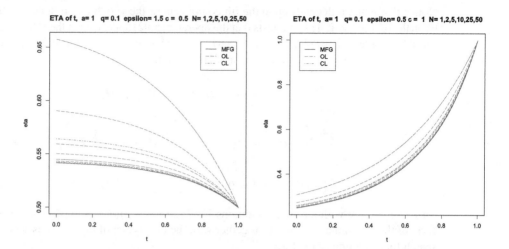

Figure 5.1. *Plot of the solution η_t of the Riccati equations (5.72), (5.80), and (5.81) for several values of the parameters and numbers of players N increasing from 1 to 50.*

In fact, it seems from the plots that as N increases, the functions η_t decrease to their common limit as $N \to \infty$. In the limit of large games ($N \to \infty$), the open loop and the closed loop (Markovian) Nash equilibriums found with the Pontryagin stochastic maximum principle coincide. The fact that the differences between open and closed loop equilibriums vanishes in the limit of large games is well-known. It is part of the *game theory folklore*. We will elaborate further on the limit $N \to \infty$ in the next chapter when we discuss mean-field games (MFGs), and identify the Riccati equation (5.81) to the solution of the MFG formulation of the model.

Solving for a Markovian Nash Equilibrium by the PDE Approach

Recall that the PDE which appears for player i in the construction of his/her best response is

$$\partial_t V^i + L^{*i}(x, \partial_x V^i(t,x), \partial^2_{xx} V^i(t,x)) = 0, \tag{5.82}$$

where the operator symbol L^{*i} is the minimized operator symbol defined by

$$L^{*i}(x, y, z) = \inf_{\alpha \in A^i} L^i(x, y, z, \alpha),$$

and in the present situation, the operator symbol L^i entering this HJB equation is given by

$$L^i(x, y, z, \alpha) = \sum_{j=1, j \neq i}^{N} [a(\overline{x} - x^j) + \alpha^{*j}(t,x)] y^j + [a(\overline{x} - x^i) + \alpha] y^i$$
$$+ \text{trace}[\sigma(t, x, (\alpha^{*-i}(t,x), \alpha))^{\dagger} z] + \frac{1}{2} \alpha^2 - q\alpha(\overline{x} - x^i) + \frac{\epsilon}{2}(\overline{x} - x^i)^2.$$

Two remarks are in order at this stage. First, because of the fact that we allow for a common noise, the inner product between the volatility matrix and the symbol z needs some special attention. Second, this quantity is quadratic in α, so the value of α^i at which the minimum is attained when all the other variables are held fixed is easy to compute. It is given by

$$\hat{\alpha}^i = \hat{\alpha}^i(x, y) = -y^i + q(\overline{x} - x^i). \tag{5.83}$$

Notice that while it depends upon the ith component of the variable y, it does not depend upon the $\hat{\alpha}^j$ for $j \neq i$. This suggests that the operator (5.82) appearing in the HJB equations should be obtained by plugging $\alpha = -\partial_{x^i} V^i(t,x) + q(\bar{x} - x^i)$ into the definition of $L^i(x, y, z, \alpha)$. Consequently, the system of HJB equations reads

$$\partial_t V^i(t,x) + \sum_{j=1}^{N}[(a+q)(\bar{x} - x^j) - \partial_{x^j} V^j(t,x)]\partial_{x^j} V^j(t,x)$$

$$+ \frac{\sigma^2}{2} \sum_{j,k=1}^{N} [\rho^2 + \delta_{j,k}((1-\rho^2)]\partial^2_{x^j x^k} V^i(t,x) + \frac{1}{2}(\epsilon - q^2)(\bar{x} - x^i)$$

$$+ \frac{1}{2}(\partial_{x^i} V^i(t,x))^2 = 0 \qquad (5.84)$$

for $i = 1, \cdots, N$ with terminal conditions $V^i(T, x) = (c/2)(\bar{x} - x^i)^2$. Given the special form of these terminal conditions, together with the discussion of the previous section, we search for a solution of the form

$$V^i(t,x) = \frac{1}{2}\eta_t(\bar{x} - x^i)^2 + \mu_t, \qquad i = 1, \cdots, N,$$

for some scalar deterministic functions $t \hookrightarrow \eta_t$ and $t \hookrightarrow \mu_t$ to be determined. Taking derivatives and plugging this ansatz into the system (5.84), we deduce the following necessary conditions (which end up being sufficient as well):

$$\begin{cases} \dot{\eta}_t = 2(a+q)\eta_t + (1 - \frac{1}{N^2})\eta_t^2 + q^2 - \epsilon, & \eta_T = c, \\ \dot{\mu}_t = -\mu_t = \sigma^2(1-\rho^2)(1-\frac{1}{N})\eta_t, & \mu_T = 0. \end{cases} \qquad (5.85)$$

The second equation can be integrated once one knows η_t. It gives

$$\mu_t = \sigma^2(1-\rho^2)\left(1 - \frac{1}{N}\right)\int_t^T \eta_s ds,$$

while the first equation is the same Riccati equation (5.80) as in the case of the search for Markovian Nash equilibriums via the Pontryagin maximum principle approach. So the Markov equilibrium identified with the HJB approach is the same as with the Pontryagin approach. After all, we did everything we did for this to be the case!

5.6 ▪ A Predatory Trading Game Model

Predatory trading occurs when some agents trade in anticipation of a forced liquidation by a large investor. The predators trade strategically to maximize their profits and destabilize the large investor: they initially trade in the same direction as the liquidating agent, reducing liquidity and causing prices to overshoot. In this section, we present a simplified version of a predatory trading model where a distressed trader (*victim*) needs to liquidate a position of $x_0^2 > 0$ units of a risky asset over a short period of time ending at time T. We capture the main features of the situation with several forms of the game depending upon the constraints we impose on the traders (e.g., terminal inventory penalty for the distressed trader), or the assumptions we make on the strategies the distressed trader is allowed to use (e.g., he/she can only sell in the market). The other strategic traders (*predators*) are assumed to be completely solvent, they can go either long or short in the risky asset during the period $[0, T]$, and in some forms of the model, they even have a longer trading horizon $\tilde{T} > T$ to unload unwanted inventory. Even though many of the results can be extended to many-player models, we focus on duopolies of one distressed trader and one predatory trader. This assumption is only made for the sake of exposition and tractability reasons.

5.6.1 ▪ Mathematical Setup

We consider a single period $[0, T]$, assuming that both agents trade in accordance with the restriction we impose on the trading strategies (see the definitions of $\mathbb{A}^1$ and $\mathbb{A}^2$ below). The model reads

$$\begin{cases} dX_t^0 = \Gamma(\alpha_t)dt + \sigma dW_t, \\ dX_t^1 = \alpha_t^1 dt, \\ dX_t^2 = \alpha_t^2 dt, \end{cases} \tag{5.86}$$

where the first equation gives the dynamics of a price X_t^0, which can be thought of as the *midprice* of the order book, and not necessarily the price at which transactions (buys and sells) take place in the market, $\Gamma(\alpha) = \Gamma(\alpha^1, \alpha^2)$ represents the permanent price impact due to the trading intensities α^1 and α^2 of the two traders, and X_t^1 and X_t^2 the positions at time t of the predator and the distressed trader, respectively. We denote by $X_t \in \mathbb{R}^3$ the vector whose components are X_t^0, X_t^1, and X_t^2, and $\alpha_t = (\alpha_t^1, \alpha_t^2)$. For the sake of the present discussion we use the simple form

$$\Gamma(\alpha) = \Gamma(\alpha^1, \alpha^2) = \gamma(\alpha^1 + \alpha^2)$$

for a constant $\gamma > 0$. This simple form of linear permanent impact is often used in practice. It corresponds to the simplistic model of a *flat order book*, or equivalently, to quadratic transaction costs.

Notice that even before we specify the optimization part of the game, namely the cost/reward functions, we can see that model (5.86) is not amenable to the weak formulation presented in Section 5.3 since the volatility structure is degenerate: the state of the system is three-dimensional while the noise shocks are only one-dimensional. However, since the volatility of the dynamics of the state as given in (5.86) is constant (and hence does not depend upon the control) we shall use reduced Hamiltonians throughout.

The running cost/reward functions will be the same in all versions of the game considered below. Both players will be assumed to be risk neutral in the sense that they try to maximize their expected wealths (or minimize their expected costs) at the end of the trading period. So, given a strategy profile $\boldsymbol{\alpha} = (\alpha^1, \alpha^2)$ for both players, the objective of player i is to maximize the expectation of his revenue from trading, namely, the expectation of $-\int_0^T P_t dX_t^i$ to which he/she may have to add the cost of a penalty levied at the terminal time T. Here P_t is the price at which the transaction takes place at time t. In addition to the permanent price impact already appearing in the dynamics of X_t^0, it also includes a temporary impact $\Lambda(\alpha)$ depending upon the rates of trading of the two players. For the sake of simplicity, we choose to use a linear function

$$\Lambda(\alpha) = \lambda(\alpha^1 + \alpha^2)$$

for some $\lambda > 0$. The goal of the predator (player 1) is to minimize

$$J^1(\boldsymbol{\alpha}) = \mathbb{E}\left[\int_0^T f^1(t, X_t, \alpha_t)dt + g^1(X_T)\right],$$

where the running cost function f^1 is given by

$$f^1(t, x, \alpha) = \alpha^1(x^0 + \Lambda(\alpha)) = \alpha^1(x^0 + \lambda(\alpha^1 + \alpha^2)),$$

so his/her (reduced) Hamiltonian reads

$$\begin{aligned} \tilde{H}^1(t, x, y^1, \alpha) &= \gamma(\alpha^1 + \alpha^2)y^{1,0} + \alpha^1 y^{1,1} + \alpha^2 y^{1,2} + \alpha^1(x^0 + \lambda(\alpha^1 + \alpha^2)) \\ &= \lambda(\alpha^1)^2 + (x^0 + \lambda\alpha^2 + y^{1,1} + \gamma y^{1,0})\alpha^1 + (\gamma y^{1,0} + y^{1,2})\alpha^2. \end{aligned}$$

At the same time, the victim tries to minimize

$$J^2(\boldsymbol{\alpha}) = \mathbb{E}\left[\int_0^{\tilde{T}} f^2(t, X_t, \alpha_t)dt + g^2(X_T)\right],$$

where the running cost function f^2 is defined by

$$f^2(t, x, \alpha) = \alpha^2(x^0 + \lambda(\alpha^1 + \alpha^2)),$$

so his/her (reduced) Hamiltonian reads

$$\tilde{H}^2(t, x, y^2, \alpha) = \gamma(\alpha^1 + \alpha^2)y^{2,0} + \alpha^1 y^{2,1} + \alpha^2 y^{2,2} + \alpha^2(x^0 + \lambda(\alpha^1 + \alpha^2))$$
$$= \lambda(\alpha^2)^2 + (x^0 + \lambda\alpha^1 + y^{2,2} + \gamma y^{2,0})\alpha^2 + \gamma y^{2,0}\alpha^1 + \alpha^1 y^{2,1}.$$

5.6.2 ▪ One Period, No Hard Constraint

We first consider a single period $[0, T]$, and the only (soft) constraint we impose is an inventory penalty. We assume that

$$g^1(\mathbf{x}) \equiv \frac{c_1}{2}(x^1)^2 \qquad \text{and} \qquad g^2(\mathbf{x}) = \frac{c_2}{2}(x^2)^2$$

for some nonnegative constants c_1 and c_2, these constants being 0 in the absence of inventory penalties and large when the inventory penalty is intended to be binding. We assume that the sets $\mathbb{A}^1$ and $\mathbb{A}^2$ of admissible strategies for the two players are given by

$$\mathbb{A}^i = \mathcal{H}^2_{[0,T]} = \left\{\boldsymbol{\alpha}^i = (\alpha_t^i)_{0 \leq t \leq \tilde{T}}; \text{ adapted process such that } \mathbb{E}\int_0^T |\alpha_t^i| dt < \infty\right\}$$

for $i = 1, 2$. In this version of the game, the natural candidates for the equilibrium strategies are given by the necessary condition of the stochastic maximum principle and Isaacs conditions. The (unrestricted) minimization over α^1 of the reduced Hamiltonian H^1 requires that the first-order condition

$$2\lambda\alpha^1 + \lambda\alpha^2 = -x^0 - \gamma y^{1,0} - y^{1,1} \tag{5.87}$$

be satisfied, while the (unrestricted) minimization over α^2 of the reduced Hamiltonian H^2 requires that the first-order condition

$$2\lambda\alpha^2 + \lambda\alpha^1 = -x^0 - y^{2,2} - \gamma y^{2,0} \tag{5.88}$$

be satisfied. Solving the linear system (5.87)–(5.88) gives

$$\begin{cases} \hat{\alpha}^1(t, x, (y^1, y^2)) = -\frac{1}{3\lambda}[x^0 + 2\gamma y^{1,0} + 2y^{1,1} - y^{2,2} - \gamma y^{2,0}], \\ \hat{\alpha}^2(t, x, (y^1, y^2)) = -\frac{1}{3\lambda}[x^0 - \gamma y^{1,0} - y^{1,1} + 2y^{2,2} + 2\gamma y^{2,0}], \end{cases} \tag{5.89}$$

and it is plain to check that these candidates satisfy the Isaacs conditions.

Open Loop Equilibriums by the Stochastic Maximum Approach

If we denote by $(\boldsymbol{Y}^1, \boldsymbol{Z}^1)$ and $(\boldsymbol{Y}^2, \boldsymbol{Z}^2)$ the adjoint processes, using the solution $\hat{\alpha}$ above for the system given by the Isaacs conditions, an open loop Nash equilibrium can be derived from the solution of the FBSDE:

$$\begin{cases} dX_t^0 = \gamma[\hat{\alpha}^1(t, X_t, Y_t^1, Y_t^2) + \hat{\alpha}^2(t, X_t, Y_t^1, Y_t^2)]dt + \sigma dW_t, \\ dX_t^1 = \hat{\alpha}^1(t, X_t, Y_t^1, Y_t^2)dt, \\ dX_t^2 = \hat{\alpha}^2(t, X_t, Y_t^1, Y_t^2)dt, \\ dY_t^{1,0} = -\hat{\alpha}^1(t, X_t, Y_t^1, Y_t^2)dt + Z_t^{1,0}dW_t, \\ dY_t^{1,1} = Z_t^{1,1}dW_t, \\ dY_t^{1,2} = Z_t^{1,2}dW_t, \\ dY_t^{2,0} = -\hat{\alpha}^2(t, X_t, Y_t^1, Y_t^2)dt + Z_t^{2,0}dW_t, \\ dY_t^{2,1} = Z_t^{2,1}dW_t, \\ dY_t^{2,2} = Z_t^{2,2}dW_t. \end{cases} \quad (5.90)$$

The backward equations for $Y_t^{1,1}, Y_t^{1,2}, Y_t^{2,1}$, and $Y_t^{2,2}$ are particularly simple because the Hamiltonians H^1 and H^2 do not depend upon x^1 and x^2. Our choices for the inventory penalties force the terminal conditions to be $Y_T^{1,0} = \partial_{x^0}g^1(X_T) = Y_T^{1,2} = \partial_{x^2}g^1(X_T) = 0$ and $Y_T^{1,1} = \partial_{x^1}g^1(X_T) = c_1 X_T^1$ for the adjoint processes of player 1. Similarly, $Y_T^{2,0} = \partial_{x^0}g^2(X_T) = 0$, $Y_T^{2,1} = \partial_{x^1}g^2(X_T) = 0$ and $Y_T^{2,2} = \partial_{x^2}g^2(X_T) = c_2 X_T^2$. These terminal conditions imply that $Y_t^{1,2} \equiv Y_t^{2,1} \equiv 0$. Consequently, the above FBSDE rewrites as

$$\begin{cases} dX_t^0 = -\frac{\gamma}{3\lambda}[2X_t^0 + \gamma Y_t^{1,0} + Y_t^{1,1} + Y_t^{2,2} + \gamma Y_t^{2,0}]dt + \sigma dW_t, & X_0^0 = x_0^0, \\ dX_t^1 = -\frac{1}{3\lambda}[X_t^0 + 2\gamma Y_t^{1,0} + 2Y_t^{1,1} - Y_t^{2,2} - \gamma Y_t^{2,0}]dt, & X_T^1 = x_0^1, \\ dX_t^2 = -\frac{1}{3\lambda}[X_t^0 - \gamma Y_t^{1,0} - Y_t^{1,1} + 2Y_t^{2,2} + 2\gamma Y_t^{2,0}]dt, & X_T^2 = x_0^2, \\ dY_t^{1,0} = \frac{1}{3\lambda}[X_t^0 + 2\gamma Y_t^{1,0} + 2Y_t^{1,1} - Y_t^{2,2} - \gamma Y_t^{2,0}]dt + Z_t^{1,0}dW_t, & Y_T^{1,0} = 0, \\ dY_t^{1,1} = Z_t^{1,1}dW_t, & Y_T^{1,1} = c_1 X_T^1, \\ dY_t^{2,0} = \frac{1}{3\lambda}[X_t^0 - \gamma Y_t^{1,0} - Y_t^{1,1} + 2Y_t^{2,2} + 2\gamma Y_t^{2,0}]dt + Z_t^{2,0}dW_t, & Y_T^{2,0} = 0, \\ dY_t^{2,2} = Z_t^{2,2}dW_t, & Y_T^{2,2} = c_2 X_T^2, \end{cases} \quad (5.91)$$

which is an FBSDE where the forward equation is three-dimensional and the backward equation is four-dimensional. It can be solved as a linear system because if we set

$$\overline{A} = -\frac{1}{3\lambda}\begin{bmatrix} 2\gamma & 0 & 0 \\ 1 & 0 & 0 \\ 1 & 0 & 0 \end{bmatrix}, \quad \overline{B} = -\frac{1}{3\lambda}\begin{bmatrix} \gamma^2 & \gamma & \gamma^2 & \gamma \\ 2\gamma & 2 & -\gamma & -1 \\ -\gamma & -1 & 2\gamma & 2 \end{bmatrix}, \quad \text{and} \quad \Sigma = \begin{bmatrix} \sigma \\ 0 \\ 0 \end{bmatrix},$$

then the forward component of the FBSDE (5.91) can be rewritten as

$$dX_t = [\overline{A}X_t + \overline{B}Y_t]dt + \Sigma dW_t \quad (5.92)$$

with the appropriate initial condition, provided we set $Y_t = [Y_t^{1,0}, Y_t^{1,1}, Y_t^{2,0}, Y_t^{2,2}]^\dagger$. Now, if we also set

$$C = \frac{1}{3\lambda}\begin{bmatrix} 1 & 0 & 0 \\ 0 & 0 & 0 \\ 1 & 0 & 0 \\ 0 & 0 & 0 \end{bmatrix}, \quad \overline{D} = \frac{1}{3\lambda}\begin{bmatrix} 2\gamma & 2 & -\gamma & -1 \\ 0 & 0 & 0 & 0 \\ -\gamma & -1 & 2\gamma & 2 \\ 0 & 0 & 0 & 0 \end{bmatrix}, \quad \text{and} \quad Z_t = \begin{bmatrix} Z_t^{1,0} \\ Z_t^{1,1} \\ Z_t^{2,0} \\ Z_t^{2,2} \end{bmatrix},$$

then the backward component of the FBSDE (5.91) can be rewritten as

$$dY_t = [\overline{C}X_t + \overline{D}Y_t]dt + Z_t dW_t \tag{5.93}$$

with the appropriate terminal condition. We learned in Chapter 2 that we can search for a solution in the form

$$Y_t = P_t X_t + p_t, \qquad 0 \le t \le T,$$

for some deterministic 4×3 and 4×1 matrix-valued differentiable functions P_t and p_t, to be determined later on. Computing dY_t from this expression, using (5.92), and identifying with (5.93), we find that P_t must satisfy the matrix Riccati equation

$$\dot{P}_t = -P_t \overline{B} P_t + [\overline{D}P_t - P_t \overline{A}] + \overline{C}, \qquad P_T = \begin{bmatrix} 0 & 0 & 0 \\ 0 & c_1 & 0 \\ 0 & 0 & 0 \\ 0 & 0 & c_2 \end{bmatrix}. \tag{5.94}$$

We momentarily assume existence and uniqueness of a continuous function $[0,T] \ni t \hookrightarrow P_t$ solving (5.94). Once this equation is solved, we choose $Z_t = P_t \Sigma$ for the martingale part, and the four-dimensional vector p_t can be obtained by solving the three-dimensional ordinary differential equation

$$\dot{p}_t = [-P_t \overline{B} + \overline{D}]p_t dt, \qquad p_T = \begin{bmatrix} 0 \\ 0 \\ 0 \\ 0 \end{bmatrix}. \tag{5.95}$$

Since this equation is linear and the terminal condition is 0, we conclude that $p_t \equiv 0$. We denote by $[U(t,s)]_{0 \le s \le t \le T}$ the propagator of the equation (5.92) with $\Sigma = 0$. In other words, $U(t,s)$ is a 4×4 matrix satisfying the ordinary differential equation

$$\frac{d}{dt}U(t,s) = [\overline{A} + \overline{B}P_t]U(t,s)$$

over the interval $[s,T]$ with initial condition $U(s,s) = I_3$, where as usual, I_3 denotes the 3×3 identity matrix. The unique solution of (5.92) can be expressed in terms of this propagator via the formula

$$X_t = U(t,0)x_0 + \sigma \int_0^t U(t,s) \begin{bmatrix} 1 \\ 0 \\ 0 \end{bmatrix} dW_t,$$

which shows that in equilibrium, the state X_t is Gaussian. Moreover, it also shows that for $i = 1, 2$, the variance v_t^i of the random variable X_t^i is given by

$$v_t^i = \sigma^2 \int_0^t [U(t,s)_{i+1,1}]^2 ds$$

if we denote by $U(t,s)_{k,l}$ the entry of the matrix $U(t,s)$ on the kth row and lth column. When viewed as a function of s for t fixed, say $t = T$ for the sake of the present discussion, $U(t,s)$ solves the adjoint equation

$$\frac{d}{ds}U(t,s) = U(t,s)[\overline{A} + \overline{B}P_s]$$

over the interval $[0,t]$ with terminal condition $U(t,t) = I_3$.

5.6. A Predatory Trading Game Model

Remark 5.24. *Plugging $Y_t = P_t X_t$ into (5.89) shows that the equilibrium rates of trading are in closed loop feedback form since they are affine functions of the state X_t. Accordingly, they are Gaussian processes.*

Remark 5.25. *It is reasonable to expect that, whenever c_1 and c_2 increase unboundedly toward $+\infty$, one should end up with $X_T^1 = X_T^2 = 0$ since the terminal inventory penalties become prohibitive, forcing traders not to hold any remaining inventory at time T. However, the random variables X_T^1 and X_T^2 cannot be deterministic for finite c_1 and c_2. Indeed, if X_T^2 is deterministic (for example), $v_T^2 = 0$ and consequently, $U(T,s)_{3,1} = 0$ for almost every $s \in [0,T]$, and in fact for every $s \in [0,T]$ by continuity of $s \hookrightarrow U(T,s)$. So, for every s,*

$$0 = \frac{d}{ds} U(T,s)_{3,1} = \left[U(T,s)[\overline{A} + \overline{B}P_s] \right]_{3,1}$$
$$= \lim_{s \nearrow T} \left[U(T,s)[\overline{A} + \overline{B}P_s] \right]_{3,1} = [\overline{A} + \overline{B}P_T]_{3,1} = -\frac{1}{3\lambda},$$

which is impossible. However, we shall show below in Subsection 5.6.3 that a direct analysis can be conducted in the case $X_T^1 = X_T^2 = 0$, which intuitively corresponds to the case $c_1 = c_2 = \infty$.

Some of the practical consequences of the above theoretical results are summarized in the following bullet points. They are illustrated graphically in Figures 5.2, 5.3, and 5.4.

- Figures 5.2 and 5.3 illustrate the fact that the prey will try to sell and lower its position, the predator trading first in the same direction to increase the price impact, and then in the opposite direction by buying back to take advantage of the lower prices, and to lessen his/her terminal inventory penalty. This expected behavior is quite obvious in the case of plastic markets ($\lambda << \gamma$) for which the permanent impact dominates the temporary impact.

- Figures 5.2 and 5.3 also point to the fact that the larger c_1 and c_2, the more likely it is that the traders will have little or no inventory at time $t = T$, and the more deterministic (and smoother) the paths of their positions.

- Figure 5.4 shows that in equilibrium, the distributions of the traders' positions are nontrivial and unimodal, and that they could very well be Gaussian as we argued in the text. The density estimates superimposed on the histograms confirm that larger values of c_1 and c_2 increase the inventory penalty, and lead to tighter distributions showing fewer fluctuations, especially at the end of the trading period.

The Case of the No-Buy Constraint

We now assume that the distressed trader is not allowed to buy the asset, or in other words, we require that the condition "$\alpha_t^2 \leq 0$ for $0 \leq t \leq T$" be added to the definition of the set $\mathbb{A}^2$ of admissible controls. In this case, the minimization of $H^1(t,x,y^2,\alpha)$ is done over the domain $\alpha^2 \in (-\infty, 0]$, and the minimizer must satisfy

$$\hat{\alpha}^2 = -\frac{1}{2\lambda} [\lambda \hat{\alpha}^1 + x^0 + \gamma y^{2,0} + y^{2,2}]^+ \tag{5.96}$$

instead of (5.87). If we define the domain

$$D = \{(x, y^1, y^2); \ x^0 + 2\gamma y^{1,0} + 2y^{1,1} - y^{2,2} - \gamma y^{2,0} > 0\}, \tag{5.97}$$

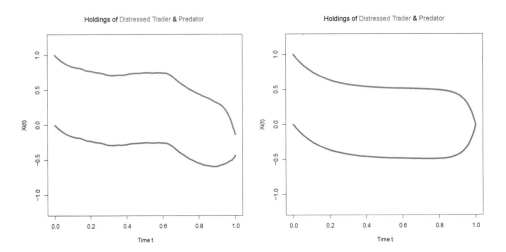

Figure 5.2. *Plots of the holdings X_t^1 (blue) and X_t^2 (red) of the predator and the prey in equilibrium in the case of a plastic market ($\lambda = 0.05\gamma$) with inventory penalty coefficients $c_1 = c_2 = 25$ (left), and stiff inventory penalty coefficients $c_1 = c_2 = 2500$ (right).*

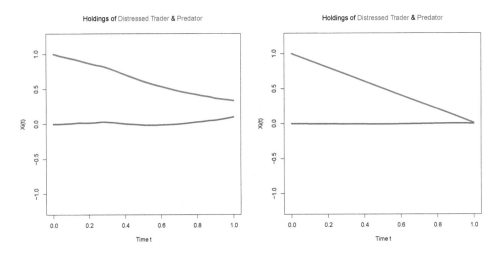

Figure 5.3. *Plots of the holdings X_t^1 (blue) and X_t^2 (red) of the predator and the prey in equilibrium in the case of an elastic market ($\lambda = 15\gamma$) with inventory penalty coefficients $c_1 = c_2 = 25$ (left), and $c_1 = c_2 = 2500$ (right).*

the function $(t, x, y^1, y^2) \hookrightarrow (\hat{\alpha}^1(t, x, y^1, y^2), \hat{\alpha}^2(t, x, y^1, y^2))$ satisfying the Isaacs condition is given by

$$\hat{\alpha}^1(t,x,y^1,y^2) = -\frac{1}{3\lambda}[x^0 + 2\gamma y^{1,0} + 2y^{1,1} - y^{2,2} - \gamma y^{2,0}]\mathbf{1}_D(x,y^1,y^2) \qquad (5.98)$$

and

$$\hat{\alpha}^2(t,x,y^1,y^2) = -\frac{1}{3\lambda}[x^0 - \gamma y^{1,0} - y^{1,1} + 2y^{2,2} + 2\gamma y^{2,0}]\mathbf{1}_D(x,y^1,y^2) \quad (5.99)$$

$$-\frac{1}{2\lambda}[x^0 + \gamma y^{2,0} + y^{2,2})\mathbf{1}_{D^c}(x,y^1,y^2)$$

5.6. A Predatory Trading Game Model

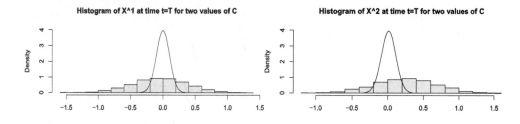

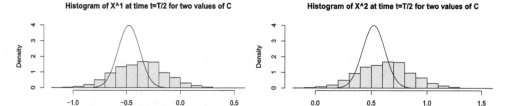

Figure 5.4. *Histograms (constructed from $N = 2500$ Monte Carlo samples) of X_t^1 (left) and X_t^2 (right) of the holdings in equilibrium, of the predator and the prey. The light blue histograms are computed from the samples of X_t^i half way for $t = T/2$ (bottom) and at the end for $t = T$ (top), the simulations being done for the value $c_1 = c_2 = 50$ of the inventory penalty. Over the top of each histogram, we superimposed an estimate of the density of the same quantities X_t^i from Monte Carlo simulations done with the value $c_1 = c_2 = 2500$.*

instead of (5.89). Plugging these equations into (5.91) gives an FBSDE which is no longer affine. We cannot solve it with the methods used earlier in the nonconstrained case.

5.6.3 ▪ Direct Analysis of a Constrained Model

In this subsection, we require that both traders liquidate their positions by time T. In this case, the set of admissible strategies $\mathbb{A}^1$ for player 1 (predator) is

$$\mathbb{A}^1 = \left\{ \boldsymbol{\alpha}^1 = (\alpha_t^1)_{0 \leq t \leq \tilde{T}} \in \mathcal{H}^2_{[0,\tilde{T}]}; \; X^1_{\tilde{T}} = x_0^1 + \int_0^{\tilde{T}} \alpha_t^1 dt = 0 \right\},$$

and similarly, the set of admissible strategies for the prey reads

$$\mathbb{A}^2 = \left\{ \boldsymbol{\alpha}^2 = (\alpha_t^1)_{0 \leq t \leq \tilde{T}} \in \mathcal{H}^2_{[0,\tilde{T}]}; \; X^2_T = x_0^2 + \int_0^T \alpha_t^2 dt = 0 \right\}.$$

The following computations mimic our analysis of the constrained optimal control solved explicitly in Chapter 3. Their goal is to rewrite the objective functions in the presence of the added constraints. In particular, $\boldsymbol{X}^1$ and $\boldsymbol{X}^2$ are in $\mathcal{H}^2_{[0,T]}$ whenever the controls $\boldsymbol{\alpha}^1$ and $\boldsymbol{\alpha}^2$ are admissible. We first rewrite the cost to the predator in a form which will turn out to be convenient for the optimization. Using integration by parts, the fact that $X^1_{\tilde{T}} = 0$, and the fact that $\boldsymbol{X}^1 \in \mathcal{H}^2_{[0,T]}$, we get

$$J^1(\boldsymbol{\alpha}^1, \boldsymbol{\alpha}^2) = \mathbb{E}\left[\int_0^T [X_t^0 + \lambda(\alpha_t^1 + \alpha_t^2)]\alpha_t^1 \, dt \right]$$
$$= \lambda \mathbb{E} \int_0^T (\alpha_t^1)^2 dt + \lambda \mathbb{E} \int_0^T \alpha_t^1 \alpha_t^2 dt + \mathbb{E} \int_0^T X_t^0 \alpha_t^1 dt$$
$$= (i) + (ii) + (iii).$$

Moreover,

$$(iii) = \mathbb{E}\int_0^T X_t^0 dX_t^1 = -x_0^0 x_0^1 - \mathbb{E}\int_0^T X_t^1 dX_t^0$$

$$= -x_0^0 x_0^1 - \gamma \mathbb{E}\int_0^T X_t^1 \alpha_t^1 dt - \gamma \mathbb{E}\int_0^T X_t^1 \alpha_t^2 dt$$

$$= -x_0^0 x_0^1 + \frac{\gamma}{2}(x_0^1)^2 + \gamma x_0^1 x_0^2 + \gamma \mathbb{E}\int_0^T X_t^2 \alpha_t^1 dt,$$

which implies that

$$J^1(\boldsymbol{\alpha}^1, \boldsymbol{\alpha}^2) = \lambda \mathbb{E}\int_0^T (\alpha_t^1)^2 dt + \mathbb{E}\int_0^T \alpha_t^1 \beta_t^2 dt + x_0^1\left(\frac{\gamma}{2}x_0^1 - x_0^0\right) + \gamma x_0^1 x_0^2$$

for

$$\beta_t^2 = \lambda \alpha_t^2 + \gamma\left(x_0^2 + \int_0^t \alpha_s^2 ds\right). \tag{5.100}$$

Using the notation $\langle \cdot, \cdot \rangle$ for the inner product of the Hilbert space $\mathcal{H}^2_{[0,T]}$, we have

$$J^1(\boldsymbol{\alpha}^1, \boldsymbol{\alpha}^2) = \lambda\langle \boldsymbol{\alpha}^1, \boldsymbol{\alpha}^1\rangle + \langle \boldsymbol{\beta}^2, \boldsymbol{\alpha}^1\rangle + x_0^1\left(\frac{\gamma}{2}x_0^1 - x_0^0\right) + \gamma x_0^1 x_0^2,$$

which proves that, for $\boldsymbol{\alpha}^2$ fixed, $\boldsymbol{\alpha} \hookrightarrow J^1(\boldsymbol{\alpha}, \boldsymbol{\alpha}^2)$ is strictly convex and will have a unique minimum (if any) over the convex set $\mathbb{A}^1$. With the goal of constructing the best response to $\boldsymbol{\alpha}^2$, we derive a necessary condition (as a function of $\boldsymbol{\alpha}^2$) for this minimum by introducing a Lagrange multiplier θ as a square integrable random variable, and minimizing the Lagrangian

$$L(\boldsymbol{\alpha}, \theta) = J^1(\boldsymbol{\alpha}, \boldsymbol{\alpha}^2) - \theta(\langle \mathbf{1}, \boldsymbol{\alpha}\rangle - x_0^1),$$

where $\mathbf{1}$ denotes the function identically equal to 1 on the interval $[0, \tilde{T}]$. The first-order conditions give

$$\theta = 2\lambda \alpha_t + \beta_t^2 \quad \text{or in other words} \quad \alpha_t = \frac{1}{2\lambda}[\theta - \beta_t^2].$$

Choosing θ in order to satisfy the constraint gives

$$\theta = \frac{1}{\tilde{T}}\int_0^{\tilde{T}} \beta_t^2 dt - \frac{2\lambda}{\tilde{T}}x_0^1,$$

and finally we get

$$\hat{\alpha}_t^1 = -\frac{1}{2\lambda}\left[\beta_t^2 - \frac{1}{\tilde{T}}\int_0^{\tilde{T}} \beta_s^2 ds\right] - \frac{1}{\tilde{T}}x_0^1 \tag{5.101}$$

for the best response to $\boldsymbol{\alpha}^2$. Similarly, we rewrite the cost to the distressed player as

$$J^2(\boldsymbol{\alpha}^1, \boldsymbol{\alpha}^2) = \mathbb{E}\left[\int_0^T [X_t^0 + \lambda(\alpha_t^1 + \alpha_t^2)]\alpha_t^2 \, dt\right]$$

$$= \lambda \mathbb{E}\int_0^T (\alpha_t^2)^2 dt + \lambda \mathbb{E}\int_0^T \alpha_t^1 \alpha_t^2 dt + \mathbb{E}\int_0^T X_t^0 \alpha_t^2 dt$$

$$= (i) + (ii) + (iii).$$

5.6. A Predatory Trading Game Model

Using integration by parts, the fact that $X_T^2 = 0$, and the fact that $\boldsymbol{X}^2 \in \mathcal{H}^2_{[0,T]}$, we get

$$(iii) = \mathbb{E}\int_0^{\tilde T} X_t^0 dX_t^2 = \mathbb{E}[X_T^0 X_T^2 - X_0^0 X_0^2] - \mathbb{E}\int_0^T X_t^2 dX_t^0$$

$$= -x_0^0 x_0^2 - \gamma \mathbb{E}\int_0^T X_t^2 \alpha_t^1 dt - \gamma \mathbb{E}\int_0^T X_t^2 dX_t^2$$

$$= x_0^2\left(\frac{\gamma}{2} x_0^2 - x_0^0\right) - \gamma \mathbb{E}\int_0^T X_t^2 dX_t^1$$

$$= x_0^2\left(\frac{\gamma}{2} x_0^2 - x_0^0\right) + \gamma x_0^2 x_0^1 + \gamma \mathbb{E}\int_0^T X_t^1 \alpha_t^2 dt.$$

Consequently,

$$J^2(\boldsymbol{\alpha}^1, \boldsymbol{\alpha}^2) = \lambda \mathbb{E}\int_0^T (\alpha_t^2)^2 dt + \mathbb{E}\int_0^T \beta_t^1 \alpha_t^2 dt + x_0^2\left(\frac{\gamma}{2} x_0^2 - x_0^0\right) + \gamma x_0^2 x_0^1 \quad (5.102)$$

with

$$\beta_t^1 = \lambda \alpha_t^1 + \gamma\left(x_0^1 + \int_0^t \alpha_s^1 ds\right). \quad (5.103)$$

With the same notation $\langle \cdot, \cdot \rangle$ as before, we have

$$J^2(\boldsymbol{\alpha}^1, \boldsymbol{\alpha}^2) = \lambda \langle \boldsymbol{\alpha}^2, \boldsymbol{\alpha}^2\rangle + \langle \boldsymbol{\beta}^1, \boldsymbol{\alpha}^2\rangle + x_0^2\left(\frac{\gamma}{2} x_0^2 - x_0^0\right) + \gamma x_0^2 x_0^1,$$

which proves that, for $\boldsymbol{\alpha}^1$ fixed, $\boldsymbol{\alpha} \hookrightarrow J^2(\boldsymbol{\alpha}^1, \boldsymbol{\alpha})$ is strictly convex and will have a unique minimum (if any) over the convex set $\mathbb{A}^2$. Trying to find the best response to $\boldsymbol{\alpha}^1$, we introduce as before a Lagrange multiplier θ as a square integrable random variable, and minimize the Lagrangian

$$L(\boldsymbol{\alpha}, \theta) = J^2(\boldsymbol{\alpha}^1, \boldsymbol{\alpha}) - \theta(\langle \boldsymbol{1}, \boldsymbol{\alpha}\rangle - x_0^2).$$

The first-order conditions give

$$\theta = 2\lambda \alpha_t^2 + \beta_t^1 \quad \text{or in other words} \quad \alpha_t^2 = \frac{1}{2\lambda}[\theta - \beta_t^1].$$

Choosing θ in order to satisfy the constraint gives

$$\theta = \frac{1}{T}\int_0^T \beta_t^1 dt - \frac{2\lambda}{T} x_0^2,$$

and finally we find that

$$\hat\alpha_t^2 = -\frac{1}{2\lambda}\left[\beta_t^1 - \frac{1}{T}\int_0^T \beta_s^1 ds\right] - \frac{1}{T} x_0^2 \quad (5.104)$$

is the best response to $\boldsymbol{\alpha}^1$.

Plugging the expressions (5.103) and (5.100) of β_t^1 and β_t^2 into the formulas (5.104) and (5.101), giving $\hat\alpha_t^1$ and $\hat\alpha_t^2$, we get the following expression for the best response map $(\boldsymbol{\alpha}^1, \boldsymbol{\alpha}^2) \hookrightarrow (\hat{\boldsymbol{\alpha}}^1, \hat{\boldsymbol{\alpha}}^2)$:

$$\begin{cases} \hat\alpha_t^1 = -\frac{1}{2}\alpha_t^2 - \frac{\gamma}{2\lambda}\left(X_t^2 - \frac{1}{T}\int_0^T X_s^2 ds\right) - \frac{1}{2T} x_0^1 - \frac{1}{2T}(x_0^1 + x_0^2), \\ \hat\alpha_t^2 = -\frac{1}{2}\alpha_t^1 - \frac{\gamma}{2\lambda}\left(X_t^1 - \frac{1}{T}\int_0^T X_s^1 ds\right) - \frac{1}{2T} x_0^2 - \frac{1}{2T}(x_0^1 + x_0^2). \end{cases} \quad (5.105)$$

The analogs of the Isaacs conditions are obtained by stating that (α^1, α^2) is a fixed point for the above best response function. To solve for such a fixed point, we set

$$Y_t^1 = X_t^1 - \frac{1}{T}\int_0^T X_s^1 ds \quad \text{and} \quad Y_t^2 = X_t^2 - \frac{1}{T}\int_0^T X_s^2 ds$$

so that the fixed-point conditions become

$$\begin{cases} \dot{Y}_t^1 + \frac{1}{2}\dot{Y}_t^2 = -\frac{\gamma}{2\lambda}Y_t^2 - \frac{1}{2T}x_0^1 - \frac{1}{2T}(x_0^1 + x_0^2), \\ \frac{1}{2}\dot{Y}_t^1 + \dot{Y}_t^2 = -\frac{\gamma}{2\lambda}Y_t^1 - \frac{1}{2T}x_0^2 - \frac{1}{2T}(x_0^1 + x_0^2). \end{cases} \quad (5.106)$$

Adding these two conditions gives an ordinary differential equation for $Y_t^1 + Y_t^2$:

$$\frac{d}{dt}(Y_t^1 + Y_t^2) = -\frac{\gamma}{3\lambda}(Y_t^1 + Y_t^2) - \frac{3}{2T}(x_0^1 + x_0^2),$$

from which we get

$$Y_t^1 + Y_t^2 = -(x_0^1 + x_0^2)\left(e^{-\gamma T/(3\lambda)} - 1\right)^{-1} e^{-\gamma t/(3\lambda)} - \frac{3\lambda}{\gamma T}(x_0^1 + x_0^2), \quad (5.107)$$

since $Y_T^1 + Y_T^2 - (Y_0^1 + Y_0^2) = X_T^1 - X_0^1 + X_T^2 - X_0^2 = -(x_0^1 + x_0^2)$. Similarly, if we subtract the second condition in (5.106) from the first one, we get an ordinary differential equation for $Y_t^1 - Y_t^2$:

$$\frac{d}{dt}(Y_t^1 - Y_t^2) = \frac{\gamma}{\lambda}(Y_t^1 - Y_t^2) + \frac{1}{T}(x_0^2 - x_0^1)$$

from which we get

$$Y_t^1 - Y_t^2 = (x_0^2 - x_0^1)\left(e^{-\gamma T/\lambda} - 1\right)^{-1} e^{\gamma t/\lambda} + \frac{\lambda}{\gamma T}(x_0^2 - x_0^1), \quad (5.108)$$

since $Y_T^1 - Y_T^2 - (Y_0^1 - Y_0^2) = X_T^1 - X_0^1 - (X_T^2 - X_0^2) = x_0^2 - x_0^1$. Now, from (5.107) and (5.108), we derive expressions for Y_t^1 and Y_t^2, and after differentiating these expressions with respect to t, we finally get

$$\begin{cases} \hat{\alpha}_t^1 = ae^{-\gamma t/(3\lambda)} + be^{\gamma t/\lambda}, \\ \hat{\alpha}_t^2 = ae^{-\gamma t/(3\lambda)} - be^{\gamma t/\lambda} \end{cases} \quad (5.109)$$

with

$$a = -\frac{\gamma}{3\lambda}\left(1 - e^{-\gamma T/(3\lambda)}\right)^{-1}\frac{x_0^1 + x_0^2}{2}, \quad (5.110)$$

$$b = \frac{\gamma}{\lambda}\left(e^{\gamma T/\lambda} - 1\right)^{-1}\frac{x_0^2 - x_0^1}{2}. \quad (5.111)$$

Notice that both open loop equilibrium controls $\hat{\alpha}_t^1$ and $\hat{\alpha}_t^2$ are deterministic!

5.6.4 ▪ Search for Markovian Nash Equilibriums

In this subsection, we assume as before that the prey needs to liquidate his/her initial position x_0^2 by the end of the period $[0, T]$, while the predator can still trade over the period $[T, \tilde{T}]$, and we search for Markovian Nash equilibriums. Assuming that each agent can see the transaction price, his/her own position (obviously), but also the other player's position (this is highly unrealistic in many cases, but bear with me for a moment), the search for a Markovian Nash equilibrium in closed loop form reduces to the search for two functions ϕ^1 and ϕ^2 on $[0, T] \times [0, \infty) \times [0, x_0] \times \mathbb{R}$ and $[0, \tilde{T}] \times [0, \infty) \times [0, x_0] \times \mathbb{R}$ such that, if the dynamics of the state $\boldsymbol{X}_t$ of the system are given by

$$\begin{cases} dX_t^0 = \gamma\phi(t, \boldsymbol{X}_t)dt + \gamma\phi^2(t, \boldsymbol{X}_t)dt + \sigma dW_t, \\ dX_t^1 = \phi(t, \boldsymbol{X}_t)dt, \\ dX_t^2 = \phi^2(t, \boldsymbol{X}_t)dt \end{cases} \quad (5.112)$$

over the period $[0, T]$, then

$$\phi^1 \in \arg\inf_\phi J^1(\phi, \phi^2),$$

where we use the obvious notation $J^1(\phi, \phi^2)$ for $J^1(\boldsymbol{\alpha}, \boldsymbol{\alpha}^2)$ for the Markov controls $\alpha_t = \phi(t, \boldsymbol{X}_t)$ and $\alpha_t^2 = \phi^2(t, \boldsymbol{X}_t)$. So the Markov strategy ϕ^1 of the distressed player appears as the Markov optimal control for the Markovian optimal stochastic control problem (remember that we momentarily assume that the function ϕ^2 is given)

$$\inf_\phi \mathbb{E}\left[\int_0^T \left(\lambda\phi(t, \boldsymbol{X}_t)^2 + (X_t^0 + \lambda\phi^2(t, \boldsymbol{X}))\phi(t, \boldsymbol{X}_t)\right)dt\right],$$

which depends upon the choice of ϕ^2. The operator symbol reads

$$L^1(t, \mathbf{x}, y, \mathbf{z}, \alpha) = \frac{\sigma^2}{2}z^{00} + \gamma[\phi^2(t, \mathbf{x}) + \alpha]y^0 + \alpha y^1 + \phi^2(t, \mathbf{x})y^2 + \alpha(x_0 + \lambda(\alpha + \phi^2(t, \mathbf{x})), \quad (5.113)$$

and because this expression is quadratic in α, one can perform the minimization over α explicitly. We consider the two models identified earlier.

The objective function of the predator over the first period $[0, T]$ is

$$J^2(\boldsymbol{\alpha}) = \mathbb{E}\left[\int_0^T f^2(t, \boldsymbol{X}_t, \alpha_t)dt + g(\boldsymbol{X}_T)\right],$$

where the running cost function f^2 is defined by

$$f^2(t, \mathbf{x}, \alpha) = \alpha^2(x^0 + \lambda(\alpha^1 + \alpha^2))$$

and the terminal cost function g can be interpreted as a *scrap value*. We shall not discuss the choice of the function g here.

If we assume that the minimization is unconstrained, the minimum is attained for

$$\alpha^* = -\frac{1}{2\lambda}\left(x_0 + \lambda\phi^2(t, \mathbf{x}) + \gamma y_0 + y_1\right), \quad (5.114)$$

and plugging this value back into (5.113), we obtain for the corresponding minimized operator symbol

$$L^{1*}(t, \mathbf{x}, y, \mathbf{z}) = \frac{\sigma^2}{2}z_{0,0} + \gamma\phi^2(t, \mathbf{x})y^0 + \phi^2(t, \mathbf{x})y_2 - \frac{1}{4\lambda}[x_0 + \lambda\phi^2(t, \mathbf{x}) + \gamma y_0 + y_1]^2.$$

Consequently, the HJB value function V^1 for this stochastic control problem reads as

$$\partial_t V^1(t,\mathbf{x}) + \frac{\sigma^2}{2}\partial^2_{x_0 x_0} V^1(t,\mathbf{x}) + \gamma\phi^2(t,\mathbf{x})\partial_{x_0} V^1(t,\mathbf{x}) \tag{5.115}$$
$$+ \phi^2(t,\mathbf{x})\partial_{x_2} V^1(t,\mathbf{x}) - \frac{1}{4\lambda}[x_0 + \lambda\phi^2(t,\mathbf{x}) + \gamma\partial_{x_0} V^1(t,\mathbf{x}) + \partial_{x_1} V^1(t,\mathbf{x})]^2.$$

with terminal condition $V^1(T,\mathbf{x}) \equiv 0$.

5.7 • Notes & Complements

The formulation of the Isaacs condition as given in Definition 5.8 is credited to Isaacs in the case of two-player ($N = 2$) zero-sum games, and to Friedman in the general case of noncooperative N-player games. Generalizations of the stochastic maximum principle to games with stochastic dynamics including jumps have been considered by several authors. We refer the interested reader to [101] and the references therein.

Our discussion of the zero-sum games in the text is minimal. Despite the fact that most of the mathematical technology used to study these games is specific to this special form of model, and usually does not extend to more general game setups, we wanted to introduce them for the sake of completeness. The published literature on zero-sum stochastic differential games is vast. It is mostly focused on the existence of the value of the game and the existence of a Nash equilibrium in the form of a saddle point. The existence of the value of the game is often proven by showing first that the lower and upper values of the game can be defined as functions of t and x, and as such, appear as upper and lower viscosity solutions of a specific PDE. Appealing to comparison theorems from the theory of viscosity solutions of PDEs is the usual way to conclude the proof of the existence of the value of the game. The reader interested in learning more about the subject may consult [59], [24], [29, 30], [66], and [52]. The example used to warn about the differences between open and closed loop equilibriums is due to Buckdahn. We learned it from Jianfeng Zhang; see [107].

In the notation of the text, deterministic LQ games correspond to $C_t \equiv 0$, $D_t \equiv 0$, and $\gamma_t \equiv 0$. They were studied by Friedman [61], Lukes and Russel [88], Bensoussan [14], and Eisele [51]. Hamadène constructed an open loop Nash equilibrium in the case of volatility independent of the control in [65], and Peng and Wu generalized his result in [105].

Our presentation of the systemic risk toy model is based on the paper by Carmona, Fouque, and Sun [39]. Clearly, our interest in this model is in the fact that its open and closed loop forms can be solved explicitly, and the large game limit appears effortlessly. So in this sense, it is a perfect introduction to the discussion of mean-field games (MFGs), hence our decision to present it in full detail, despite its shortcomings when viewed as a model for systemic risk in an interbank system. Indeed, according to the model, banks can borrow from each other without having to repay their debts, and without the liabilities of the banks being included in the model, the rationale for the threshold crossing bankruptcy is rather weak.

The semilinear LQ game model discussed in the text was first studied by Bensoussan and Frehse in the Markovian case of deterministic coefficients [16], and generalized to nonanticipative coefficients (possibly depending upon the past of the random scenario $\omega \in \Omega$) by Lepeltier, Wu, and Yu in [83].

The predatory trading model used in the text is based on the paper of Carlin, Lobo, and Viswanathan [31]. These authors extend to the multiplayer game setting an earlier model for optimal execution proposed by Bertsimas and Lo [21], Almgren and Chriss [5], and Almgren [4]. They consider only deterministic strategy profiles and derived the Nash equilibrium given by (5.109), (5.110) and (5.111). The linear structure of the problem makes it possible to show that this set of deterministic controls is also an equilibrium for the open loop problem as long as the constraints $X^1_T = X^2_T = 0$ are imposed. The following argument can be used to prove this fact. We learned of it from a private conversation with A. Schied. We define the corresponding sets of admissible strategies $\mathbb{A}^1_d$ and $\mathbb{A}^2_d$ by

$$\mathbb{A}^1_d = \left\{ \boldsymbol{\alpha}^1 = (\alpha^1_t)_{0 \leq t \leq \tilde{T}}; \text{ deterministic}, X^1_{\tilde{T}} = x^1_0 + \int_0^{\tilde{T}} \alpha^1_t dt = 0 \right\}$$

5.7. Notes & Complements

and

$$\mathbb{A}_d^2 = \left\{ \boldsymbol{\alpha}^2 = (\alpha_t^1)_{0 \leq t \leq \tilde{T}}; \text{ deterministic, } X_T^2 = x_0^2 + \int_0^T \alpha_t^2 dt = 0 \text{ and } \alpha_t^2 \equiv 0 \text{ for } T \leq t \leq \tilde{T} \right\}.$$

This argument shows that if $(\boldsymbol{\alpha}^1, \boldsymbol{\alpha}^2) \in \mathbb{A}^1 \times \mathbb{A}^2$ is a deterministic Nash equilibrium in $\mathbb{A}_d = \mathbb{A}_d^1 \times \mathbb{A}_d^2$ among the deterministic admissible strategies, then it is also an open loop Nash equilibrium. Indeed, if we fix $\boldsymbol{\alpha}^1$, then for any $\boldsymbol{\alpha} \in \mathbb{A}^2$, using (5.102), Jensen's inequality, and the fact that α_t^1 and X_t^1 and β^1 are deterministic, we get

$$J^2(\boldsymbol{\alpha}^1, \boldsymbol{\alpha}) = \lambda \mathbb{E} \int_0^{\tilde{T}} (\alpha_t)^2 dt + \lambda \mathbb{E} \int_0^{\tilde{T}} \beta_t^1 \alpha_t dt + x_0^2 \left(\frac{\gamma}{2} x_0^2 - x_0^0 \right) + \gamma x_0^2 x_0^1$$

$$\geq \lambda \int_0^T (\overline{\alpha}_t)^2 dt + \int_0^{\tilde{T}} \beta_t^1 \alpha_t dt + x_0^2 \left(\frac{\gamma}{2} x_0^2 - x_0^0 \right) + \gamma x_0^2 x_0^1 = J^2(\boldsymbol{\alpha}^1, \overline{\boldsymbol{\alpha}}),$$

where we used the notation $\overline{\alpha}_t = \mathbb{E}[\alpha_t]$ for the expectation of α_t. Notice that $\overline{\boldsymbol{\alpha}}$ is deterministic and admissible, so using the fact that $(\boldsymbol{\alpha}^1, \boldsymbol{\alpha}^2)$ is a Nash equilibrium among the deterministic admissible strategies, the above inequality gives

$$J^2(\boldsymbol{\alpha}^1, \boldsymbol{\alpha}) \geq J^2(\boldsymbol{\alpha}^1, \overline{\boldsymbol{\alpha}}) \geq J^2(\boldsymbol{\alpha}^1, \boldsymbol{\alpha}^2),$$

proving that $\boldsymbol{\alpha}^2$ is the *best response* to $\boldsymbol{\alpha}^1$, even in the larger set of nondeterministic adapted controls in $\mathbb{A}^2$. Similarly, one shows that $\boldsymbol{\alpha}^1$ is the *best response* to $\boldsymbol{\alpha}^2$ in $\mathbb{A}^1$, concluding the proof that $(\boldsymbol{\alpha}^1, \boldsymbol{\alpha}^2)$ is also the unique open loop Nash equilibrium in the constrained set $\mathbb{A}^1 \times \mathbb{A}^2$.

The extension to the two-period model in which the prey is not allowed to trade during the second period can be found in a paper by Schöneborn and Schied [111]. The short introduction to the problem of existence and characterization of Markovian Nash equilibriums with control strategies in closed loop form is based on a paper by Carmona and Yang [41] in which details can be found.

Chapter 6
Mean-Field Games

This chapter is devoted to a soft introduction to a special class of large games called mean-field games. In these games, the individual players have statistically similar behaviors, and their mutual interactions are through average quantities, hence the terminology *mean-field*, borrowed from statistical physics. The goal of the theory is to derive effective equations for the optimal behavior of any single player when the size of the population grows unboundedly. Despite the fact that the formulation and the solution of these models require rather restrictive assumptions, practical applications abound, and we describe some of them, even if we do not always have complete answers to the mathematical and numerical challenges they raise.

Most of the tools presented in the previous chapters can be brought to bear in the analysis of these game models. We use them to highlight a probabilistic approach to the difficult problem of existence of Nash equilibriums. We also present uniqueness results.

6.1 ▪ Introduction and First Definitions

While mean-field games have often been presented in the context of models with a continuum of players, our presentation will focus on games with a finite, though large, number of players, the models with a continuum of players being viewed as a mathematical abstraction of the limit of the game models when the number of players tends to infinity.

6.1.1 ▪ Games with Mean-Field Interactions

In this subsection, we recall some of the notation introduced in Section 1.3 of Chapter 1 for the analysis of SDEs of McKean–Vlasov type, their particle approximations, and the propagation of chaos. In the present context, we consider a special class of N-player stochastic differential games satisfying a strong symmetry property (forcing the players to be statistically identical), in which each player's influence on the whole system vanishes as the number of players grows unboundedly. Mathematically, this last assumption forces the interaction between the players to occur through empirical distribution of the individual states of the players. As a result, the mathematical formalism will borrow from the analysis of stochastic differential equations of McKean–Vlasov type (recall Section 1.3) and mean-field BSDEs (recall Section 2.2).

As a result of the desire for the players to be (statistically) identical, we assume that

$$d_1 = d_2 = \cdots = d_N$$

and we denote this common dimension by d_0 so that $d = N d_0$. Also, we assume that

$$A^1 = A^2 = \cdots = A^N,$$

whose common value we denote A^0. A simple way to have the drifts of the state dynamics (as well as the volatility or the running costs) satisfy the symmetry condition would be to assume that they are of the form

$$b^i(t,(x^1,\cdots,x^N),(\alpha^1,\cdots,\alpha^N)) = \frac{1}{N}\sum_{j=1}^N \hat{b}(t,x^i,x^j,\alpha^i_t) = \int \hat{b}(t,x^i,y,\alpha^i)d\mu_x^N(dy) \tag{6.1}$$

where, as before, we denote by μ_x^N the empirical distribution of $x = (x^1,\cdots,x^N)$ defined as the probability measure

$$\mu_x^N(dx') = \frac{1}{N}\sum_{i=1}^N \delta_{x^i}(dx').$$

As always, we use the notation δ_x for the unit mass (Dirac measure) at x. Here,

$$\hat{b}: [0,T] \times \mathbb{R}^{d_0} \times \mathbb{R}^{d_0} \times A_0 \hookrightarrow \mathbb{R}^{d_0}$$

is assumed to be deterministic, and to satisfy regularity assumptions which will determine the properties of the coefficients b^i in (6.1), and eventually the properties we require of the general drift b as given in (6.2). In general mean-field game models, it is assumed that the drift b^i, the volatility σ^i of the stochastic dynamics of the private state of player i, and the corresponding running cost function f^i of the ith player are of the form

$$(b^i, \sigma^i, f^i)(t, X_t, \alpha_t) = (b^0, \sigma^0, f^0)(t, X^i_t, \mu^N_{X_t}, \alpha^i_t) \tag{6.2}$$

for some deterministic function (b^0, σ^0, f^0) defined on $[0,T] \times \mathbb{R}^{d_0} \times \mathcal{P}(\mathbb{R}^{d_0}) \times A^0$ and with values in $\mathbb{R}^{d_0} \times \mathbb{R}^{d_0 \times m_0} \times \mathbb{R}$. For most mean-field game models, namely those without a common noise, we assume that all the Wiener processes are independent and have the same dimensions $m_1 = \cdots = m_N$, say m_0. Moreover, in most of the examples we consider, we assume that the volatilities are uncontrolled in the sense that $\sigma^i(t,x,\alpha) = \sigma(t,x^i)$ for some *nice* function $\sigma : [0,T] \times \mathbb{R}^{d_0} \hookrightarrow \mathbb{R}^{d_0 \times m_0}$. In fact, most often, this common volatility function will be assumed to be constant for the sake of simplicity.

6.1.2 ▪ Mean-Field Games in a Random Environment

In addition to the assumptions of the introductory subsection above, we now assume the existence of a mean-zero Gaussian measure $\boldsymbol{W}^0 = (W^0(A,B))_{A,B}$ parameterized by the Borel subsets of a space Ξ and $[0,T]$ such that

$$\mathbb{E}[W^0(A,B)W^0(A',B')] = \nu(A \cap A')|B \cap B'|,$$

where we used the notation $|B|$ for the Lebesgue measure of a set. Here ν is a nonnegative measure on Ξ, called the spatial intensity of $\boldsymbol{W}$. Most often, we shall take $\Xi = \mathbb{R}^d$. We now assume that the dynamics of the private state of player $i \in \{1,\cdots,N\}$ are given by

$$dX^i_t = b(t, X^i_t, \overline{\mu}^N_t, \alpha^i_t)dt + \sigma(t, X^i_t, \overline{\mu}^N_t, \alpha^i_t)dW^i_t + \int_\Xi c(t, X^i_t, \overline{\mu}^N_t, \alpha^i_t, \xi)W^0(d\xi, dt)$$

for some $\mathbb{R}^d$-valued function c defined on $[0,T] \times \mathbb{R}^d \times \mathcal{P}_2(\mathbb{R}^d) \times A^i \times \mathbb{R}^d$. We shall often refer to $\boldsymbol{W}^0$ as a common noise since it affects the private states of all the players. We can think of $W^0(d\xi, dt)$ as a common random noise which is white in time (to provide the time derivative of a Brownian motion) and colored in space (the spectrum of the color being given by the Fourier transform of ν), and the motivating example we should keep in mind is a function c of the form $c(t, x, \mu, \alpha, \xi) \sim c(t, x, \mu) \delta(x - \xi)$ (or a mollified version of the delta function), in which case the integration with respect to the random measure $\boldsymbol{W}^0$ gives

$$\int_{\mathbb{R}^d} c(X_t^i, \overline{\mu}_t^N, \alpha_t^i, \xi) W^0(d\xi, dt) = c(t, X_t^i, \overline{\mu}_t^N) W^0(X_t^i, dt).$$

This formula says that at time t, the private state of player i is subject to several sources of random shocks: its own idiosyncratic noise W_t^i, but also an extra shock at the value of the private state. This setup is especially realistic in the case of the Cucker–Smale flocking model presented in Subsection 6.4.2. Indeed, in that model, $\boldsymbol{W}^0$ represents the noise of the environment in which the birds are flying. The same applies to models of a school of fish or pedestrian crowds. It will also be used in the discussion of games with major players in Subsection 6.4.4, and the systemic risk toy model revisited in Subsection 6.4.1 below.

The simplest example of a random environment corresponds to a coefficient c independent of ξ. In this case, the random measure W^0 may as well be independent of the spatial component. In other words, we can assume that $W^0(d\xi, dt) = W^0(dt)$ is independent of the space location ξ, and represents an extra noise term *common* to all the players. In this case, $W^0(dt)$ can be viewed as the Itô differential dW_t^0 of a Wiener process $\boldsymbol{W}^0$ independent of the idiosyncratic Wiener processes $\boldsymbol{W}^i$ for $i = 1, \cdots, N$.

6.1.3 • Asymptotics of the Empirical Distribution $\overline{\mu}_t^N$

The rationale for the MFG approach to the search for approximate Nash equilibriums is based on the asymptotic behavior as $N \to \infty$ of the empirical distribution $\overline{\mu}_t^N$ of $X_t = (X_t^1, \cdots, X_t^N)$ appearing in the state dynamics. We give an informal derivation of a characterization of this limiting empirical distribution where the players use distributed Markovian strategy profiles. This means that the admissible controls α_t^i of player i should be given by deterministic functions of time and the states of the players. Moreover, the symmetry requirement suggests that all these functions should be the same. In other words, we assume that $\alpha_t^i = \alpha(t, X_t^i)$ for some deterministic function $(t, x) \hookrightarrow \alpha(t, x)$.

In order to understand this limiting behavior, we fix a smooth test function ϕ with compact support in $\mathbb{R}^d$, and using Itô's formula we compute

$$d\left\langle \phi, \frac{1}{N} \sum_{j=1}^N \delta_{X_t^j} \right\rangle = \frac{1}{N} \sum_{j=1}^N d\phi(X_t^j)$$

$$= \frac{1}{N} \sum_{j=1}^N \left(\nabla \phi(X_t^j) dX_t^j + \frac{1}{2} \text{trace}\{\nabla^2 \phi(X_t^j) d[X^j, X^j]_t\} \right)$$

$$= \frac{1}{N} \sum_{j=1}^N \nabla \phi(X_t^j) \sigma(t, X_t^j, \overline{\mu}_t^N, \alpha(t, X_t^j)) dW_t^j$$

$$+ \frac{1}{N} \sum_{j=1}^N \nabla \phi(X_t^j) b(t, X_t^j, \overline{\mu}_t^N, \alpha(t, X_t^j)) dt$$

$$+ \frac{1}{N}\sum_{j=1}^{N}\nabla\phi(X_t^j)\int_{\Xi} c(t, X_t^j, \overline{\mu}_t^N, \alpha(t, X_t^j), \xi)W^0(d\xi, dt)$$

$$+ \frac{1}{N}\sum_{j=1}^{N}\text{trace}\bigg\{\bigg(\frac{1}{2}[\sigma\sigma^\dagger](t, X_t^j, \overline{\mu}_t^N, \alpha(t, X_t^j))$$

$$+ \int_{\Xi}[cc^\dagger](t, X_t^j, \overline{\mu}_t^N, \alpha(t, X_t^j), \xi)\nu(d\xi)\bigg)\nabla^2\phi(t, X_t^j)\bigg\}dt.$$

Our goal is to take the limit as $N \to \infty$ in this expression. Using the definition of the measures $\overline{\mu}_t^N$, we can rewrite the above equality as

$$d\langle\phi, \overline{\mu}_t^N\rangle = O(N^{-1/2}) + \langle(\nabla\phi)b(t, \cdot, \overline{\mu}_t^N, \alpha(t, \cdot)), \overline{\mu}_t^N\rangle dt$$

$$+ \bigg\langle(\nabla\phi)\int_{\Xi}c(t, \cdot, \overline{\mu}_t^N, \alpha(t, \cdot), \xi)W^0(d\xi, dt), \overline{\mu}_t^N\bigg\rangle$$

$$\bigg\langle\text{trace}\bigg\{\bigg(\frac{1}{2}[\sigma\sigma^\dagger](t, \cdot, \overline{\mu}_t^N, \alpha(t, \cdot)) + \int_{\Xi}[cc^\dagger](t, \cdot, \overline{\mu}_t^N, \alpha(t, \cdot), \xi)\nu(d\xi)\bigg)\nabla^2\phi\bigg\}, \overline{\mu}_t^N\bigg\rangle dt,$$

which shows that, by integration by parts, in the limit $N \to \infty$,

$$\mu_t = \lim_{N\to\infty}\overline{\mu}_t^N$$

appears as the solution of the stochastic partial differential equation (SPDE)

$$d\mu_t = -\nabla[b(t, \cdot, \mu_t, \alpha(t, \cdot))\mu_t]dt - \nabla\bigg(\int_{\Xi}c(t, \cdot, \overline{\mu}_t^N, \alpha(t, \cdot), \xi)W^0(d\xi, dt)\mu_t\bigg) \quad (6.3)$$

$$+ \text{trace}\bigg[\nabla^2\bigg(\frac{1}{2}[\sigma\sigma^\dagger](t, \cdot, \mu_t, \alpha(t, \cdot)) + \int_{\Xi}[cc^\dagger](t, \cdot, \mu_t, \alpha(t, \cdot), \xi)\nu(d\xi)\bigg)\mu_t\bigg]dt.$$

Note that, when the ambient noise is not present (i.e., either $c \equiv 0$ or $\boldsymbol{W}^0 = 0$), then this SPDE reduces to a deterministic PDE, and that this PDE is the Kolmogorov equation giving the forward dynamics of the distribution at time t of the diffusion process with drift b and volatility σ.

6.1.4 ▪ Solution Strategy for Mean-Field Games

In the search for a Nash equilibrium α, one assumes that all the players j but one keep the same strategy profile α, and the remaining player deviates from this strategy in the hope of being better off. If the number of players is large (think $N \to \infty$), one expects that the empirical measure $\overline{\mu}_t^N$ will not be much affected by the deviation of one single player, and for all practical purposes, one should be able to assume that the empirical measure $\overline{\mu}_t^N$ is approximately equal to its limit μ_t. If one remembers the definition of the best response map, in the case of large symmetric games, the search for approximate Nash equilibriums could be done through the solution of the optimization problem of one single player interacting with the other players through the empirical distribution $\overline{\mu}_t^N$. In the asymptotic regime of large games, the empirical measures $\overline{\mu}_t^N$ can be replaced by the solution μ_t of the SPDE (6.3). Effective equations appear in this limiting regime, and if we can implement a consistency condition corresponding to the fixed-point argument following the construction of the best response map in the construction of Nash equilibriums, then we should be endowed with approximate Nash equilibriums. Indeed, we shall show that, assigning to

6.1. Introduction and First Definitions

each player the strategy $\boldsymbol{\alpha}$, provided by the solution of the effective equation in the limit $N \to \infty$, provides an approximate Nash equilibrium.

The implementation of this strategy can be broken down into two steps for pedagogical reasons.

(i) For each arbitrary measure-valued stochastic process $\boldsymbol{\mu} = (\mu_t)_{0 \leq t \leq T}$ adapted to the filtration generated by the random measure $\boldsymbol{W}^0$, solve the (standard) stochastic control problem (with random coefficients)

$$\inf_{\alpha \in \mathbb{A}} \mathbb{E}\left[\int_0^T f(t, X_t, \mu_t, \alpha_t)dt + g(X_T, \mu_T)\right]$$

subject to the dynamic constraint

$$dX_t = b(t, X_t, \mu_t, \alpha_t)dt + \sigma(t, X_t, \mu_t, \alpha_t)dW_t + \int_\Xi c(t, X_t^i, \mu_t, \alpha_t^i, \xi) W^0(d\xi, dt)$$

with initial condition $X_0 = x_0$ that are over controls in feedback form Markovian in $\boldsymbol{X}$, and conditional on the past of the random measure $\boldsymbol{\mu}$.

(ii) Determine the measure-valued stochastic process $\boldsymbol{\mu} = (\mu_t)_{0 \leq t \leq T}$ so that the solution of the SPDE (6.3) in $\mathcal{P}_2(\mathbb{R}^d)$ (computed from the optimally controlled state and the optimal feedback function $\alpha(t, x)$) coincides with $\boldsymbol{\mu} = (\mu_t)_{0 \leq t \leq T}$.

Clearly, this second point requires the solution of a fixed-point problem in an infinite-dimensional space, while the first one involves the solution of an optimization problem in a space of stochastic processes. The following remarks provide specific forms of this set of three prescriptions in several important particular cases.

Remarks

1. In the absence of the ambient random field noise term $\boldsymbol{W}^0$, the measure-valued adapted stochastic process $(\mu_t)_{0 \leq t \leq T}$ can be taken as a deterministic function $[0, T] \ni t \hookrightarrow \mu_t \in \mathcal{P}_2(\mathbb{R}^d)$, and the control problem in item (ii) is a standard Markovian control problem. Moreover, as we already noticed, the SPDE becomes a deterministic PDE, the Kolmogorov equation giving the dynamics of the marginal distribution at time t of the diffusion process associated to a given Markov control α. Consequently, the fixed-point item (ii) reduces to the search for a deterministic flow of measures $[0, T] \ni t \hookrightarrow \mu_t \in \mathcal{P}_2(\mathbb{R}^d)$ such that the law of the optimally controlled process (resulting from the solution of the first point) is in fact μ_t, i.e.,

$$\forall t \in [0, T], \quad \mathcal{L}(X_t) = \mu_t.$$

2. As we shall see later in the chapter, the models introduced for the purpose of the analysis of systemic risk and games with major players are special cases of the general games in a random environment when the random shocks are provided by a single Wiener process instead of a spatial random field. In these cases, the solutions of the SPDEs can be identified to the forward Kolmogorov equations giving the time evolution of the marginal conditional distribution of the diffusion.

As explained in the introduction, our goal is to solve the fixed-point problem and to show that indeed, its solution provides approximate Nash equilibriums for large games, and in the process, to quantify the nature of this approximation. This program is presented in Section 6.3 below in the absence of the ambient noise random field.

6.1.5 • The Case of Linear-Quadratic (LQ) Mean-Field Games

The first example we shall discuss in detail concerns the linear-quadratic (LQ) models for which the mean-field nature of the interaction requires revisiting the notation used to describe these games. The state of the system is the aggregation $X_t = [X_t^1, \cdots, X_t^N]^\dagger$ of N individual private states $X_t^i \in \mathbb{R}^{d'}$ so that $d = Nd'$ and $m = Nd'$. Their dynamics are given by the linear SDEs

$$dX_t^i = [\tilde{A}_t X_t^i + \tilde{\tilde{A}}_t \overline{X}_t + \tilde{B}_t \alpha_t^i + \tilde{\beta}_t]dt + \sigma dW_t^i, \qquad i = 1, \cdots, N, \qquad (6.4)$$

where the W^i are N-independent d'-dimensional Wiener processes, and where we used the notation $\overline{X}_t$ for the (empirical) sample average

$$\overline{X}_t = \frac{1}{N}\sum_{i=1}^N X_t^i$$

of the individual player private states X_t^i. Notice that in this subsection, we only consider models without a common noise. This is only for the sake of simplicity. As we saw in Section 5.5 of Chapter 5, a common noise term can easily be added without overly increasing the level of technicality of the discussion. In the case of MFGs, the matrix A_t appearing in the drift of the dynamics of the state of an LQ game is assumed to be of the form

$$A_t = \begin{bmatrix} \tilde{A}_t + \tilde{\tilde{A}}_t/N & \tilde{\tilde{A}}_t/N & \cdots & \tilde{\tilde{A}}_t/N & \tilde{\tilde{A}}_t/N \\ \tilde{\tilde{A}}_t/N & \tilde{A}_t + \tilde{\tilde{A}}_t/N & \cdots & \tilde{\tilde{A}}_t/N & \tilde{\tilde{A}}_t/N \\ \vdots & \vdots & & \vdots & \vdots \\ \tilde{\tilde{A}}_t/N & \tilde{\tilde{A}}_t/N & \cdots & \tilde{A}_t + \tilde{\tilde{A}}_t/N & \tilde{\tilde{A}}_t/N \\ \tilde{\tilde{A}}_t/N & \tilde{\tilde{A}}_t/N & \cdots & \tilde{\tilde{A}}_t/N & \tilde{A}_t + \tilde{\tilde{A}}_t/N \end{bmatrix}.$$

In other words, A_t is a sum of a block-diagonal matrix with $d' \times d'$ diagonal blocks equal to $\tilde{A}_t$, and a full matrix of $d' \times d'$ blocks equal to $\tilde{\tilde{A}}_t$,

$$A_t = \begin{bmatrix} \tilde{A}_t & 0 & \cdots & 0 & 0 \\ 0 & \tilde{A}_t & \cdots & 0 & 0 \\ \vdots & \vdots & & \vdots & \vdots \\ 0 & 0 & \cdots & \tilde{A}_t & 0 \\ 0 & 0 & \cdots & 0 & \tilde{A}_t \end{bmatrix} + \frac{1}{N}\begin{bmatrix} \tilde{\tilde{A}}_t & \tilde{\tilde{A}}_t & \cdots & \tilde{\tilde{A}}_t & \tilde{\tilde{A}}_t \\ \tilde{\tilde{A}}_t & \tilde{\tilde{A}}_t & \cdots & \tilde{\tilde{A}}_t & \tilde{\tilde{A}}_t \\ \vdots & \vdots & & \vdots & \vdots \\ \tilde{\tilde{A}}_t & \tilde{\tilde{A}}_t & \cdots & \tilde{\tilde{A}}_t & \tilde{\tilde{A}}_t \\ \tilde{\tilde{A}}_t & \tilde{\tilde{A}}_t & \cdots & \tilde{\tilde{A}}_t & \tilde{\tilde{A}}_t \end{bmatrix}.$$

Notice from (6.4) that the action α^i of player i is the only action (among all the actions α^j) entering the dynamics of the private state of player i. Since $k^1 = \cdots = k^N = k'$ for symmetry reasons, the equation for the dynamics of the state of the game can be rewritten as

$$dX_t = [A_t X_t + B_t \alpha_t + \beta_t]dt + \sigma dW_t, \qquad i = 1, \cdots, N, \qquad (6.5)$$

6.1. Introduction and First Definitions

where $\sigma > 0$ and the matrix B_t is a block-diagonal matrix with $d' \times k'$ diagonal blocks equal to $\tilde{B}_t$, and we have

$$B_t = \begin{bmatrix} \tilde{B}_t & 0 & \cdots & 0 & 0 \\ 0 & \tilde{B}_t & \cdots & 0 & 0 \\ \vdots & \vdots & & \vdots & \vdots \\ 0 & 0 & \cdots & \tilde{B}_t & 0 \\ 0 & 0 & \cdots & 0 & \tilde{B}_t \end{bmatrix} \quad \text{and} \quad \beta_t = \begin{bmatrix} \tilde{\beta}_t \\ \tilde{\beta}_t \\ \vdots \\ \tilde{\beta}_t \\ \tilde{\beta}_t \end{bmatrix}.$$

To connect with the notation used so far for LQ games,

$$B_t = [B_t^1, \cdots, B_t^N]$$

so that B_t^i is the $d \times k'$ matrix obtained by piling up N successive $d' \times k'$ blocks of zeros, except for the ith block which is taken to be $\tilde{B}_t$ instead of a block of zeros. In this way, if $y \in \mathbb{R}^d$, $B_t^{i\dagger} y = \tilde{B}^\dagger y^i$ where y^i is the ith block when y is viewed as $[y^1, \cdots, y^N]^\dagger$ with each y^j in $\mathbb{R}^{d'}$.

In an LQ game with mean-field interactions, the function f^i giving the running cost of player i is often assumed to be of the form

$$f^i(t, x, \alpha) = \frac{1}{2} \|\tilde{M}_t x^i + \tilde{\tilde{M}}_t \bar{x}\|^2 + \frac{1}{2} \alpha^{i\dagger} N_t \alpha^i,$$

where $\tilde{M}_t$ and $\tilde{\tilde{M}}_t$ are $d' \times d'$ matrices and N_t is a $k' \times k'$ matrix, while the terminal cost is assumed to be given an expression such as

$$g^i(x) = \frac{1}{2} \|\tilde{Q} x^i + \tilde{\tilde{Q}} \bar{x}\|^2.$$

Notice that the matrices M_t^i, N_t^i, and Q^i are independent of i for symmetry reasons. Consequently, when computing the $\hat{\alpha}$ satisfying the Isaacs condition on all the adjoint processes, we get

$$\hat{\alpha}_t^i = \hat{\alpha}^i(t, X_t, (Y_t^1, \cdots, Y_t^N), (Z_t^1, \cdots, Z_t^N)) = -[N_t]^{-1} B_t^i Y_t^i = -[N_t]^{-1} \tilde{B}_t Y_t^{ii},$$

where Y_t^{ii} is the ith block of length d' in the $d = Nd'$-dimensional vector Y_t^i. With this form of the model, it is possible to use the methods introduced in Subsection 5.2.2 of Chapter 5 to search for Nash equilibriums of the game. This is exactly what we did in the case of the systemic risk toy model completely analyzed in Section 5.5.

However, since finite player games are rather difficult to solve, even in the particular case of LQ models, the mean-field formulation may be of interest since, as we shall see in Section 6.3 below, its solution may provide approximate Nash equilibriums. Given the above discussion, the LQ MFG model reads as follows.

(i) For each arbitrary deterministic function $[0, T] \ni t \hookrightarrow m_t$, solve the (standard) LQ stochastic control problem

$$\inf_{\alpha \in \mathbf{A}} \mathbb{E} \left[\frac{1}{2} \int_0^T \left(\|\tilde{M}_t X_t + \tilde{\tilde{M}}_t m_t\|^2 + \frac{1}{2} \alpha^\dagger N_t \alpha_t \right) dt + \frac{1}{2} \|\tilde{Q} X_T + \tilde{\tilde{Q}} m_T\|^2 \right]$$

subject to the linear dynamic constraint

$$dX_t = [\tilde{A}_t X_t + \tilde{\tilde{A}}_t m_t + \tilde{B}_t \alpha_t^i + \tilde{\beta}_t] dt + \sigma dW_t^i$$

with initial condition $X_0 = x_0$.

(ii) Determine the function $(m_t)_{0 \leq t \leq T}$ so that at each time $t \in [0,T]$, the solution of the mean of the marginal distribution of the optimal state coincides with m_t, i.e., $m_t = \mathbb{E}X_t$ for all $t \in [0,T]$.

Given what we learned in Chapter 3, this problem seems tractable. Indeed, for each fixed flow of means $(m_t)_{0 \leq t \leq T}$, the control problem of point (i) is a standard LQ control problem and its solution reduces to the solution of a matrix Riccati equation. While matrix Riccati equations are typically difficult to solve in general, the strict convexity of the cost function guarantees the existence and uniqueness of an optimal control, and hence of a solution of this matrix Riccati equation. Because of the Markovian nature of the dynamics, requiring that point (ii) holds forces m_t to be the solution of a linear ODE, whose solution can be substituted back into the Riccati equation. So, as in the case of the control of McKean–Vlasov LQ systems, the whole problem reduces to the solution of a specific matrix Riccati equation. We shall give some examples below.

In most cases, and especially when we try to solve the N-player game, we assume $d' = 1$ and $k' = 1$ in order to simplify the notation. In other words, we assume that all the player private states as well as the individual actions are scalar. In that particular case, the Riccati equation is scalar and necessary, and sufficient conditions for its well-posedness are readily available.

6.2 ▪ A Full Solution Without the Common Noise

The goal of this section is to solve the mean-field game formulation of the problem under a set of reasonably general conditions. Later on in the chapter, we shall show that the solution provides approximate Nash equilibriums for large games, and in the process, quantify the nature of this approximation.

6.2.1 ▪ The Hamiltonian and the Stochastic Maximum Principle

For the sake of simplicity, we assume that the set A of actions α which can be taken by a given player at a fixed time is the whole Euclidean space $\mathbb{R}^k$. Moreover, in order to avoid technicalities, we assume that the volatility is uncontrolled, and that it is in fact a constant $\sigma > 0$.

The Hamiltonian

The fact that the volatility is uncontrolled allows us to use the reduced Hamiltonian

$$H(t, x, \mu, y, \alpha) = b(t, x, \mu, \alpha) \cdot y + f(t, x, \mu, \alpha) \tag{6.6}$$

for $t \in [0,T]$, $x, y \in \mathbb{R}^d$, $\alpha \in \mathbb{R}^k$, and $\mu \in \mathcal{P}_2(\mathbb{R}^d)$. Our objective is to use the stochastic maximum principle to fully solve the problem, so we first try to minimize the Hamiltonian with respect to the control parameter, and understand how minimizers depend upon the other variables of the Hamiltonian. In order to do so, it is convenient to assume that H is strictly convex with respect to α. To prove existence and uniqueness of a minimizer $\hat{\alpha}(t, x, y, \mu)$, and to control its dependence with respect to the arguments (x, y, μ), we assume that there exist two constants λ and c_L such that

(C.1) the drift b is an affine function of α in the sense that it is of the form

$$b(t, x, \mu, \alpha) = b_1(t, x, \mu) + b_2(t)\alpha, \tag{6.7}$$

where the mapping $[0,T] \ni t \hookrightarrow b_2(t) \in \mathbb{R}^d$ is bounded;

6.2. A Full Solution Without the Common Noise

(C.2) for any $t \in [0, T]$ and $\mu \in \mathcal{P}_2(\mathbb{R}^d)$, the function $\mathbb{R}^d \times \mathbb{R}^k \ni (x, \alpha) \hookrightarrow f(t, x, \mu, \alpha)$ is twice continuously differentiable (i.e., it is C^2), with partial derivatives which are Lipschitz continuous in the variables x, μ, and α uniformly in t; moreover, the second-order partial derivatives of f with respect to x and α satisfy

$$\begin{pmatrix} \partial^2_{xx} f(t, x, \mu, \alpha) & \partial^2_{x\alpha} f(t, x, \mu, \alpha) \\ \partial^2_{x\alpha} f(t, x, \mu, \alpha) & \partial^2_{\alpha\alpha} f(t, x, \mu, \alpha) \end{pmatrix} \geq \begin{pmatrix} 0 & 0 \\ 0 & \lambda I_k \end{pmatrix} \quad (6.8)$$

for some positive constant $\lambda > 0$ where I_k denotes the identity matrix of dimension k.

Note that assumption **(C.1)** used in the first part of the discussion will be subsumed by assumptions **(MFG.1)** made later on in the section. The minimization of the Hamiltonian is taken care of by the following result.

Existence of a Lipschitz Minimizer

Lemma 6.1. *Under assumptions* **(C.1)**–**(C.2)**, *for all* $(t, x, \mu, y) \in [0, T] \times \mathbb{R}^d \times \mathcal{P}_2(\mathbb{R}^d) \times \mathbb{R}^d$, *there exists a unique minimizer* $\hat{\alpha}(t, x, \mu, y)$ *of the Hamiltonian*

$$\hat{\alpha}(t, x, \mu, y) = \mathrm{argmin}_{\alpha \in \mathbb{R}^d} H(t, x, \mu, y, \alpha), \quad t \in [0, T], \; x, y \in \mathbb{R}^d, \; \mu \in \mathcal{P}(\mathbb{R}^d).$$

Moreover, the function $[0, T] \ni t \hookrightarrow \hat{\alpha}(t, 0, 0, 0)$ *is bounded and the function* $[0, T] \times \mathbb{R}^d \times \mathcal{P}_2(\mathbb{R}^d) \times \mathbb{R}^d \hookrightarrow \hat{\alpha}(t, x, \mu, y)$ *is Lipschitz continuous with respect to* (x, μ, y) *uniformly in* $t \in [0, T]$, *the Lipschitz constant depending only upon* λ, *the supremum norm of* b_2, *and the Lipschitz constant of* $\partial_\alpha f$ *in the variables* x *and* μ.

Proof. For any given (t, x, y, μ), the function $\mathbb{R}^k \ni \alpha \hookrightarrow H(t, x, \mu, y, \alpha)$ is twice continuously differentiable. Since its second partial derivative $\partial^2_{\alpha\alpha} f(t, x, \mu, \alpha)$ is invertible and the norm of its inverse is uniformly bounded by (6.8), the implicit function theorem implies the existence and uniqueness of the minimizer $\hat{\alpha}(t, x, y, \mu)$, together with its boundedness on bounded subsets. Recall that a bounded subset of $\mathcal{P}_2(\mathbb{R}^d)$ is a set of probability measures with uniformly bounded second moments. Since $\hat{\alpha}(t, x, y, \mu)$ appears as the unique solution of the equation

$$\partial_\alpha H\bigl(t, x, \mu, y, \hat{\alpha}(t, x, y, \mu)\bigr) = 0,$$

the implicit function theorem implies that $\hat{\alpha}$ is differentiable with respect to (x, y) with a continuous (hence locally bounded) derivative. By uniform boundedness of b_2 and $\partial^2_{x,\alpha} f$, the derivatives must be globally bounded. The regularity of $\hat{\alpha}$ with respect to μ follows from the following remark. If $(t, x, y) \in [0, T] \times \mathbb{R}^d \times \mathbb{R}^d$ is fixed and μ, μ' are generic elements in $\mathcal{P}_2(\mathbb{R}^d)$, $\hat{\alpha}$ and $\hat{\alpha}'$ denoting the associated minimizers, we deduce from the convexity assumption (6.8)

$$\begin{aligned} \lambda |\hat{\alpha}' - \hat{\alpha}|^2 &\leq \langle \hat{\alpha}' - \hat{\alpha}, \partial_\alpha f(t, x, \mu, \hat{\alpha}') - \partial_\alpha f(t, x, \mu, \hat{\alpha}) \rangle \\ &= \langle \hat{\alpha}' - \hat{\alpha}, \partial_\alpha H(t, x, \mu, y, \hat{\alpha}') - \partial_\alpha H(t, x, \mu, y, \hat{\alpha}) \rangle \\ &= \langle \hat{\alpha}' - \hat{\alpha}, \partial_\alpha H(t, x, \mu, y, \hat{\alpha}') - \partial_\alpha H(t, x, \mu', y, \hat{\alpha}') \rangle \quad (6.9) \\ &= \langle \hat{\alpha}' - \hat{\alpha}, \partial_\alpha f(t, x, \mu, \hat{\alpha}') - \partial_\alpha f(t, x, \mu', \hat{\alpha}') \rangle \\ &\leq C |\hat{\alpha}' - \hat{\alpha}| \, W_2(\mu', \mu), \end{aligned}$$

and we easily complete the proof from there. $\square$

Convenient Forms of the Stochastic Maximum Principle

Going back to the program (i)–(ii) outlined in Subsection 6.1.4, we understand that the first step consists in solving a standard stochastic control problem each time the candidates for the marginal distributions $\boldsymbol{\mu} = (\mu_t)_{0 \le t \le T}$ are held fixed. One way to do so is to express the value function of the optimization problem as the solution of the corresponding Hamilton–Jacobi–Bellman equation. This is the cornerstone of the analytic approach to the MFG theory originally advocated by Lasry and Lions, the matching problem (ii) being resolved by coupling the Hamilton–Jacobi–Bellman partial differential equation with a Kolmogorov equation intended to identify the $\boldsymbol{\mu} = (\mu_t)_{0 \le t \le T}$ with the marginal distributions of the optimal state of the problem. Instead, we choose a probabilistic approach based on a probabilistic description of the optimal states of the optimization problem (i) as provided by the stochastic maximum principle.

We shall use the sufficient part of the stochastic maximum principle presented in Chapter 4 to prove the existence of solutions to the problem (i) stated in Subsection 6.1.4. In addition to **(C.1)**–**(C.2)** we shall also assume the following.

(C.3) The function $[0,T] \ni t \hookrightarrow b_1(t,x,\mu)$ is affine in x in the sense that it is of the form $[0,T] \ni t \hookrightarrow b_0(t,\mu) + b_1(t,\mu)x$, where b_0 and b_1 are $\mathbb{R}^d$- and $\mathbb{R}^{d \times d}$-valued, respectively, and are bounded on bounded subsets of their respective domains.

In fact, we shall rely on a slightly stronger version of the sufficient part of the stochastic maximum principle. This will allow us to handle the dependence with respect to the fixed flow of marginal distributions and solve the fixed-point step (ii).

Lemma 6.2. *Under assumptions* **(C.1)**–**(C.3)**, *if the mapping $[0,T] \ni t \hookrightarrow \mu_t \in \mathcal{P}_2(\mathbb{R}^d)$ is measurable and bounded, and the cost functional J is defined by*

$$J(\boldsymbol{\beta}, (\mu_t)_{0 \le t \le T}) = \mathbb{E}\left[g(U_T, \mu_T) + \int_0^T f(t, U_t, \mu_t, \beta_t) dt\right]$$

for any progressively measurable process $\boldsymbol{\beta} = (\beta_t)_{0 \le t \le T}$ in $\mathbb{H}^{2,k}$, where $\mathbf{U} = (U_t)_{0 \le t \le T}$ is the controlled diffusion process

$$U_t = x_0 + \int_0^t b(s, U_s, \mu_s, \beta_s) ds + \sigma W_t, \quad t \in [0,T]$$

for some $x_0 \in \mathbb{R}^d$, and if the forward-backward system

$$\begin{cases} dX_t = b\bigl(t, X_t, \mu_t, \hat{\alpha}(t, X_t, Y_t, \mu_t)\bigr) dt + \sigma dW_t, & X_0 = x_0, \\ dY_t = -\partial_x H(t, X_t, \mu_t, Y_t, \hat{\alpha}(t, X_t, Y_t, \mu_t)) + Z_t dW_t, & Y_T = \partial_x g(X_T, \mu_T), \end{cases} \quad (6.10)$$

has a solution $(X_t, Y_t, Z_t)_{0 \le t \le T}$ such that

$$\mathbb{E}\left[\sup_{0 \le t \le T}\bigl(|X_t|^2 + |Y_t|^2\bigr) + \int_0^T |Z_t|^2 dt\right] < +\infty, \quad (6.11)$$

then it holds that

$$J(\hat{\boldsymbol{\alpha}}; (\mu_t)_{0 \le t \le T}) + \frac{\lambda}{2} \mathbb{E}\int_0^T |\beta_t - \hat{\alpha}_t|^2 dt \le J(\boldsymbol{\beta}; (\mu_t)_{0 \le t \le T}),$$

where $\hat{\boldsymbol{\alpha}} = (\hat{\alpha}_t)_{0 \le t \le T}$ is defined by $\hat{\alpha}_t = \hat{\alpha}(t, X_t, \mu_t, Y_t)$.

6.2. A Full Solution Without the Common Noise

Proof. We first notice that by Lemma 6.1, $\hat{\alpha} = (\hat{\alpha}_t)_{0 \le t \le T} \in \mathbb{H}^{2,k}$. Following the proof of the stochastic maximum principle given in Chapter 4, we get

$$J(\beta; (\mu_t)_{0 \le t \le T}) \ge J(\hat{\alpha}; (\mu_t)_{0 \le t \le T}) + \mathbb{E} \int_0^T \Big[H(t, U_t, \mu_t, Y_t, \beta_t) - H(t, X_t, \mu_t, Y_t, \hat{\alpha}_t)$$
$$- \langle U_t - X_t, \partial_x H(t, X_t, \mu_t, Y_t, \hat{\alpha}_t) \rangle - \langle \beta_t - \hat{\alpha}_t, \partial_\alpha H(t, X_t, \mu_t, Y_t, \hat{\alpha}_t) \rangle \Big] dt.$$

By linearity of b, the Hessian of H satisfies (6.8) and the required convexity assumption is satisfied. The result easily follows. □

Remark 6.3. *Lemma 6.2 has interesting consequences. First, it says that the optimal control, if it exists, must be unique. Second, it also implies that given two solutions (X, Y, Z) and (X', Y', Z') to (6.10), $d\mathbb{P} \otimes dt$-a.e., it holds that*

$$\hat{\alpha}(t, X_t, \mu_t, Y_t) = \hat{\alpha}(t, X'_t, \mu_t, Y'_t),$$

so that X and X' coincide by the Lipschitz property of the coefficients of the forward equation. As a consequence, (Y, Z) and (Y', Z') coincide as well.

Notice that the bound provided by Lemma 6.2 above is sharp within convex models, as shown, for example, by the following slight variation on the same theme.

Proposition 6.4. *Under the same assumptions and notation as in Lemma 6.2 above, if we consider in addition another measurable and bounded mapping $[0, T] \ni t \hookrightarrow \mu'_t \in \mathcal{P}_2(\mathbb{R}^k)$ and the controlled diffusion process $\mathbf{U}' = (U'_t)_{0 \le t \le T}$ defined by*

$$U'_t = x'_0 + \int_0^t b(s, U'_s, \mu'_s, \beta_s) ds + \sigma W_t, \quad t \in [0, T],$$

for an initial condition $x'_0 \in \mathbb{R}^d$ possibly different from x_0, then

$$J(\hat{\alpha}; (\mu_t)_{0 \le t \le T}) + \langle x'_0 - x_0, Y_0 \rangle + \frac{\lambda}{2} \mathbb{E} \int_0^T |\beta_t - \hat{\alpha}_t|^2 dt$$
$$\le J(\beta; (\mu'_t)_{0 \le t \le T}) + \mathbb{E} \bigg[\int_0^T \langle b_0(t, \mu'_t) - b(t, \mu_t), Y_t \rangle dt \bigg]. \tag{6.12}$$

Proof. As before, we go back to the strategy of the original proof of the stochastic maximum principle and we expand

$$\bigg(\langle U'_t - X_t, Y_t \rangle + \int_0^t \big[f(s, U'_s, \mu'_s, \beta_s) - f(s, X_s, \mu_s, \hat{\alpha}_s) \big] ds \bigg)_{0 \le t \le T}$$

using Itô's formula. Since the initial conditions x_0 and x'_0 are possibly different, we get the additional term $\langle x'_0 - x_0, Y_0 \rangle$ in the left-hand side of (6.12). Similarly, since the drift of $\mathbf{U}'$ is driven by $(\mu'_t)_{0 \le t \le T}$, we get the additional difference of the drifts in order to account for the fact that the drifts are driven by the different flows of probability measures. □

6.2.2 ▪ The Mean-Field FBSDE

As we explained earlier, our strategy is to solve the standard stochastic control problem (i) using the stochastic maximum principle and to reduce its solution to the solution of an FBSDE. But solving this FBSDE would not solve the entire mean-field game problem. Indeed, we still have to take care of part (ii), which requires finding a fixed point for the flow $\mu = (\mu_t)_t$ of marginal probabilities, requiring that $\mu_t = \mathcal{L}(X_t)$ for all t. Substituting $\mathcal{L}(X_t)$ for μ_t into the FBSDE changes the nature of the FBSDE since the marginal distribution of the solution appears in the coefficients.

In a nutshell, the probabilistic approach to the solution of the mean-field game problem results in the solution of an FBSDE of the McKean–Vlasov type

$$\begin{cases} dX_t = b\big(t, X_t, \mathcal{L}(X_t), \hat{\alpha}(t, X_t, \mathcal{L}(X_t), Y_t)\big)dt + \sigma dW_t, \\ dY_t = -\partial_x H\big(t, X_t, \mathcal{L}(X_t), Y_t, \hat{\alpha}(t, X_t, \mathcal{L}(X_t), Y_t)\big)dt + Z_t dW_t \end{cases} \quad (6.13)$$

with initial condition $X_0 = x_0 \in \mathbb{R}^d$ and terminal condition $Y_T = \partial_x g(X_T, \mathcal{L}(X_t))$.

This discussion, and especially the above FBSDE, are reminiscent of our analysis of Section 4.4 on the control of McKean–Vlasov dynamics. Indeed, FBSDE (6.13) is very similar, though simpler than (4.83). It is simpler because, since we are dealing with a classical control problem, partial derivatives of the Hamiltonian and the terminal cost function with respect to their measure argument do not have to appear in the backward part of the FBSDE. Section 4.4 of Chapter 4 concluded with an existence and uniqueness result (recall Theorem 4.27) which we stated without proof because of the highly technical nature of the arguments. Here, we prove the existence of a solution by taking advantage of the strong convexity assumption on the coefficients using the well-known fact that in optimization theory, convexity often generates compactness. Our goal is to use compactness to solve the matching problem (ii) by applying Schauder's fixed-point theorem in an appropriate space of finite measures on $\mathcal{C}([0,T], \mathbb{R}^d)$.

Standing Assumptions and Main Result

Assumptions **(MFG.1)**–**(MFG.6)** used in the proof of the main result of this section subsume assumptions **(C.1)**–**(C.3)** used earlier. We assume that there exist two positive constants λ and c_L such that we have the following assumptions.

(MFG.1). Smoothness. For each fixed $t \in [0,T]$ and $\mu \in \mathcal{P}_2(\mathbb{R}^d)$, the running cost f (resp., terminal cost g) is continuously differentiable with respect to (x, α) (resp., x), with c_L Lipschitz continuous derivatives in the arguments x, μ, and α. The Lipschitz property with respect to μ is understood in the sense of the 2-Wasserstein distance W_2.

(MFG.2). Local Lipschitz continuity. The functions $[0,T] \ni t \hookrightarrow f(t, 0, \delta_0, 0)$, $[0,T] \ni t \hookrightarrow \partial_x f(t, 0, \delta_0, 0)$, $[0,T] \ni t \hookrightarrow \partial_\mu f(t, 0, \delta_0, 0)$, and $[0,T] \ni t \hookrightarrow \partial_\alpha f(t, 0, \delta_0, 0)$ are bounded by c_L. Moreover,

$$|f(t, x', \mu', \alpha') - f(t, x, \mu, \alpha)| + |g(x', \mu') - g(x, \mu)|$$
$$\leq c_L \left[1 + |x'| + |x| + |\alpha'| + |\alpha| + \left(\int_{\mathbb{R}^d} |y|^2 d(\mu + \mu')(y)\right)^{1/2}\right]$$
$$[|(x', \alpha') - (x, \alpha)| + W_2(\mu', \mu)]$$

for all $t \in [0,T]$, $x, x' \in \mathbb{R}^d$, $\alpha, \alpha' \in \mathbb{R}^k$, and $\mu, \mu' \in \mathcal{P}_2(\mathbb{R}^d)$.

(MFG.3). Convexity. For (t, μ) fixed, the function f is convex in (x, α) in the sense that

$$f(t, x', \mu, \alpha') - f(t, x, \mu, \alpha) - \langle (x' - x, \alpha' - \alpha), \partial_{(x,\alpha)} f(t, x, \mu, \alpha) \rangle \geq \lambda |\alpha' - \alpha|^2$$

for all $t \in [0,T]$, $x, x' \in \mathbb{R}^d$, $\alpha, \alpha' \in \mathbb{R}^k$, and $\mu \in \mathcal{P}_2(\mathbb{R}^d)$. Similarly, for μ fixed, the function g is convex in x.

6.2. A Full Solution Without the Common Noise

(MFG.4). Uniform smoothness in α. For all $t \in [0,T]$, $x \in \mathbb{R}^d$, and $\mu \in \mathcal{P}_2(\mathbb{R}^d)$, $|\partial_\alpha f(t, x', \mu, 0)| \leq c_L$.

(MFG.5). Weak mean-reverting. The two functions $\mathbb{R}^d \ni x \hookrightarrow x \cdot \partial_x g(0, \delta_x)$ and $[0,T] \times \mathbb{R}^d \ni (t,x) \hookrightarrow x \cdot \partial_x f(t, 0, \delta_x, 0)$ take nonnegative values.

(MFG.6). Linear drift. The coefficient b is linear in x and α when μ is given; specifically,

$$b(t, x, \mu, \alpha) = b_0(t, \mu) + b_1(t)x + b_2(t)\alpha, \tag{6.14}$$

where b_1 and b_2 are continuous functions of $t \in [0,T]$ with values in $\mathbb{R}^{d \times d}$ and $\mathbb{R}^{d \times k}$, respectively, with Euclidean norms bounded by c_L, and the function $b_0 : [0,T] \times \mathcal{P}(\mathbb{R}^d) \ni (t, \mu) \hookrightarrow b_0(t, \mu) \in \mathbb{R}^d$ is continuous and also bounded by c_L for any $t \in [0,T]$ and any $\mu, \mu' \in \mathcal{P}_2(\mathbb{R}^d)$:

$$|b_0(t, \mu') - b_0(t, \mu)| \leq c_L W_2(\mu, \mu').$$

In this framework, we shall prove the following.

Theorem 6.5. *Under* **(MFG.1)–(MFG.6)**, *the forward-backward system* (6.13) *has a solution.*

Remark 6.6. *Assumptions* **(MFG.1)–(MFG.2)** *are standard Lipschitz properties. In this setting, condition* **(MFG.4)** *is a way of controlling the smoothness of the running cost f with respect to α uniformly in the other variables. Similarly,* **(MFG.3)** *and* **(MFG.6)** *are quite standard in optimization theory as they guarantee that the Hamiltonian H is convex in the variables (x, α).*

Condition **(MFG.5)** *is unusual. It looks like a mean-reverting condition for recurrent diffusion processes. Its role is to control the expectation of the forward equation in* (6.13) *and to establish an a priori bound which plays a crucial role in the proof of compactness. We used the adjective "weak" to emphasize the fact that the expectation has some boundedness property but is not expected to converge with time.*

A typical example we may have in mind is the so-called linear-quadratic setting, when b_0, f, and g have the form

$$b_0(t, \mu) = b_0(t) \int_{\mathbb{R}^d} x d\mu(x),$$

$$g(x, \mu) = \frac{1}{2} \Big| qx + \bar{q} \int_{\mathbb{R}^d} x d\mu(x) \Big|^2,$$

$$f(t, x, \mu, \alpha) = \frac{1}{2} \Big| m(t)x + \bar{m}(t) \int_{\mathbb{R}^d} x d\mu(x) \Big|^2 + \frac{1}{2} |n(t)\alpha|^2,$$

where q, $\bar{q}$, $m(t)$, and $\bar{m}(t)$ are elements of $\mathbb{R}^{d \times d}$ and $n(t)$ is an element of $\mathbb{R}^{k \times k}$. In this framework, **(MFG.5)** *reads*

$$\forall x \in \mathbb{R}^d, \quad \forall t \in [0,T], \quad \langle qx, \bar{q}x \rangle \geq 0, \quad \langle m(t)x, \bar{m}(t)x \rangle \geq 0.$$

In particular, in the one-dimensional setting $d = 1$, **(MFG.5)** *says that $q\bar{q}$ and $m(t)\bar{m}(t)$ must be nonnegative. As shown in [37], this condition is not optimal for existence, as the conditions $q(q + \bar{q}) \geq 0$ and $m(t)(m(t) + \bar{m}(t)) \geq 0$ are then sufficient to guarantee the solvability of* (6.13). *Obviously, the gap between the two is the price one pays to treat general systems within a single framework.*

Stronger Form of the Matching Problem

The proof of Theorem 6.5 is divided into three main steps. The first one is intended to make precise the statement of the matching problem embedded in the fixed point step (ii). We use the strong form of the convexity assumption on the running cost function, together with the consequences of the stochastic maximum principle identified above, to give a more quantitative version of the existence and uniqueness result which we were able to obtain directly by applying Theorem 2.29 of Chapter 2.

Lemma 6.7. *For each measure μ on $\mathcal{C}([0,T],\mathbb{R}^d)$ satisfying*

$$\int_{\mathbb{R}^d} \sup_{0 \leq t \leq T} |\omega_t|^2 d\mu(\omega) < +\infty \tag{6.15}$$

with $(\mu_t)_{0 \leq t \leq T}$ as marginal distributions, the FBSDE (6.10) is uniquely solvable. If we denote by $(X_t^{x_0;\mu}, Y_t^{x_0;\mu}, Z_t^{x_0;\mu})_{0 \leq t \leq T}$ the unique solution, there exists a constant c depending only upon the constants in (**MFG.1**)–(**MFG.6**) *such that for all $x_0, x_0' \in \mathbb{R}^d$,*

$$|Y_0^{x_0';\mu} - Y_0^{x_0;\mu}| \leq c|x_0' - x_0|.$$

Notice that the random variables $Y_0^{x_0;\mu}$ and $Y_0^{x_0';\mu}$ are deterministic because of the zero-one law.

Proof. Since $\partial_x H(t,x,\mu,\alpha) = b_1(t)y + \partial_x f(t,x,\mu,\alpha)$, Lemma 6.1 implies that the driver $[0,T] \times \mathbb{R}^d \times \mathbb{R}^k \ni (t,x,y) \hookrightarrow \partial_x H(t,x,\mu_t,\hat{\alpha}(t,x,\mu_t,y))$ of the backward part of FBSDE (6.10) is Lipschitz continuous in the variables (x,y) uniformly in t. So Theorem 2.23 of Subsection 2.5.3 in Chapter 2 implies existence and uniqueness when T is small enough. Equivalently, when T is arbitrary, there exists $\delta > 0$, depending on the Lipschitz constant of the coefficients in the variables x and y such that unique solvability holds on $[T-\delta, T]$, that is, when the initial condition x_0 of the forward process is prescribed at some time $t_0 \in [T-\delta, T]$. The solution is then denoted by $(X_t^{t_0,x_0}, Y_t^{t_0,x_0}, Z_t^{t_0,x_0})_{t_0 \leq t \leq T}$. Following Delarue [49], existence and uniqueness hold on the whole $[0,T]$, provided

$$\forall x_0, x_0' \in \mathbb{R}^d, \quad |Y_{t_0}^{t_0,x_0} - Y_{t_0}^{t_0,x_0'}|^2 \leq c|x_0 - x_0'|^2 \tag{6.16}$$

for some constant c independent of t_0 and δ. By (6.12), we have

$$\hat{J}^{t_0,x_0} + \langle x_0' - x_0, Y_{t_0}^{t_0,x_0} \rangle + \lambda \mathbb{E} \int_{t_0}^T |\hat{\alpha}_t^{t_0,x_0} - \hat{\alpha}_t^{t_0,x_0'}|^2 dt \leq \hat{J}^{t_0,x_0'}, \tag{6.17}$$

where

$$\hat{J}^{t_0,x_0} = J\big((\hat{\alpha}(t, X_t^{t_0,x_0}, \mu_t, Y_t^{t_0,x_0}))_{t_0 \leq t \leq T}; (\mu_t)_{t_0 \leq t \leq T}\big),$$
$$\hat{J}^{t_0,x_0'} = J\big((\hat{\alpha}(t, X_t^{t_0,x_0'}, \mu_t, Y_t^{t_0,x_0'}))_{t_0 \leq t \leq T}; (\mu_t)_{t_0 \leq t \leq T}\big),$$
$$\hat{\alpha}_t^{t_0,x_0} = \hat{\alpha}(t, X_t^{t_0,x_0}, \mu_t, Y_t^{t_0,x_0}),$$
$$\hat{\alpha}_t^{t_0,x_0'} = \hat{\alpha}(t, X_t^{t_0,x_0'}, \mu_t, Y_t^{t_0,x_0'}).$$

Exchanging the roles of x_0 and x_0', in a similar way we get

$$\hat{J}^{t_0,x_0'} + \langle x_0 - x_0', Y_{t_0}^{t_0,x_0'} \rangle + \lambda \mathbb{E} \int_{t_0}^T |\hat{\alpha}_t^{t_0,x_0} - \hat{\alpha}_t^{t_0,x_0'}|^2 dt \leq \hat{J}^{t_0,x_0}. \tag{6.18}$$

6.2. A Full Solution Without the Common Noise

Adding (6.17) and (6.18), we deduce that

$$2\lambda \mathbb{E} \int_{t_0}^{T} |\hat{\alpha}_t^{t_0,x_0} - \hat{\alpha}_t^{t_0,x_0'}|^2 dt \leq \langle x_0' - x_0, Y_{t_0}^{t_0,x_0'} - Y_{t_0}^{t_0,x_0} \rangle. \quad (6.19)$$

By standard SDE estimates, there exists a constant c (the value of which may vary from line to line), independent of t_0 and δ, such that

$$\mathbb{E}\left[\sup_{t_0 \leq t \leq T} |X_t^{t_0,x_0} - X_t^{t_0,x_0'}|^2 \right] \leq c\mathbb{E} \int_{t_0}^{T} |\hat{\alpha}_t^{t_0,x_0} - \hat{\alpha}_t^{t_0,x_0'}|^2 dt.$$

Similarly, by standard BSDE estimates

$$\mathbb{E}[|Y_{t_0}^{t_0,x_0} - Y_{t_0}^{t_0,x_0'}|^2] \leq c\left(\mathbb{E}\left[\sup_{t_0 \leq t \leq T} |X_t^{t_0,x_0} - X_t^{t_0,x_0'}|^2 \right] + \mathbb{E} \int_{t_0}^{T} |\hat{\alpha}_t^{t_0,x_0} - \hat{\alpha}_t^{t_0,x_0'}|^2 dt \right)$$

$$\leq c\mathbb{E} \int_{t_0}^{T} |\hat{\alpha}_t^{t_0,x_0} - \hat{\alpha}_t^{t_0,x_0'}|^2 dt. \quad (6.20)$$

Plugging (6.20) into (6.19), we complete the proof of (6.16). □

We shall work with the following stronger version of the matching problem (ii).

Definition 6.8. *To each probability measure μ on $\mathcal{C}([0,T],\mathbb{R}^d)$ of order 2, i.e., satisfying (6.15) with $(\mu_t)_{0 \leq t \leq T}$ as marginal distributions, we associate the law $\mathcal{L}(X^{x_0;\mu})$ of the forward component of the solution to the FBSDE (6.10) with x_0 as initial condition. The resulting mapping is denoted by*

$$\Phi : \mathcal{P}_2\big(\mathcal{C}([0,T],\mathbb{R}^d)\big) \ni \mu \hookrightarrow \mathbb{P}_{X^{x_0;\mu}} \in \mathcal{P}_2\big(\mathcal{C}([0,T],\mathbb{R}^d)\big).$$

We then call a solution to the matching problem (ii) any fixed point μ of Φ. Clearly, $X^{x_0;\mu}$ satisfies (6.13) for each such μ.

Existence under Additional Boundedness Conditions

Proposition 6.9. *The system (6.13) is solvable if, in addition to **(MFG.1)–(MGF.6)**, we also assume that $\partial_x f$ and $\partial_x g$ are uniformly bounded, i.e., there exists a constant c_B such that*

$$\forall t \in [0,T], \ x \in \mathbb{R}^d, \mu \in \mathcal{P}_2(\mathbb{R}^d), \ \alpha \in \mathbb{R}^k, \quad |\partial_x g(x,\mu)|, |\partial_x f(t,x,\mu,\alpha)| \leq c_B.$$

Proof. The coefficients $\partial_x f$ and $\partial_x g$ being bounded, the terminal condition in (6.10) is bounded and the growth of the driver is of the form

$$|\partial_x H\big(t,x,\mu_t,y,\hat{\alpha}(t,x,\mu_t,y)\big)| \leq c_B + c_L |y|.$$

Standard BSDE estimates imply that there exists a constant c depending only upon c_B and c_L such that, for any $\mu \in \mathcal{P}_2(\mathcal{C}([0,T],\mathbb{R}^d))$,

$$\forall t \in [0,T], \quad |Y_t^{x_0;\mu}| \leq c, \quad (6.21)$$

almost surely. Plugging this bound into Lemma 6.1, we deduce that

$$\forall t \in [0,T], \quad \hat{\alpha}(t, X_t^{x_0;\boldsymbol{\mu}}, \mu_t, Y_t^{x_0;\boldsymbol{\mu}}) \leq c\left[1 + |X_t^{x_0;\boldsymbol{\mu}}| + \left(\int_{\mathbb{R}^d} |x|^2 d\mu_t(x)\right)^{1/2}\right] \quad (6.22)$$

with possibly a different constant c. Using now the forward equation in (6.10), we get

$$\mathbb{E}\left[\sup_{0 \leq t \leq T} |X_t^{x_0;\boldsymbol{\mu}}|^2\right] \leq c + \frac{1}{2} \sup_{0 \leq t \leq T}\left(\int_{\mathbb{R}^d} |x|^2 d\mu_t(x)\right).$$

In particular,

$$\int_{\mathcal{C}([0,T],\mathbb{R}^d)} \sup_{0 \leq t \leq T} |\omega_t|^2 d\boldsymbol{\mu}(\omega) \leq 2c \quad \Rightarrow \quad \mathbb{E}\left[\sup_{0 \leq t \leq T} |X_t^{x_0;\boldsymbol{\mu}}|^2\right] \leq 2c. \quad (6.23)$$

Below, we thus consider the restriction of Φ to the subset

$$\mathcal{E} = \left\{\boldsymbol{\mu} \in \mathcal{P}_2(\mathcal{C}([0,T],\mathbb{R}^d)) : \int_{\mathcal{C}([0,T],\mathbb{R}^d)} \sup_{0 \leq t \leq T} |\omega_t|^2 d\boldsymbol{\mu}(\omega) \leq 2c\right\}.$$

Clearly, Φ maps $\mathcal{E}$ into itself. Notice that the family of processes $((X_t^{x_0;\boldsymbol{\mu}})_{0 \leq t \leq T})_{\boldsymbol{\mu} \in \mathcal{E}}$ is tight in $\mathcal{C}([0,T],\mathbb{R}^d)$. Indeed, for any $\boldsymbol{\mu} \in \mathcal{E}$ and $0 \leq s \leq t \leq T$,

$$|X_t^{x_0;\boldsymbol{\mu}} - X_s^{x_0;\boldsymbol{\mu}}| \leq c_L(t-s)^{1/2}\left[\sup_{0 \leq r \leq T} |X_r^{x_0;\boldsymbol{\mu}}| + \left(\int_{\mathcal{C}([0,T],\mathbb{R}^d)} \sup_{0 \leq t \leq T} |\omega_t|^2 d\boldsymbol{\mu}(\omega)\right)^{1/2}\right],$$

and tightness follows from (6.23).

In fact, the proof of (6.23) shows that

$$\sup_{\boldsymbol{\mu} \in \mathcal{E}}\left[\mathbb{E}\left[\sup_{0 \leq t \leq T} |X_t^{x_0;\boldsymbol{\mu}}|^4\right]\right] < +\infty, \quad (6.24)$$

implying that $\Phi(\mathcal{E})$ is relatively compact for the 2-Wasserstein distance on $\mathcal{C}([0,T],\mathbb{R}^d)$. Indeed, tightness says that it is relatively compact for the topology of weak convergence of measures and (6.24) says that any weakly convergent sequence $(\mathcal{L}(X^{x_0;\boldsymbol{\mu}_n}))_{n \geq 1}$ with $\boldsymbol{\mu}_n \in \mathcal{E}$ for any $n \geq 1$ is convergent for the 2-Wasserstein distance.

We now use the modified versions of the stochastic maximum principle to show that Φ is continuous on $\mathcal{E}$. Given another measure $\boldsymbol{\mu}' \in \mathcal{E}$, we deduce from (6.12) in Proposition 6.4 that

$$J(\hat{\alpha}; (\mu_t)_{0 \leq t \leq T}) + 2\lambda \mathbb{E}\int_0^T |\hat{\alpha}'_t - \hat{\alpha}_t|^2 dt$$
$$\leq J(\hat{\alpha}'; (\mu_t)_{0 \leq t \leq T}) + \mathbb{E}\int_0^T \langle b_0(t, \mu'_t) - b_0(t, \mu_t), Y_t\rangle dt. \quad (6.25)$$

with the shortened notations

$$\hat{\alpha}_t = \hat{\alpha}(t, X_t^{x_0;\boldsymbol{\mu}}, \mu_t, Y_t^{x_0;\boldsymbol{\mu}}), \quad \hat{\alpha}'_t = \hat{\alpha}(t, X_t^{x_0;\boldsymbol{\mu}'}, \mu'_t, Y_t^{x_0;\boldsymbol{\mu}'}), \quad t \in [0,T].$$

By optimality of $\hat{\alpha}'$ for the cost functional $J(\cdot; (\mu'_t)_{0 \leq t \leq T})$, we claim that

$$J(\hat{\alpha}'; (\mu_t)_{0 \leq t \leq T}) \leq J(\hat{\alpha}'; (\mu'_t)_{0 \leq t \leq T}) + J(\hat{\alpha}'; (\mu_t)_{0 \leq t \leq T}) - J(\hat{\alpha}'; (\mu'_t)_{0 \leq t \leq T})$$
$$\leq J(\hat{\alpha}; (\mu'_t)_{0 \leq t \leq T}) + J(\hat{\alpha}'; (\mu_t)_{0 \leq t \leq T}) - J(\hat{\alpha}'; (\mu'_t)_{0 \leq t \leq T}).$$

6.2. A Full Solution Without the Common Noise

Since μ and μ' are in $\mathcal{E}$, we deduce from **(MFG.1)**, (6.22), and (6.23) that

$$J(\hat{\alpha}; (\mu_t)_{0 \le t \le T}) - J(\hat{\alpha}; (\mu'_t)_{0 \le t \le T}) \le c \int_0^T W_2(\mu, \mu')dt,$$

$$J(\hat{\alpha}'; (\mu_t)_{0 \le t \le T}) - J(\hat{\alpha}'; (\mu'_t)_{0 \le t \le T}) \le c \int_0^T W_2(\mu, \mu'_t)dt$$

so that, from (6.25) and (6.21) again

$$\mathbb{E}\int_0^T |\hat{\alpha}'_t - \hat{\alpha}_t|^2 dt \le c \int_0^T W_2(\mu, \mu'_t)dt.$$

We can now conclude the existence of a fixed point using Schauder's theorem. $\square$

Approximation Procedure

The conjunction of the convexity of f and g, and the boundedness of $\partial_x f$ and $\partial_x g$, puts a heavy damper on the applicability of the above existence result. For instance, boundedness of $\partial_x f$ and $\partial_x g$ fails when f and g are quadratic with respect to x. In order to overcome this limitation, we approximate the cost functions f and g by two sequences $(f^n)_{n \ge 1}$ and $(g^n)_{n \ge 1}$ satisfying **(MFG.1)**–**(MFG.6)** uniformly with respect to $n \ge 1$, and such that, for any $n \ge 1$, $\partial_x f^n$ and $\partial_x g^n$ are bounded.

Lemma 6.10. *There exist two positive constants λ' and c'_L, depending only upon λ and c_L, and two sequences of functions $(f^n)_{n \ge 1}$ and $(g^n)_{n \ge 1}$ such that*

*(i) for any $n \ge 1$, f^n and g^n satisfy **(MFG.1)**–**(MFG.5)** with respect to the parameters λ' and $c_{L'}$ and $\partial_x f^n$ and $\partial_x g^n$ are bounded,*

(ii) the sequences $(f^n, \partial_x f^n, \partial_\alpha f^n)_{n \ge 1}$ and $(g^n, \partial_x g^n, \partial_\alpha g^n)_{n \ge 1}$ converge towards $(f, \partial_x f, \partial_\alpha f)$ and $(g, \partial_x g, \partial_\alpha g)$ uniformly on bounded subsets of $[0, T] \times \mathbb{R}^d \times \mathcal{P}_2(\mathbb{R}^d) \times \mathbb{R}^d$ and $\mathbb{R}^d \times \mathcal{P}_2(\mathbb{R}^d)$, respectively.

Recall that a subset of $\mathcal{P}_2(\mathbb{R}^d)$ is bounded if it is a set of probability measures with uniformly bounded second-order moments. Notice that such bounded sets are tight and hence relatively compact for the metric W_2.

Proof. We only give the construction of the approximation of the running cost f, the procedure also applying to the approximation of the terminal cost g. To lighten notation, we also forget the dependence of f upon t.

For any $n \ge 1$, we define f_n as a truncated Legendre transform

$$f_n(x, \mu, \alpha) = \sup_{|y| \le n} \inf_{z \in \mathbb{R}^d} [y \cdot (x - z) + f(z, \mu, \alpha)] \tag{6.26}$$

for $(x, \alpha) \in \mathbb{R}^d \times \mathbb{R}^k$ and $\mu \in \mathcal{P}_2(\mathbb{R}^d)$. By standard properties of the Legendre transform of convex functions,

$$f_n(x, \mu, \alpha) \le \sup_{y \in \mathbb{R}^d} \inf_{z \in \mathbb{R}^d} [y \cdot (x - z) + f(z, \mu, \alpha)] = f(x, \mu, \alpha). \tag{6.27}$$

Moreover, by strict convexity of f in x,

$$f_n(x, \mu, \alpha) \ge \inf_{z \in \mathbb{R}^d} [f(z, \mu, \alpha)] \ge \inf_{z \in \mathbb{R}^d} [\gamma |z|^2 + \partial_x f(0, \mu, \alpha) \cdot z] + f(0, \mu, \alpha)$$
$$\ge -\frac{1}{4\gamma} |\partial_x f(0, \mu, \alpha)|^2 + f(0, \mu, \alpha), \tag{6.28}$$

proving that f_n takes finite real values. Clearly, f_n is also n-Lipschitz continuous in x. The proof that these approximate cost functions satisfy properties (i) and (ii) in the statement of the lemma is long and technically involved, so we shall not give it here. The interested reader can find the details in [32]. □

Given the result of Lemma 6.10, for each $n \geq 1$, we can consider the approximated Hamiltonian
$$H^n(t, x, \mu, y, \alpha) = b(t, x, \mu, \alpha) \cdot y + f^n(t, x, \mu, y, \alpha)$$
also defined for $t \in [0, T]$, $x, y \in \mathbb{R}^d$, $\alpha \in \mathbb{R}^k$, and $\mu \in \mathcal{P}(\mathbb{R}^d)$, together with the approximated minimizer
$$\hat{\alpha}^n(t, x, y, \mu) = \text{argmin}_\alpha H^n(t, x, \mu, y, \alpha).$$

By Lemma 6.1, we know that the functions $(\hat{\alpha}^n)_{n\geq 1}$ are Lipschitz continuous with respect to x, y, and μ uniformly in $n \geq 1$ (i.e., with a Lipschitz constant independent of n), proving equicontinuity. By λ'-convexity of H^n with respect to α, it is plain to prove that the sequence $(\hat{\alpha}^n(t, x, y, \mu))_{n\geq 1}$ is bounded uniformly in (t, x, y, μ) restricted to bounded subsets. Following (6.9), we deduce that the sequence $(\hat{\alpha}^n)_{n\geq 1}$ converges towards $\hat{\alpha}$ as $n \to +\infty$ uniformly in (t, x, y, μ) restricted to bounded subsets.

From Proposition 6.9, we learn that for any $n \geq 1$, (6.13), with $(\partial_x f, \partial_x g)$ replaced by $(\partial_x f^n, \partial_x g^n)$, has a solution $(\boldsymbol{X}^n, \boldsymbol{Y}^n, \boldsymbol{Z}^n)$. It is relatively easy to prove that the sequence $(\boldsymbol{X}^n)_{n\geq 1}$ is tight. The next step is to prove that the sequence $(u^n)_{n\geq 1}$ of decoupling fields is equicontinuous on compact subsets of $[0, T] \times \mathbb{R}^d$. Finally, extracting a convergent subsequence $(\mathcal{L}(\boldsymbol{X}^{n_p}), u^{n_p})_{p\geq 1}$, it is possible to construct from the limit (μ, u) a triple of processes $(\boldsymbol{X}, \boldsymbol{Y}, \boldsymbol{Z})$ solving the mean-field FBSDE (6.13), concluding the proof of Theorem 6.5. Again, while natural, this strategy of proof requires technical arguments, and we refer the interested reader to [32] for details.

Remark 6.11. *The boundedness property of the drift with respect to the measure parameter can be restrictive. Typical examples of when this condition can be relaxed include linear-quadratic models. When the optimization problem is driven by quadratic coefficients and the dimension d is equal to 1, we know from Bensoussan et al. [19] and Carmona, Delarue, and Lachapelle [37] that the adjoint equations are solvable when the drift b is linear with respect to the expectation of the measure parameter. In fact, it is possible to generalize this result to the case when the cost functions satisfy the same assumptions as in* **(MFG.1)–(MFG.5)** *and b is of the form*

$$b(t, x, \mu, \alpha) = a_t x + a'_t \int_\mathbb{R} x' d\mu(x') + \alpha, \quad x, \alpha \in \mathbb{R}, \quad \mu \in \mathcal{P}(\mathbb{R}), \tag{6.29}$$

for continuous functions $[0, T] \ni t \hookrightarrow a_t \in \mathbb{R}$ and $[0, T] \ni t \hookrightarrow a'_t \in \mathbb{R}$.

6.3 · Propagation of Chaos and Approximate Nash Equilibriums

While the rationale for the formulation of mean-field games articulated in Subsection 6.1.4 is clear given the nature of Nash equilibriums (as opposed to other forms of optimization suggesting the optimal control of stochastic dynamics of the McKean–Vlasov type, as studied in Chapter 4), it may not be obvious how the solution of the FBSDE introduced and solved in the previous sections relates to the existence of Nash equilibriums for large

6.3. Propagation of Chaos and Approximate Nash Equilibriums

games. In this section we prove that the solution of the mean-field FBSDE derived from the mean-field game problem actually provides ε-Nash equilibriums when N is large enough.

The proof relies on the fact that the FBSDE value function is Lipschitz continuous and, standard arguments in the theory of the propagation of chaos.

6.3.1 ▪ Decoupling Field of the Mean-Field FBSDE

In this subsection, we prove that the decoupling field (value function) of FBSDE (6.13) is Lipschitz continuous. Recall that we assume that assumptions **(MFG.1)–(MFG.6)** hold.

Lemma 6.12. *For any solution* $(\boldsymbol{X}, \boldsymbol{Y}, \boldsymbol{Z}) = (X_t, Y_t, Z_t)_{0 \leq t \leq T}$ *to* (6.13), *there exists a function* $u : [0, T] \times \mathbb{R}^d \hookrightarrow \mathbb{R}^d$ *satisfying the growth and Lipschitz properties*

$$\forall t \in [0, T], \quad \forall x, x' \in \mathbb{R}^d, \quad \begin{cases} |u(t, x)| \leq c(1 + |x|), \\ |u(t, x) - u(t, x')| \leq c|x - x'| \end{cases} \quad (6.30)$$

for some constant $c \geq 0$, *and such that* $\mathbb{P}$-*a.s. for all* $t \in [0, T]$, $Y_t = u(t, X_t)$. *In particular, for any* $\ell \geq 1$, $\mathbb{E}[\sup_{0 \leq t \leq T} |X_t|^\ell] < +\infty$.

Proof. This result follows directly from the result of Lemma 6.7, from the fact that the constant c appearing in the statement of this lemma depends only upon the constants in assumptions **(MFG.1)–(MFG.6)**, and from the way we constructed the solution of the mean-field FBSDE (6.13) by a limiting argument which does not affect the estimates (6.30). □

We shall also need a control on the rate of convergence of empirical measures. Even though sharper estimates exist (see, for example, [35] for precise statements and complete proofs), we shall use an old result of Horowitz et al. (see for example Section 10 in [109]) which we state as a lemma for future reference.

Lemma 6.13. *Given* $\mu \in \mathcal{P}_{d+5}(\mathbb{R}^d)$, *i.e., a probability measure* μ *on* $\mathbb{R}^d$ *such that*

$$\int_{\mathbb{R}^d} |x|^{d+5} \mu(dx) < +\infty,$$

there exists a constant C, *depending only upon* d *and the above* $(d + 5)$*th moment, such that*

$$\mathbb{E}[W_2^2(\bar{\mu}^N, \mu)] \leq C N^{-2/(d+4)},$$

where $\bar{\mu}^N$ *stands for the empirical measure of any sample of size* N *from the distribution* μ.

6.3.2 ▪ Existence of Approximate Nash Equilibriums

Again, we assume that assumptions **(MFG.1)–(MFG.6)** hold and we consider the N-player stochastic differential game where the dynamics of the private state $\boldsymbol{U}^i$ of player $i \in \{1, \cdots, N\}$ are given by

$$dU_t^i = b(t, U_t^i, \bar{\nu}_N, \beta_t^i)dt + \sigma dW_t^i, \quad 0 \leq t \leq T, \quad U_0^i = x_0, \quad 1 \leq i \leq N, \quad (6.31)$$

where $(\boldsymbol{W}^i = (W_t^i)_{0 \leq t \leq T})_{1 \leq i \leq N}$ are N-independent d-dimensional Brownian motions, $(\boldsymbol{\beta}^i = (\beta_t^i)_{0 \leq t \leq T})_{1 \leq i \leq N}$ are N square integrable $\mathbb{R}^k$-valued processes that are progressively measurable with respect to the filtration generated by $(\boldsymbol{W}^1, \ldots, \boldsymbol{W}^N)$, and $\bar{\boldsymbol{\nu}}^N =$

$(\bar{\nu}_t^N)_{0 \le t \le T}$ are the empirical measures

$$\bar{\nu}_t^N = \frac{1}{N} \sum_{j=1}^{N} \delta_{U_t^j}, \quad 0 \le t \le T. \tag{6.32}$$

We switched from the notation X_t^i to U_t^i for the private state of a generic player because we want to use the notation X_t^i for the private state associated to a (distributed) strategy providing an approximate Nash equilibrium. See the statement of the theorem below. For each $1 \le i \le N$, we denote by

$$\bar{J}^{N,i}(\boldsymbol{\beta}^1, \ldots, \boldsymbol{\beta}^N) = \mathbb{E}\left[g(U_T^i, \bar{\nu}_T^N) + \int_0^T f(t, U_t^i, \bar{\nu}_t^N, \beta_t^i) dt\right] \tag{6.33}$$

the cost to the ith player. Our goal is to construct an ϵ-Nash equilibrium for the N-player game from a solution of the mean-field FBSDE (6.13). The main result is the following.

Theorem 6.14. *Assume that* **(MFG.1)–(MGF.6)** *are in force, let* $(\boldsymbol{X}, \boldsymbol{Y}, \boldsymbol{Z}) = (X_t, Y_t, Z_t)_{0 \le t \le T}$ *be a solution of the mean-field FBSDE (6.13), let u be the associated decoupling field, set $\mu_t = \mathcal{L}(X_t)$ for $0 \le t \le T$, and define the associated cost by*

$$J = \mathbb{E}\left[g(X_T, \mu_T) + \int_0^T f(t, X_t, \mu_t, \hat{\alpha}(t, X_t, \mu_t, Y_t)) dt\right], \tag{6.34}$$

where $\hat{\alpha}$ is the minimizer function constructed in Lemma 6.1. Next, we consider the solution $(X_t^1, \ldots, X_t^N)_{0 \le t \le T}$ of the system of N stochastic differential equations

$$dX_t^i = b\bigl(t, X_t^i, \bar{\mu}_t^N, \hat{\alpha}(t, X_t^i, \mu_t, u(t, X_t^i))\bigr) dt + \sigma dW_t^i, \quad 0 \le t \le T, \tag{6.35}$$

where

$$\bar{\mu}_t^N = \frac{1}{N} \sum_{j=1}^{N} \delta_{X_t^j} \tag{6.36}$$

is the empirical distribution of the X_t^i.

Then, the set $(\alpha_t^{N,i} = \hat{\alpha}(t, X_t^i, \mu_t, u(t, X_t^i)))_{1 \le i \le N, 0 \le t \le T}$ of strategies is an approximate Nash equilibrium of the N-player game (6.31)–(6.33) in the sense that

(i) there exists an integer N_0 such that for any $N \ge N_0$ and $A > 0$, there exists a constant C_A such that for any admissible strategy $\boldsymbol{\beta}^1 = (\beta_t^1)_{0 \le t \le T}$ for the first player

$$\mathbb{E}\int_0^T |\beta_t^1|^2 dt \ge C_A \implies \bar{J}^{N,1}(\boldsymbol{\beta}^1, \boldsymbol{\alpha}^{N,2}, \ldots, \boldsymbol{\alpha}^{N,N}) \ge J + A; \tag{6.37}$$

(ii) for any $A > 0$, there exists a sequence of positive real numbers $(\epsilon_N)_{N \ge 1}$ converging toward 0, such that for any admissible strategy $\boldsymbol{\beta}^1 = (\beta_t^1)_{0 \le t \le T}$ for the first player

$$\mathbb{E}\int_0^T |\beta_t^1|^2 dt \le A \implies \begin{cases} \bar{J}^{N,1}(\boldsymbol{\beta}^1, \boldsymbol{\alpha}^{N,2}, \ldots, \boldsymbol{\alpha}^{N,N}) \ge J - \varepsilon_N, \\ \bar{J}^{N,i}(\boldsymbol{\beta}^1, \boldsymbol{\alpha}^{N,2}, \ldots, \boldsymbol{\alpha}^{N,N}) \ge J - \varepsilon_N, \quad 2 \le i \le N. \end{cases} \tag{6.38}$$

Proof. Let $\boldsymbol{\beta}^1 = (\beta_t^1)_{0 \le t \le T}$ be a progressively measurable process satisfying $\mathbb{E}\int_0^T |\beta_t^1|^2 dt < \infty$, and let us use the quantities defined in (6.31), (6.32), and (6.33) with $\beta_t^i = \boldsymbol{\alpha}_t^{N,i}$ for

6.3. Propagation of Chaos and Approximate Nash Equilibriums

$i = 2, \cdots, N$. The convexity of f and g with respect to (x, α) around $x = 0$, $\alpha = 0$, and the lower bound (6.8) implies that

$$\bar{J}^{N,1}(\beta^1, \alpha^{N,2}, \ldots, \alpha^{N,N}) \geq \mathbb{E}\bigg[g(0, \bar{\nu}_T^N) + \int_0^T f(t, 0, \bar{\nu}_t^N, 0)dt\bigg] + \lambda \mathbb{E}\int_0^T |\beta_t^1|^2 dt$$
$$+ \mathbb{E}\bigg[\langle U_T^1, \partial_x g(0, \bar{\nu}_T^N)\rangle + \int_0^T \langle U_t^1, \partial_x f(t, 0, \bar{\nu}_t^N, 0)\rangle dt\bigg].$$

The Lipschitz assumption with respect to the Wasserstein distance and the definition of the latter imply that for any $t \in [0, T]$,

$$|f(t, 0, \bar{\nu}_t^N, 0) - f(t, 0, \delta_0, 0)| \leq C\bigg[\int_{\mathbb{R}^d} |x|^2 d\bar{\nu}_t^N(x)\bigg]^{1/2} = C\bigg[\frac{1}{N}\sum_{i=1}^N |U_t^i|^2\bigg]^{1/2}$$

with a similar inequality for g. As usual, the value of the constant C may vary from line to line. From this, we get

$$\bar{J}^{N,1}(\beta^1, \alpha^{N,2}, \ldots, \alpha^{N,N}) \geq \mathbb{E}\bigg[g(0, \delta_0) + \int_0^T f(t, 0, \delta_0, 0)dt\bigg] + \lambda \mathbb{E}\int_0^T |\beta_t^1|^2 dt$$
$$+ \mathbb{E}\bigg[\langle U_T^1, \partial_x g(0, \bar{\nu}_T^N)\rangle + \int_0^T \langle U_t^1, \partial_x f(t, 0, \bar{\nu}_t^N, 0)\rangle dt\bigg]$$
$$- C\bigg[\frac{1}{N}\sum_{i=1}^N \sup_{0 \leq t \leq T} \mathbb{E}|U_t^i|^2\bigg]^{1/2}.$$

Using the Lipschitz property of $\partial_x g$ and $\partial_x f$ with respect to the Wasserstein distance in a similar way, we see that for any $\delta > 0$, there exists a constant C_δ such that

$$\bar{J}^{N,1}(\beta^1, \alpha^{N,2}, \ldots, \alpha^{N,N}) \geq \mathbb{E}\bigg[g(0, \delta_0) + \int_0^T f(t, 0, \delta_0, 0)dt\bigg] + \lambda \mathbb{E}\int_0^T |\beta_t^1|^2 dt$$
$$- \delta \sup_{0 \leq t \leq T} \mathbb{E}[|U_t^1|^2] - C_\delta\bigg(1 + \frac{1}{N}\sum_{i=1}^N \sup_{0 \leq t \leq T} \mathbb{E}[|U_t^i|^2]\bigg). \quad (6.39)$$

Because of the special structure of the drift b, standard estimates for solutions of stochastic differential equations imply the existence of a constant C such that

$$\mathbb{E}\bigg(\sup_{0 \leq t \leq T} |U_t^1|^2\bigg) \leq C\bigg(1 + \mathbb{E}\int_0^T |\beta_t^1|^2 dt + \frac{1}{N}\sum_{j=1}^N \mathbb{E}\int_0^T |U_t^j|^2 dt\bigg)$$

$$\mathbb{E}\bigg(\sup_{0 \leq t \leq T} |U_t^i|^2\bigg) \leq C\bigg(1 + \frac{1}{N}\sum_{j=1}^N \mathbb{E}\int_0^T |U_t^j|^2\bigg), \quad 2 \leq i \leq N,$$

$$\frac{1}{N} \sup_{0 \leq t \leq T} \mathbb{E}\bigg[\sum_{j=1}^N |U_t^j|^2 dt\bigg] \leq C\bigg(1 + \frac{1}{N}\mathbb{E}\int_0^T |\beta_t^1|^2 dt\bigg)$$

so that

$$\mathbb{E}\bigg(\sup_{0 \leq t \leq T} |U_t^1|^2\bigg) \leq C\bigg(1 + \mathbb{E}\int_0^T |\beta_t^1|^2 dt\bigg).$$

Therefore, if δ is chosen small enough in (6.39), we can find a constant C such that

$$\bar{J}^{N,1}(\boldsymbol{\beta}^1, \boldsymbol{\alpha}^{N,2}, \ldots, \boldsymbol{\alpha}^{N,N}) \geq C + \left(\frac{\lambda}{2} - \frac{C}{N}\right) \mathbb{E} \int_0^T |\beta_t^1|^2 dt.$$

This proves that there exists an integer N_0 such that for $N \geq N_0$ and $A > 0$, there exists a constant C_A such that

$$\mathbb{E} \int_0^T |\beta_t^1|^2 dt \geq C_A \implies \bar{J}^{N,1}(\boldsymbol{\beta}^1, \boldsymbol{\alpha}^{N,2}, \ldots, \boldsymbol{\alpha}^{N,N}) \geq A,$$

because of the definition (6.34) of the cost J. Then, we can choose A such that

$$\mathbb{E} \int_0^T |\beta_t^1|^2 dt \geq C_A \implies \bar{J}^{N,1}(\boldsymbol{\beta}^1, \boldsymbol{\alpha}^{N,2}, \ldots, \boldsymbol{\alpha}^{N,N}) \geq J + A. \quad (6.40)$$

Below, we will choose $\boldsymbol{\beta}^1$ satisfying (6.40).

For the next step of the proof we introduce the system of decoupled independent and identically distributed states

$$d\overline{X}_t^i = b\bigl(t, \overline{X}_t^i, \mu_t, \hat{\alpha}(t, \overline{X}_t^i, \mu_t, u(t, \overline{X}_t^i))\bigr) dt + \sigma dW_t^i, \quad 0 \leq t \leq T,$$

so that the common law of the processes $\overline{\boldsymbol{X}}^i$ is $\mathcal{L}(\boldsymbol{X})$. In particular, $\mathcal{L}(X_t^i) = \mu_t$ for any $t \in [0,T]$ and $i \in \{1, \cdots, N\}$.

Using the regularity of the FBSDE decoupling field derived in Lemma 6.12 and the estimate recalled in Lemma 6.13, we can follow Sznitman's proof [113] (see also Theorem 1.3 of [72]), and get

$$\forall 1 \leq i \leq N, \quad \sup_{0 \leq t \leq T} \mathbb{E}\bigl[|X_t^i - \bar{X}_t^i|^2\bigr] \leq CN^{-2/(d+4)}$$

(recall that $(X^1, \ldots, X^N)$ solves (6.35)–(6.36), which implies

$$\forall 1 \leq i \leq N, \quad \sup_{0 \leq t \leq T} \mathbb{E}\bigl[W_2^2(\bar{\mu}_t^N, \mu_t)\bigr] \leq CN^{-2/(d+4)}.$$

Using the Lipschitz regularity of the coefficients g and f assumed in (**MFG.2**), of the minimizer $\hat{\alpha}$ proven in Lemma 6.1, and the Cauchy–Schwarz inequality, we get

$$J - \bar{J}^{N,i}(\boldsymbol{\alpha}^{N,1}, \ldots, \boldsymbol{\alpha}^{N,N})$$
$$= \mathbb{E}\biggl[g(X_T, \mu_T) + \int_0^T f(t, X_t, \mu_t, \hat{\alpha}(t, X_t, \mu_t, u(t, X_t))) dt$$
$$\quad - g(X_T^i, \bar{\mu}_T^N) - \int_0^T f(t, X_t^i, \bar{\mu}_t^N, \hat{\alpha}(t, X_t^i, \mu_t, u(t, X_t^i))) dt\biggr]$$
$$\leq C\mathbb{E}\biggl[|X_T - X_T^i| + W_2(\mu_T, \bar{\mu}_T^N) + \int_0^T \bigl(|X_t - X_t^i| + W_2(\mu_t, \bar{\mu}_T^N)$$
$$\quad + |\hat{\alpha}(t, X_t, \mu_t, u(t, X_t)) + \hat{\alpha}(t, X_t^i, \mu_t, u(t, X_t^i))|\bigr) dt\biggr]$$
$$\leq C\mathbb{E}\biggl[|X_T - X_T^i| + W_2(\mu_T, \bar{\mu}_T^N) + \int_0^T \bigl(|X_t - X_t^i| + W_2(\mu_t, \bar{\mu}_T^N)\bigr) dt\biggr]$$

6.3. Propagation of Chaos and Approximate Nash Equilibriums

for some constant $C > 0$. This proves that for any $1 \leq i \leq N$,

$$\bar{J}^{N,i}(\boldsymbol{\alpha}^{1,N}, \ldots, \boldsymbol{\alpha}^{N,N}) \geq J - CN^{-2/(d+4)}. \tag{6.41}$$

By standard stability arguments for SDEs, we have

$$\sup_{0 \leq t \leq T} \mathbb{E}\big[|U_t^1 - X_t^1|^2\big] \leq \frac{C}{N} \sum_{j=1}^N \mathbb{E}\int_0^T |U_t^j - X_t^j|^2 dt + C\mathbb{E}\int_0^T |\beta_t^1 - \boldsymbol{\alpha}_t^{N,1}|^2 dt,$$

$$\sup_{0 \leq t \leq T} \mathbb{E}\big[|U_t^i - X_t^i|^2\big] \leq \frac{C}{N} \sum_{j=1}^N \mathbb{E}\int_0^T |U_t^j - X_t^j|^2 dt, \quad 2 \leq i \leq N.$$

Therefore,

$$\frac{1}{N}\sum_{j=1}^N \mathbb{E}\int_0^T |U_t^j - X_t^j|^2 dt \leq \frac{C}{N}\mathbb{E}\int_0^T |\beta_t^1 - \boldsymbol{\alpha}_t^{N,1}|^2 dt,$$

so that

$$\sup_{0 \leq t \leq T} \mathbb{E}\big[|U_t^i - X_t^i|^2\big] \leq \frac{C}{N}\mathbb{E}\int_0^T |\beta_t^1 - \boldsymbol{\alpha}_t^{N,1}|^2 dt, \quad 2 \leq i \leq N.$$

As a by-product, by (6.41), for any $A > 0$, there exists a constant C_A such that

$$\mathbb{E}\int_0^T |\beta_t^1|^2 dt \leq A$$

implies

$$\sup_{0 \leq t \leq T} \mathbb{E}\big[|U_t^i - \overline{X}_t^i|^2\big] \leq C_A N^{-2/(d+4)}, \quad 2 \leq i \leq N,$$

so that, for any $t \in [0, T]$,

$$\mathbb{E}\big[W_2^2(\bar{\nu}_t^N, \mu_t)\big] \leq C_A N^{-2/(d+4)}. \tag{6.42}$$

As in (6.41), we deduce (up to a modification of C_A)

$$\bar{J}^{N,i}(\boldsymbol{\beta}^1, \boldsymbol{\alpha}^{N,2}, \ldots, \boldsymbol{\alpha}^{N,N}) \geq J - C_A N^{-2/(d+4)}, \quad 2 \leq i \leq N. \tag{6.43}$$

Finally we define $(\bar{U}_t^1)_{0 \leq t \leq T}$ as the solution of the SDE

$$d\bar{U}_t^1 = b(t, \bar{U}_t^1, \mu_t, \beta_t^1)dt + \sigma dW_t^1, \quad 0 \leq t \leq T.$$

By (6.42) and by the stability of SDEs, we deduce

$$\sup_{0 \leq t \leq T} \mathbb{E}\big[|U_t^1 - \bar{U}_t^1|^2\big] \leq C_A N^{-2/(d+4)}$$

so that

$$\bar{J}^{N,1}(\boldsymbol{\beta}^1, \boldsymbol{\alpha}^{N,2}, \ldots, \boldsymbol{\alpha}^{N,N}) \geq J(\boldsymbol{\beta}^1) - C_A N^{-2/(d+4)},$$

where $J(\boldsymbol{\beta}^1)$ stands for the limit cost of $\boldsymbol{\beta}^1$, namely,

$$J(\boldsymbol{\beta}^1) = \mathbb{E}\bigg[g(\bar{U}_T^1, \mu_T) + \int_0^T f\big(t, \bar{U}_t^1, \mu_t, \beta_t^1\big)dt\bigg]. \tag{6.44}$$

Since $J \leq J(\boldsymbol{\beta}^1)$ (notice that the stochastic maximum principle still applies despite the fact that $\boldsymbol{\beta}^1$ is adapted to a filtration which can be larger than the filtration generated by $\boldsymbol{W}^1$), we get in the end

$$\bar{J}^{N,1}(\boldsymbol{\beta}^1, \boldsymbol{\alpha}^{N,2}, \ldots, \boldsymbol{\alpha}^{N,N}) \geq J - C_A N^{-2/(d+4)}. \tag{6.45}$$

Putting together (6.43) and (6.45) completes the proof. □

6.3.3 ▪ Comparison with the Control of McKean–Vlasov SDEs

In both cases, the solution of the optimal control of SDEs of McKean–Vlasov type and the solution of an MFG problem reduce to an FBSDE of McKean–Vlasov type.

In both cases, the final form of the FBSDE is obtained by the conjunction of a fixed-point argument and the solution of an optimization problem, which we tackle in both cases by an appropriate form of the Pontryagin stochastic maximum principle. However, despite the strong similarities, these FBSDEs are different because these two steps, namely "fixed point" and "optimization", are not taken in the same order, leading to two different problems! We emphasize this important fact by reviewing the steps taken to solve these two problems.

In the solution of the optimal control of SDEs of McKean–Vlasov type, the fixed-point step is taken before the optimization. Indeed, the fixed-point argument is embedded in the form of the state dynamics, forcing the marginal distribution of the state to be present in the coefficients of the SDE, giving the dynamics of the state. Recall from Section 4.4 of Chapter 4 that the problem can be stated for open loop controls $\alpha = (\alpha_t)_{0 \le t \le T}$ (adapted to any specific information structure) in the form

$$\alpha^* = \arg\min_\alpha \mathbb{E}\left[\int_0^T f(t, X_t, \mathcal{L}(X_t), \alpha_t)dt + g(X_T, \mathcal{L}(X_T))\right] \quad (6.46)$$

subject to

$$dX_t = b(t, X_t, \mathcal{L}(X_t), \alpha_t)dt + \sigma(t, X_t, \mathcal{L}(X_t), \alpha_t)dW_t, \quad t \in [0, T].$$

This form of the problem is screaming for a probabilistic analysis based on an appropriate version of the Pontryagin stochastic maximum principle. We developed the tools necessary for such an analysis in Section 4.4 of Chapter 4. As in the classical theory, the adjoint equation appears as a BSDE whose driver is given by the derivative of the Hamiltonian with respect to the state variable. While the Hamiltonian is formally the same as in the case of the MFG problem, the state variable is now a couple (x, μ) accounting for the possible values of $(X_t, \mathcal{L}(X_t))$ which is the actual controlled state evolved dynamically. Reproducing (4.83) used in Chapter 4, we get

$$\begin{cases} dX_t = b(t, X_t, \mathcal{L}(X_t), \hat{\alpha}(t, X_t, \mathcal{L}(X_t), Y_t, Z_t))dt, \quad X_0 = x, \\ \quad + \sigma(t, X_t, \mathcal{L}(X_t), \hat{\alpha}(t, X_t, \mathcal{L}(X_t), Y_t, Z_t))dW_t, \\ dY_t = -\partial_x H(t, X_t, \mathcal{L}(X_t), Y_t, Z_t, \hat{\alpha}(t, X_t, \mathcal{L}(X_t), Y_t, Z_t))dt + Z_t dW_t, \\ \quad - \tilde{\mathbb{E}}[\partial_\mu H(t, \tilde{X}_t, \mathcal{L}(\tilde{X}_t), \tilde{Y}_t, \tilde{Z}_t, \hat{\alpha}(t, \tilde{X}_t, \mathcal{L}(\tilde{X}_t), \tilde{Y}_t, \tilde{Z}_t))(X_t)]dt \end{cases}$$

with terminal condition $Y_T = \partial_x g(X_T, \mathcal{L}(X_t)) + \tilde{\mathbb{E}}[\partial_\mu g(\tilde{X}_T, \mathcal{L}(\tilde{X}_t))(X_T)]$. This is in stark contrast with the MFG case for which the optimization is a control problem for which the flow of measures μ_t is frozen. For that reason, the driver of the adjoint BSDE does not involve partial derivatives of the Hamiltonian with respect to the measure argument. Indeed, recall that the FBSDE whose solution provides an MFG equilibrium reads

$$\begin{cases} dX_t = b(t, X_t, \mathcal{L}(X_t), \hat{\alpha}(t, X_t, \mathcal{L}(X_t), Y_t, Z_t))dt, \quad X_0 = x, \\ \quad + \sigma(t, X_t, \mathcal{L}(X_t), \hat{\alpha}(t, X_t, \mathcal{L}(X_t), Y_t, Z_t))dW_t, \\ dY_t = -\partial_x H(t, X_t, Y_t, Z_t, \hat{\alpha}(t, X_t, \mathcal{L}(X_t), Y_t, Z_t))dt + Z_t dW_t \end{cases}$$

with terminal condition $Y_T = \partial_x g(X_T, \mathcal{L}(X_t))$. Clearly, both the driver and the terminal condition of the adjoint equation of the control of McKean–Vlasov dynamics have an extra term which is not present in the MFG adjoint equation. This translates into differences in the FBSDEs, and as a result, there is no reason to expect the solutions to be the same. This point was driven home in [37], where simple examples were provided as illustrations.

The differences between the two problems can also be seen at the level of the construction of approximate equilibriums. We saw in the previous subsection how the solution of the MFG problem can be used to provide players of an N-player game with approximate Nash equilibriums from distributed strategies computed from the optimal control of the MFG problem. A similar result holds in the case of the control of McKean–Vlasov dynamics. If the McKean–Vlasov SDE is viewed as the limiting dynamics as $N \to \infty$ of a group of N players who act independently of each other using distributed strategies given by the same feedback function, then the solution of the optimal control of the McKean–Vlasov SDE also provides distributed strategies forming an approximate equilibrium. However, the equilibrium is not a Nash equilibrium, but more akin to a cooperative equilibrium which one would expect from *franchised* groups. Details on this result can be found in [34].

6.4 ▪ Applications and Open Problems

This final section is devoted to the discussion of a few applications in which mean-field games appear naturally. We review them because of their practical interest, even if in some cases, once the mathematical models are setup, we are still unable to solve them completely!

6.4.1 ▪ Revisiting the Simple Model for Systemic Risk

We first revisit the toy model of systemic risk introduced and fully analyzed in Section 5.5 of Chapter 5. The beauty of the model (and the main reason why I like it so much) is the fact that not only can the PDE approach be used, but the probabilistic approach can also be solved explicitly, both in its open loop and Markovian forms. So, given the fact that we already constructed explicit exact Nash equilibriums for all the finite player games, there is no real need to consider the MFG formulation if the only perk is to be able to construct approximate Nash equilibriums for large games. This is indeed the case, especially in the absence of the common noise W^0. But still, it is instructive to write down the MFG formulation and see how it appears as the limit of finite player games.

In the generality required by the presence of the common noise, the MFG formulation of the systemic risk toy model reads as follows:

(i) For each real-valued random function $t \hookrightarrow m_t$ adapted to the filtration generated by the common noise W^0, solve the optimal control problem

$$\inf_{\alpha \in \mathbb{A}} \mathbb{E}\left[\int_0^T \left[\frac{1}{2}|\alpha_t|^2 - q\alpha_t(m_t - X_t) + \frac{\epsilon}{2}(m_t - X_t)^2\right]dt + \frac{c}{2}(m_T - X_T)^2\right]$$

under the dynamic constraint $dX_t = a(m_t - X_t)\,dt + \sigma dB_t$, where $B_t = \sqrt{1-\rho^2}dW_t + \rho dW_t^0$ for some Wiener process $W = (W_t)_{0 \le t \le T}$ independent of W^0.

(ii) Solve the fixed-point problem

$$m_t = \mathbb{E}[X_t | \mathcal{F}^{W^0}], \qquad 0 \le t \le T.$$

Remark 6.15. *In the absence of the common noise W^0 (in other words, when $\rho = 0$), the function $(m_t)_{0 \le t \le T}$ is deterministic and the fixed-point condition simply means that in equilibrium, m_t should be the expectation of the state X_t. In this form, the problem is* *a particular case of the LQ mean-field game models discussed in Subsection 6.1.5. The solution of this MFG formulation reduces to the solution of a Riccati equation which is readily identified to (5.81) via the usual analysis of the FBSDE derived from the Pontryagin stochastic maximum principle. So, given our prior discussion of the large game limit, the MFG formulation appears naturally in the asymptotic regime $N \to \infty$.*

The above formulation of the MFG problem is obtained by directly applying the strategy outlined in Subsection 6.1.4 as derived from the analysis of the limiting behavior of the empirical distribution performed in Subsection 6.1.3. While the limit of the empirical distributions $\overline{\mu}_t^N$ is expected to satisfy an SPDE of the Kolmogorov type, the mean m_t is expected to satisfy an SDE over the probability structure of the common noise $\boldsymbol{W}^0$, or at least to be an Itô process on the filtration generated by the common noise. Details can be found in [39] and [33], in which this toy model is also used to derive the master equation of the problem.

The next example comes from the analysis of large population dynamics.

6.4.2 ▪ First Cucker–Smale Model for Swarming and Flocking

We assume that the positions at time t in three-dimensional space of N birds (or fish) are given by $x^1(t), \cdots, x^N(t)$ and their velocities by $v^1(t), \cdots, v^N(t)$. The original Cucker–Smale model states that the flock (or school) evolves according to the system of ordinary differential equations

$$\begin{cases} dx^i(t) = v^i(t)dt, \\ dv^i(t) = \frac{1}{N} \sum_{j=1}^{N} w_{ij}(t)(v^j(t) - v^i(t))dt \end{cases} \quad (6.47)$$

with initial conditions $x^i(0) = x_0^i \in \mathbb{R}^d$ and $v^i(0) = v_0^i \in \mathbb{R}^d$ for $i = 1, \cdots, N$, where the weights $w_{i,j}(t)$ are defined by

$$w_{i,j}(t) = w(|x^i(t) - x^j(t)|) \quad (6.48)$$

for a single weight function chosen by Cucker and Smale to be

$$w(d) = \frac{1}{(1+d^2)^\beta}, \quad d \in \mathbb{R},$$

for some parameter $\beta \geq 0$. The authors prove that for $0 \leq \beta < 1/2$, flocking occurs in the sense that, independently of the values chosen for the initial conditions, all the velocity vectors $v^i(t)$ converge to a common value as $t \to \infty$, and the distances between individuals remain bounded in this limit. This result cannot be expected for $\beta > 1/2$ showing that $\beta = 1/2$ is a critical value for the main parameter of the model. Taking the limit $N \to \infty$ to derive a continuum model in the spirit of the derivation of the Boltzmann equation in statistical mechanics, it can be shown that the density $\varphi(t, x, v)$ at time t of individuals located at x with velocity v satisfies the equation

$$\frac{\partial \varphi}{\partial t}(t, x, v) + v \cdot \nabla_x \varphi(t, x, v) = \nabla_v([\Xi \varphi](t, x, v) \cdot \varphi(t, x, v)), \quad (6.49)$$

where the convolution operator Ξ is defined by

$$[\Xi \varphi](t, x, v) = \int \int_{\mathbb{R}^d \times \mathbb{R}^d} w(|x - x'|)(v - v')\varphi(t, x', v') \, dx' dv'. \quad (6.50)$$

6.4. Applications and Open Problems

The main reason for the popularity of this phenomenological model is the fact that the results of its analysis are consistent with some of the observations of bird flocks and fish schools. The next step in the search for a better understanding of the roots of swarming is to seek a stochastic equilibrium model whose output would produce an optimal dynamics in line with the model (6.47) postulated by Cucker and Smale. We present the proposal put forth by Caines, Malhamé, and Nourian in [95]. For each $i \in \{1, \cdots, N\}$ the stochastic dynamics of individual i are assumed to be given by equations of the form

$$\begin{cases} dx^i(t) = v^i(t)dt, \\ dv^i(t) = \alpha_t^i dt + \sigma dW_t^i \end{cases} \quad (6.51)$$

for three-dimensional adapted processes $\boldsymbol{\alpha}^i = (\alpha_t^i)_{0 \le t \le T}$ representing the controls used by the individuals in order to modify their velocities, and for independent three-dimensional Wiener processes $\boldsymbol{W}^i = (W_t^i)_{0 \le t \le T}$ and a positive constant σ. We could also assume that σ is a 3×3 matrix. Equilibrium will occur if the individuals $i \in \{1, \cdots, N\}$ manage to minimize their expected *costs* given by

$$J^i(\boldsymbol{\alpha}^1, \cdots, \boldsymbol{\alpha}^N) = \mathbb{E}\left[\int_0^T \left(\frac{1}{2}|\alpha_t^i|^2 + \left\| \frac{1}{N} \sum_{j=1}^N w(x^i(t) - x^j(t))(v^i(t) - v^j(t)) \right\|^2 \right) dt \right]. \quad (6.52)$$

In the case of bird flocks, this choice can be explained as follows: each bird tries to minimize the energy spent controlling the changes in its own velocity (hence the first term of the running cost) as well as the magnitude of the difference between its own velocity and the velocities of the other birds, as a more homogeneous flock is safer as predators will have a harder time preying on its members. Note that the further a bird is from the bulk of the flock, the less it influences the alignment of the velocities. We rewrite this model in the notation used throughout. Let us denote by $\boldsymbol{X}_t^i$ the six-dimensional vector whose first three components are given by the position $x^i(t)$ and the remaining components by the velocity vector $v^i(t)$, and view $\boldsymbol{X}_t^i$ as the value at time t of the private state of bird i. Its dynamics are linear as they are given by

$$d\boldsymbol{X}_t^i = (A\boldsymbol{X}_t^i + \alpha_t^i B)dt + \Sigma dW_t^i,$$

where the matrices A, B, and Σ are given by

$$A = \begin{bmatrix} 0 & 1 \\ 0 & 0 \end{bmatrix}, \quad B = \begin{bmatrix} 0 \\ 1 \end{bmatrix}, \quad \Sigma = \begin{bmatrix} 0 \\ \sigma \end{bmatrix}.$$

The functional it minimizes by bird i is given by

$$J^i(\boldsymbol{\alpha}^1, \cdots, \boldsymbol{\alpha}^N) = \mathbb{E}\left[\int_0^T f(t, \boldsymbol{X}_t^i, \overline{\mu}_t^N, \alpha_t^i) dt \right],$$

where as usual, $\overline{\mu}_t^N$ denotes the empirical measure of the sample $\boldsymbol{X}_t^1, \cdots, \boldsymbol{X}_t^N$ and the function f is defined by

$$f(t, \mathbf{x}, \mu, \alpha) = \frac{1}{2}|\alpha|^2 + \left\| \int\int_{\mathbb{R}^d \times \mathbb{R}^d} w(|x - x'|)(v - v') \, \mu(dx', dv') \right\|^2.$$

Recall that $\mathbf{x} = (x, v)$. Notice that, in the particular case $\beta = 0$, the weight function w is identically equal to 1, and the model reduces to an LQ game whose analysis can be reduced by considering the velocities only.

Also, if we assume that the flow $\boldsymbol{\mu} = (\mu_t)_{0 \le t \le T}$ of probability measures on $\mathbb{R}^6$ is frozen, the (reduced) Hamiltonian of the corresponding optimal control problem reads

$$H(t, \mathbf{x}, \mathbf{y}, \alpha) = A\mathbf{x} \cdot \mathbf{y} + \alpha B \cdot \mathbf{y} + f(t, \mathbf{x}, \mu_t, \alpha),$$

which is minimized (as a function of α) for $\hat{\alpha} = -B\mathbf{y}$. But despite its relative simplicity, the model has not been solved exactly (neither as a finite player game nor in its MFG formulation), except in the LQ case given by the particular choice $\beta = 0$.

Ignoring the fact that the problem is already difficult enough in the above formulation, one possible extension of the model is dictated by the following criticism. In model (6.51), the noise terms dW_t^i are independent and attached to the individuals. In other words, the random shocks depend upon the identities of the birds, and not where the birds are! A more realistic model would include random shocks depending upon the positions in space of the birds, namely something of the form $dW_t^{x^i(t)}$ where for each $x \in \mathbb{R}^3$, $\boldsymbol{W}^x$ would a Wiener process. So a more desirable equation for the dynamics of the velocities could be

$$dv^i(t) = \alpha_t^i dt + \sigma dW_t^i + \int_{\mathbb{R}^3} \delta(x^i(t) - x') W^0(dt, dx'), \tag{6.53}$$

where $\boldsymbol{W}^0 = (W^0(A \times B))_{A \in \mathcal{B}_{[0,\infty)}, B \in \mathcal{B}_{\mathbb{R}^3}}$ is a mean-zero Gaussian white noise measure independent of the Wiener processes $\boldsymbol{W}^i$, and whose intensity we denote by ν, i.e.,

$$\mathbb{E}[W^0(A \times B) W^0(A' \times B')] = |A \cap A'| \nu(B \cap B'),$$

where the notation $|C|$ is used for the Lebesgue measure of the Borel set C. The function appearing in the integral in (6.53) is given by a nonnegative smooth function $x \hookrightarrow \delta(x)$ with compact support. It can be thought of as a smoothed delta function. Indeed, should it be a delta function, the integral would reduce to

$$\int_{\mathbb{R}^3} \delta(x^i(t) - x') W^0(dt, dx') = W^0(dt, x^i(t)),$$

which is exactly the type of extra noise we wanted to add. This remark is the main reason for our introduction of the random environment in Subsection 6.1.2.

6.4.3 ▪ A Growth Model in Macroeconomics

We present a simplified version of a real business cycle model in macroeconomics. It could be viewed as a form of Aiyagari's stochastic growth model. The economy has two types of agents. The first is households, which we label by $i \in \{1, \cdots, N\}$, trying to maximize their net present value of utility of consumption

$$J^i(\mathbf{c}^1, \cdots, \mathbf{c}^N) = \mathbb{E} \int_0^\infty e^{-\rho t} U(c_t^i) dt,$$

where c_t^i denotes the consumption of household i at time t, ρ is the rate of discounting, and U is a common utility function. For the purpose of the present discussion we shall use the CRRA utility function $U(x) = x^{1-\gamma}/(1-\gamma)$. We denote by k_t^i the income of household i and we assume that, in our simplified model, it satisfies

$$dk_t^i = [w_t + k_t^i r_t - c_t^i] dt + \sigma k_t^i dW_t^i,$$

where the $\boldsymbol{W}^i = (W_t^i)_{t \ge 0}$ are independent Wiener processes. The processes $(w_t)_{t \ge 0}$ and $(r_t)_{t \ge 0}$ are exogenously given. They will be determined in equilibrium: w_t represents the

6.4. Applications and Open Problems

wages and r_t the short interest rate. A consumption strategy $\mathbf{c}^i = (c^i)_{t \geq 0}$ is admissible for household i if $c_t^i \geq 0$ and $k_t^i \geq k^*$ for all $t \geq 0$ and a fixed constant $k^* \in \mathbb{R}$ representing the borrowing limit (e.g., $k^* = 0$ would mean that the household can only save). We denote by $\overline{\mu}_t^N$ the empirical distribution of the k_t^i.

Firms form the second component of the economy. They try to maximize their profit

$$\sup_{(K_t)_t, (L_t)_t} AF(K_t, L_t) - (r_t + \delta)K_t - w_t L_t, \qquad (6.54)$$

where the constant A represents productivity, K_t and L_t the capital and labor available at time t, and $\delta > 0$ the depreciation rate, so that $r_t + \delta$ represents the cost of capital. For the purpose of this discussion, we use the Cobb–Douglas production function

$$F(K, L) = K^\alpha L^{1-\alpha} \qquad (6.55)$$

for some constant $\alpha \in (0, 1)$. The *market clearing* conditions give

$$L_t = 1, \qquad K_t = \overline{k}^N, \qquad (6.56)$$

where $\overline{k}^N$ is the mean household income given by

$$\overline{k}^N = \int k \overline{\mu}_t^N(dk), \qquad (6.57)$$

and the interest rate and wage processes are given by

$$r_t = A \partial_K F(\overline{k}^N, 1) - \delta = \alpha A (\overline{k}^N)^{\alpha-1} \qquad (6.58)$$

and

$$w_t = \partial_L F(\overline{k}^N, 1) - \delta = (1 - \alpha) A (\overline{k}^N)^\alpha, \qquad (6.59)$$

which are (essentially) the marginal cost of capital and the marginal cost of labor respectively. Plugging equations (6.58) and (6.59) into the dynamics of the household income gives the new dynamic equations

$$dk_t^i = [(1-\alpha)A(\overline{k}^N)^\alpha + \alpha A(\overline{k}^N)^{\alpha-1}k_t^i - c_t^i]dt + \sigma k_t^i dW_t^i,$$

and since the coupling between the equations is through the mean of the empirical distribution of the k_t^i, this model fits within the realm of mean-field games. Notice that the volatility σk_t^i is linear in the state (and hence vanishes) and that the drift is singular (unbounded and not Lipschitz) since it is given (in our standard notation) by the function

$$b(t, x, \mu, \alpha) = (1-\alpha)A\overline{\mu}^\alpha + \alpha A \overline{\mu}^{\alpha-1} x - \alpha,$$

where $\overline{\mu}$ denotes the mean of μ, namely $\overline{\mu} = \int x\mu(dx)$. Further discussions and elements of solutions are provided in the forthcoming book [35].

6.4.4 ▪ Games with Major Players

In many practical applications of interest, the population of players is not statistically homogeneous, and a fixed finite number of players have a different knowledge and impact on the large homogeneous population of smaller-influence players. We call the latter *minor players* and we bundle the former into one single player whom we call a *major player*. To be more specific, we consider a game with $N+1$ players $i = 0, 1, \ldots N$ whose (private) states satisfy the dynamic equations

$$dX_t^0 = b^0(t, X_t^0, \overline{\mu}_t^N, \alpha_t^0)dt + \sigma^0 dW_t^0, \tag{6.60}$$
$$dX_t^i = b(t, X_t^i, \overline{\mu}_t^N, \alpha_t^i, X_t^0)dt + \sigma dW_t^i, \qquad i = 1, \ldots, N, \tag{6.61}$$

where as usual $\overline{\mu}_t^N$ denotes the empirical measure of the private states $X_t^1, \ldots, X_t^N$ at time t of the (minor) players. Here $\boldsymbol{W}^0$ is an m^0-dimensional Wiener process independent of the N m-dimensional independent Wiener processes $\boldsymbol{W}^1, \ldots, \boldsymbol{W}^N$. σ^0 and σ could be nonnegative scalars, but they could also be $d^0 \times m^0$ and $d \times m$ matrices. We shall assume that they are deterministic for the sake of simplicity. Player 0 is called a major player because the value X_t^0 of his private state enters the dynamics of the states of players i for $i = 1, \ldots, N$, while the states of players $1, \ldots, N$ enter the dynamics of the other states (including the state of the major player) only through their empirical distribution $\overline{\mu}_t^N$. The latter are called minor players for this very reason. Obvious applications of these models involve the analysis of the interbank network where a small number of large institutions (usually considered to be too big to fail) transact with a large number of smaller institutions whose impact on the system as a whole is minimal.

Since the states of all the minor players are statistically identical, and influenced by the state of the major player in exactly the same way, it is reasonable to assume that they are exchangeable, even when the optimal strategies (in the sense of Nash equilibrium) are implemented. Recall that for any sequence of integrable exchangeable random variables $(X_i)_{i \geq 1}$, de Finetti's law of large numbers states that, almost surely,

$$\frac{1}{N} \sum_{i=1}^{N} \delta_{X_i} \Longrightarrow \mathcal{L}(X_1 | \mathcal{G})$$

for some σ-field $\mathcal{G}$, where $\Longrightarrow$ denotes convergence in distribution. We may want to apply this result for each time t to the individual states X_t^i, in which case a natural candidate for the σ-field $\mathcal{G}$ could be the element $\mathcal{F}_t^0$ of the filtration generated by the Wiener process $\boldsymbol{W}^0$ driving the dynamics of the state of the major player. This suggests that in mean-field games with major and minor players, we can proceed essentially in the same way as in the standard mean-field game theory, except for the fact that instead of fixing a *deterministic* measure flow in the first step, we fix an adapted *stochastic* measure flow, and in the last step, match this stochastic measure flow to the flow of marginal conditional distribution of the state of the representative minor player given $\mathcal{F}_t^0$. This is in accordance with intuition since, as all the minor players are influenced by the major player, they should make their decisions conditioned on the information provided by the major player. Notice that this is also consistent with the procedure used in the presence of a so-called common noise, as investigated in [38].

However, the above argument fails to apply to the major player. Indeed, no matter how many minor players are present in the game, the major player's control influences all the minor players, and in particular, the empirical distribution formed by the minor players. When we construct the limiting problem for the major player, it is thus more reasonable to allow the major player to control the stochastic measure flow, instead of fixing it a priori. This asymmetry between major and minor players was already observed in [15].

Formulation of the MFG Problem with Major and Minor Players

We use the above heuristic argument to articulate an MFG formulation for the major-minor mean-field game problem. The limiting control problem for the major player is of conditional McKean–Vlasov type with an endogenous measure flow, and the limiting control

6.4. Applications and Open Problems

problem for the representative minor player is standard, with an exogenous measure flow fixed at the beginning of the scheme. As a consequence, the limiting problem becomes a two-player stochastic differential game between the major player and a representative minor player. This is in contrast with the existing literature, where this limiting problem is usually framed as two consecutive stochastic control problems:

(i) For each $\mathbb{F}^0$-progressively measurable stochastic measure flow $\boldsymbol{\mu} = (\mu_t)_{0 \le t \le T}$, where $\mathbb{F}^0 = (\mathcal{F}_t^0)_{t \ge 0}$ denotes the filtration generated by the Wiener process $\boldsymbol{W}^0$, solve the following two-player stochastic differential game where the strategy $\boldsymbol{\alpha}^0 = (\alpha_t^0)_{0 \le t \le T}$ of the first player is assumed to be adapted to $\mathbb{F}^0$, and the strategy $\boldsymbol{\alpha} = (\alpha_t)_{0 \le t \le T}$ of the second player is assumed to be adapted to the filtration $\mathbb{F} = (\mathcal{F}_t)_{t \ge 0}$ generated by the Wiener processes $\boldsymbol{W}^0$ and $\boldsymbol{W}$, and where the controlled dynamics of the state of the system are given by

$$\begin{cases} dX_t^0 = b_0(t, X_t^0, \mathcal{L}(X_t | \mathcal{F}_t^0), \alpha_t^0) dt + \sigma_0(t, X_t^0, \mathcal{L}(X_t | \mathcal{F}_t^0), \alpha_t^0) dW_t^0, \\ dX_t = b(t, X_t, \mathcal{L}(X_t | \mathcal{F}_t^0), X_t^0, \alpha_t) dt + \sigma(t, X_t, \mathcal{L}(X_t | \mathcal{F}_t^0), X_t^0, \alpha_t) dW_t, \\ d\check{X}_t^0 = b_0(t, \check{X}_t^0, \mu_t, \alpha_t^0) dt + \sigma_0(t, \check{X}_t^0, \mu_t, \alpha_t^0) dW_t^0, \\ d\check{X}_t = b(t, \check{X}_t, \mu_t, \check{X}_t^0, \alpha_t) dt + \sigma(t, \check{X}_t, \mu_t, \check{X}_t^0, \alpha_t) dW_t \end{cases}$$
(6.62)

with initial conditions $X_0^0 = \check{X}_0^0 = x_0^0$ and $X_0 = \check{X}_0 = x_0$, and where the cost functionals for the two players are given by

$$J^0(\boldsymbol{\alpha}^0, \boldsymbol{\alpha}) = \mathbb{E}\left[\int_0^T f_0(t, X_t^0, \mathcal{L}(X_t | \mathcal{F}_t^0), \alpha_t^0) dt + g_0(X_T^0, \mathcal{L}(X_T | \mathcal{F}_T^0))\right],$$

$$J(\boldsymbol{\alpha}^0, \boldsymbol{\alpha}) = \mathbb{E}\left[\int_0^T f(t, \check{X}_t, \mu_t, \check{X}_t^0, \alpha_t) dt + g(\check{X}_T, \mu_T, \check{X}_T^0)\right],$$

where $\mathcal{L}(X_t | \mathcal{F}_t^0)$ stands for the conditional distribution of X_t given $\mathcal{F}_t^0$. By solving the game we mean *looking for Nash equilibriums* for this game.

(ii) Find a flow $\boldsymbol{\mu} = (\mu_t)_{0 \le t \le T}$ satisfying the consistency condition

$$\mu_t = \mathcal{L}(X_t | \mathcal{F}_t^0), \quad \forall t \in [0, T],$$
(6.63)

where X_t is the second component of the state (6.62) when a Nash equilibrium control couple (α^0, α) found in Step 2 is plugged into the dynamic equations of the state given in (6.62).

Notice that the above consistency condition amounts to solving a fixed-point problem in the space of stochastic measure flows. This is typical of the MFG problems with common noise. After the consistency condition (6.63) is met, (X^0, X) and $(\check{X}^0, \check{X})$ coincide, even though at the beginning of the scheme they emerge from different measure flows: (X^0, X) is defined with the endogenous measure flow $(\mathcal{L}(X_t | \mathcal{F}_t^0))_{0 \le t \le T}$, while $(\check{X}^0, \check{X})$ is defined with the exogenous measure flow $\boldsymbol{\mu} = (\mu_t)_{0 \le t \le T}$. Indeed, when computing his/her best response to the major player, a typical minor player assumes that the stochastic flow $\boldsymbol{\mu} = (\mu_t)_{0 \le t \le T}$ is fixed, as in the standard approach to mean-field games recalled at the beginning of the section. The fact that he/she is responding to a major player, who should behave in the environment given by $\boldsymbol{\mu} = (\mu_t)_{0 \le t \le T}$, is the justification for the introduction of $\check{X}_t^0$ in lieu of X_t^0 to compute his/her best response. Similarly, for the reasons given at the end of the previous subsection, the major player computes his/her best response assuming that the typical minor player uses the endogenous stochastic flow $(\mathcal{L}(X_t|\mathcal{F}_t^0))_{0\leq t\leq T}$. So, he/she is responding to the dynamics of X_t instead of the dynamics given by $\check{X}_t$. This explains this apparent doubling of the states, which disappears in equilibrium when the consistency condition is satisfied.

Notice also that even when the X_t^i are scalar, the system (6.62) describes the dynamics of a four-dimensional state driven by two independent Wiener processes. The dynamics of the first two components are of the conditional McKean–Vlasov type (because of the presence of the conditional distribution $\mathcal{L}(X_t|\mathcal{F}_t^0)$ of X_t in the coefficients), while the dynamics of the last two components are given by standard stochastic differential equations with random coefficients. In this two-player game, the cost functional J^0 of the major player is of the McKean–Vlasov type while the cost functional J of the representative minor player is of the standard type. As explained earlier, this is the main feature of our formulation of the problem. We end this subsection with the precise definition of a solution to the mean-field game described above.

Definition 6.16. *Given a tuple $(\Omega, \mathcal{F}, \mathbb{P}, \boldsymbol{W}^0, \boldsymbol{W})$, we denote by $\mathbb{F}^0$ (resp., $\mathbb{F}$) the augmented natural filtration generated by $\boldsymbol{W}^0$ (resp., $\boldsymbol{W}^0$ and $\boldsymbol{W}$), and we say that a couple of strategies $(\boldsymbol{\alpha}^0, \boldsymbol{\alpha})$, such as $\boldsymbol{\alpha}^0$ is $\mathbb{F}^0$-progressively measurable and $\boldsymbol{\alpha}$ is $\mathbb{F}$-progressively measurable, is a solution of the MFG problem with major and minor players if they form a Nash equilibrium for the two-player game defined in step* (i), *and satisfy the consistency condition in step* (ii).

The tools needed to tackle this problem were developed and presented earlier in the text. But given the hard time we had solving the MFG problem in the classical case, compounding difficulties in (1) solving optimization problems involving SDEs of McKean–Vlasov type, (2) finding Nash equilibriums, and (3) proving the existence of fixed points in spaces of stochastic measure flows, it is likely to be a daunting task. Not surprisingly, existence results are few and far between, and they deal with special cases, mostly LQ models not surprisingly, and then not even the most general of the LQ models.

Finally, we emphasize that under mild conditions, whenever a fixed point can be found in step (ii), i.e., a stochastic measure flow $\boldsymbol{\mu} = (\mu_t)_{0\leq t\leq T}$ satisfying (6.63), it is possible to use the corresponding equilibrium strategies to construct approximate Nash equilibriums for the finite-player games with major and minor players, when the number of minor players is sufficiently large. A modicum of care is needed to give a precise meaning to this statement, but like in the classical MFG case, the proof relies heavily on the theory of the propagation of chaos, although in the present situation, one needs to use a form of *conditional propagation of chaos*.

The reader interested in MFGs with major and minor players is referred to the Notes & Complements section below for some historical comments and precise references.

6.5 ▪ Notes & Complements

As stated in the abstract, the goal of this chapter was to give a soft introduction to the class of game models covered by the terminology *mean-field games*. Besides, we rely on the technology developed in the previous chapters to present the elements of what can be viewed as the probabilistic approach to mean-field games. For an in-depth analysis of the mathematical technicalities underpinning this approach, the interested reader is referred to the forthcoming book [35] by Carmona and Delarue.

Mean-field game models were introduced by Lasry and Lions around 2005–2006 in a series of papers [79, 80, 81] relying on PDE techniques. In a subsequent series of works [62, 63, 64, 77, 78] with PhD students and postdoctoral fellows, they considered applications to domains as diverse as the management of exhaustible resources, house insulation, and the analysis of pedestrian crowds.

6.5. Notes & Complements

On the other side of the pond, motivated by problems in large communication networks, Huang, Caines, and Malhamé introduced, essentially at the same time, a similar strategy which they called the Nash certainty equivalence [70]. They also studied practical applications to large populations' behavior [69]. The presentation given in this chapter of the solution of MFGs without common noise is borrowed from the work of Carmona and Delarue [32], in which complete proofs of the approximation procedure can be found. There, the authors developed the probabilistic approach based on the Pontryagin stochastic maximum principle.

A proof of Horowitz's result on the rate of convergence for empirical measures as stated in the text, can be found, for example, in Section 10 of [109]. Stronger (essential optimal) results exist. Precise statements and complete proofs can be found in Carmona and Delarue's forthcoming book [35]. Our construction of ϵ-Nash equilibriums follows the approach used by Bensoussan et al. [19] in the linear-quadratic case.

The discussion of the similarities with and the differences from the solution of the optimal control of stochastic differential equations of McKean–Vlasov type, a problem considered in Chapter 4, is modeled after the paper by Carmona, Delarue, and Lachapelle [37]. The probabilistic approach to MFG models with common noise can be found in a recent work of Carmona, Delarue, and Lacker [38]. The level of technicality required to formulate and solve the problem in this generality is far beyond the scope of this book. A different approach, more in the spirit of the reduction of the problem to the solution of FBSDEs of McKean–Vlasov type can be found in the book [35] of Carmona and Delarue.

The analysis of the systemic risk toy model as a mean-field game problem shows clearly the appearance of the mean-field equations in the limit of large games. As for the discussion of this example in Chapter 5, the material is borrowed from [39]. The limit $N \to \infty$ of large games is detailed and generalized in the discussion of the master equation index[sub]master equation provided by Carmona and Delarue in [33].

The mean-field game model of flocking was originally proposed by Nourian, Caines, and Malhamé in [95] to provide a mechanism by which flocking behavior emerges as an equilibrium, as a game counterpart of the well-known Cucker–Smale model [46]. In [95], the authors identify the mean-field limit and, *under the assumption that there exists a unique solution to the limiting mean-field game*, construct approximate Nash equilibriums for the finite-player games. Existence and approximation results for both their model and two related nearest-neighbor models were provided in the finite horizon case by Carmona and Lacker in [40].

We learned of Aiyagari's version of the macroeconomic growth model discussed in the text through a private conversation with Benjamin Moll. The discussion in the text is based on Aiyagari's original contribution [3] and the discussion of a PDE approach as explained in [1]. Further discussions and solutions under specific assumptions are provided in the forthcoming book [35] by Carmona and Delarue.

The MFG model with a major player was first introduced and studied by M. Huang in [68],t in the case of linear-quadratic models in infinite horizon. Reference [97] introduced the finite horizon counterpart of Huang's model, and [99] generalized the model to the nonlinear case. Unfortunately, the scheme proposed in [97, 99] fails to accommodate the case where the state of the major player enters the dynamics of the minor players. To be more specific, in [97, 99], the major player influences the minor players solely via their cost functionals. Reference [98] proposes a new scheme to solve the general case for LQ games in which the major player's state enters the dynamics of the minor players. The limiting control problem for the major player is solved by what the authors call *anticipative variational calculation*. In [15], the authors' take, like in [99], uses a stochastic Hamilton–Jacobi–Bellman approach and characterizes the limiting problem as a set of stochastic PDEs. The problem presented in this chapter, although very close to the framework proposed by Bensoussan, Chau, and Yam in [15], results in a two-player game involving the control of McKean–Vlasov dynamics. Its rationale is provided in detail in [42], where it is proved to be different from the models used in the previous literature and where it is solved in the linear-quadratic case.

For more practical applications of mean-field games, the interested reader is referred to the forthcoming book [35] by Carmona and Delarue.

Bibliography

[1] Y. Achdou, F. Buera, J.M. Lasry, B. Moll, and P.L. Lions. Partial differential equation models in macroeconomics. *Philosophical Transactions of the Royal Society A*, 372:20130397, 2014. (Cited on p. 251)

[2] N. Ahmed and X. Ding. Controlled McKean-Vlasov equations. *Communications in Applied Analysis*, 5(2):183–206, 2001. (Cited on p. 161)

[3] S.R. Aiyagari. Uninsured idiosyncratic risk and aggregate saving. *The Quarterly Journal of Economics*, 109(3):659–684, 1994. (Cited on p. 251)

[4] R. Almgren. Optimal execution with nonlinear impact functions and trading-enhanced risk. *Applied Mathematical Finance*, 10(1):1–18, 2003. (Cited on p. 216)

[5] R. Almgren and N. Chriss. Optimal execution of portfolio transactions. *Journal of Risk*, 3(2):5–39, 2001. (Cited on p. 216)

[6] L. Ambrosio, N. Gigli, and G. Savaré. *Gradient Flows in Metric Spaces and in the Space of Probability Measures*. Birkhäuser, Basel, 2005. (Cited on p. 143)

[7] D. Andersson and B. Djehiche. A maximum principle for SDEs of mean-field type. *Applied Mathematics & Optimization*, 63(3):341–356, 2010. (Cited on pp. 143, 145, 146, 154, 161)

[8] M. Arisawa. Ergodic problem for the Hamilton-Jacobi-Bellman equation. I. Existence of the ergodic attractor. *Annales de l'Institut Henri Poincaré: Analyse Non Linéaire*, 14:415–438, 1997. (Cited on p. 118)

[9] M. Arisawa. Ergodic problem for the Hamilton-Jacobi-Bellman equation. II. *Annales de l'Institut Henri Poincaré: Analyse Non Linéaire*, 15:1–24, 1998. (Cited on p. 118)

[10] M. Avellaneda and A. Paras. Managing the volatility risk of portfolios of derivative securities: the Lagrangian uncertain volatility model. *Applied Mathematical Finance*, 3:21–52, 1996. (Cited on p. 118)

[11] P. Barrieu and N. El Karoui. Monotone stability of quadratic semimartingales with applications to unbounded general quadratic BSDEs. *Annals of Probability*, 41:1831–1853, 2013. (Cited on pp. 59, 63)

[12] S. Benachour, B. Roynette, D. Talay, and P. Vallois. Nonlinear self-stabilizing processes—I: Existence, invariant probability, propagation of chaos. *Stochastic Processes and Their Applications*, 75:173–201, 1998. (Cited on p. 118)

[13] S. Benachour, B. Roynette, and P. Vallois. Nonlinear self-stabilizing processes—II: Convergence to invariant probability. *Stochastic Processes and Their Applications*, 75:203–224, 1998. (Cited on p. 118)

[14] A. Bensoussan. *Stochastic Control of Partially Observable Systems*. Cambridge University Press, Cambridge, UK, 1992. (Cited on pp. 160, 216)

[15] A. Bensoussan, M.H.M. Chau, and S.C.P. Yam. Mean field games with a dominating player. Technical report, 2013. (Cited on pp. 248, 251)

[16] A. Bensoussan and J. Frehse. Nonlinear elliptic systems in stochastic game theory. *Journal für die reine und angewandte Mathematik*, 350:23–67, 1984. (Cited on p. 216)

[17] A. Bensoussan and J. Frehse. On Bellman equations of ergodic type with quadratic growth Hamiltonian. In L. Cesari, editor, *Contributions to Modern Calculus of Variations*, volume 148 of *Pitman Research in Mathematics*, pages 13–26. Longman, Harlow, UK, 1987. (Cited on p. 118)

[18] A. Bensoussan, J. Frehse, and H. Nagai. Some results on risk-sensitive control with full observation. *Applied Mathematics and Optimization*, 37:1–41, 1998. (Cited on p. 117)

[19] A. Bensoussan, K.C.J. Sung, S.C.P. Yam, and S.P. Yung. Linear-quadratic mean field games. Technical report, 2014. (Cited on pp. 236, 251)

[20] D.P. Bertsekas. *Dynamic Programming and Stochastic Control*. Academic Press, New York, 1976. (Cited on p. 117)

[21] D. Bertsimas and A. Lo. Optimal control of execution costs. *Journal of Financial Markets*, 1:1–50, 1998. (Cited on p. 216)

[22] J.-M. Bismut. An introductory approach to duality in optimal stochastic control. *SIAM Review*, 20:62–78, 1978. (Cited on p. 160)

[23] R. Buckdahn, B. Djehiche, J. Li, and S. Peng. Mean-field backward stochastic differential equations: a limit approach. *Annals of Probability*, 37:1524–1565, 2009. (Cited on p. 63)

[24] R. Buckdahn and J. Li. Stochastic differential games and viscosity solutions of Hamilton–Jacobi–Bellman–Isaacs equations. *SIAM Journal on Control and Optimization*, 47:444–475, 2008. (Cited on p. 216)

[25] K. Burdzy, W. Kang, and K. Ramanan. The Skorokhod problem in a time-dependent interval. *Stochastic Processes and Their Applications*, 119:428–452, 2009. (Cited on p. 117)

[26] A. Cadenillas and I. Karatzas. The stochastic maximum principle for linear, convex optimal control with random coefficients. *SIAM Journal on Control and Optimization*, 33:590–624, 1995. (Cited on p. 160)

[27] P. Cardaliaguet. Notes from P.L. Lions' lectures at the Collège de France. Technical report, 2012. (Cited on p. 143)

[28] P. Cardaliaguet, J.M. Lasry, P.L. Lions, and A. Porretta. Long time average of mean field games. Technical report, 2012. (Cited on p. 118)

[29] P. Cardaliaguet and C. Rainer. Stochastic differential games with asymmetric information. *Applied Mathematics and Optimization*, 59:1–36, 2009. (Cited on p. 216)

[30] P. Cardaliaguet and C. Rainer. Pathwise strategies for stochastic differential games with an erratum to "Stochastic differential games with asymmetric information." *Applied Mathematics and Optimization*, 68:75–84, 2015. (Cited on p. 216)

[31] B. Carlin, M. Lobo, and S. Viswanathan. Episodic liquidity crises: Cooperative and predatory trading. *Journal of Finance*, 62:2235–2274, 2007. (Cited on p. 216)

[32] R. Carmona and F. Delarue. Probabilistic analysis of mean-field games. *SIAM Journal on Control and Optimization*, 51:2705–2734, 2013. (Cited on pp. 236, 251)

[33] R. Carmona and F. Delarue. The master equation for large population equilibriums. In *Stochastic Analysis and Applications 2014*, pages 77–128. Springer-Verlag, Berlin, 2014. (Cited on pp. 244, 251)

[34] R. Carmona and F. Delarue. Forward-backward stochastic differential equations and controlled McKean–Vlasov dynamics. *Annals of Probability*, 2015. (Cited on pp. 118, 161, 243)

[35] R. Carmona and F. Delarue. *Probabilistic Theory of Mean Field Games and Applications*. Springer-Verlag, Berlin, 2016, to appear. (Cited on pp. 26, 161, 237, 247, 250, 251)

[36] R. Carmona, F. Delarue, G-E. Espinosa, and N. Touzi. Singular forward backward stochastic differential equations and emissions derivatives. *Annals of Applied Probability*, 23:1086–1128, 2013. (Cited on p. 75)

[37] R. Carmona, F. Delarue, and A. Lachapelle. Control of McKean–Vlasov versus mean field games. *Mathematics and Financial Economics*, 7:131–166, 2013. (Cited on pp. 231, 236, 243, 251)

[38] R. Carmona, F. Delarue, and D. Lacker. Probabilistic analysis of mean field games with a common noise. Technical report, Princeton University Press, Princeton, NJ, 2014. (Cited on pp. 248, 251)

[39] R. Carmona, J.P. Fouque, and L.H. Sun. Mean field games and systemic risk: A toy model. *Communications in Mathematical Sciences*, 13:911–933, 2014. (Cited on pp. 216, 244, 251)

[40] R. Carmona and D. Lacker. A probabilistic weak formulation of mean field games and applications. *Annals of Applied Probability*, 25:1189–1231, 2015. (Cited on p. 251)

[41] R. Carmona and J. Yang. Predatory trading: A game on volatility and liquidity. *Quantitative Finance*, (under revision), 2011. (Cited on p. 217)

[42] R. Carmona and G. Zhu. A probabilistic approach to mean field games with major and minor players. *Annals of Applied Probability*, 2015, to appear. (Cited on pp. 26, 251)

[43] J. Carpenter. Does option compensation increase managerial risk appetite? *Journal of Finance*, 55:2311–2331, 2000. (Cited on p. 90)

[44] J.A. Carrillo, R.J. McCann, and C. Villani. Kinetic equilibration rates for granular media and related equations: Entropy dissipation and mass transportation estimates. *Revista Matematica Iberoamericana*, 19:971–1018, 2003. (Cited on p. 118)

[45] D. Crisan and K. Manolarakis. Solving backward stochastic differential equations using the cubature method: Application to nonlinear pricing. Technical report, 2013. (Cited on p. 63)

[46] F. Cucker and S. Smale. Emergent behavior in flocks. *IEEE Transactions on Automatic Control*, 52(5):852–862, 2007. (Cited on p. 251)

[47] J. Cvitanic, H. Pham, and N. Touzi. Super-replication in stochastic volatility models under portfolio constraints. *Journal of Applied Probability*, 36:523–545, 1999. (Cited on p. 118)

[48] M.H.A. Davis. and A.R. Norman. Portfolio selection with transaction costs. *Mathematics of Operations Research*, 15:676–713, 1990. (Cited on pp. 98, 118)

[49] F. Delarue. On the existence and uniqueness of solutions to FBSDEs in a non-degenerate case. *Stochastic Processes and Their Applications*, 99:209–286, 2002. (Cited on pp. 57, 58, 63, 232)

[50] C. Dellacherie. *Capacités et Processus Stochastiques*. Springer-Verlag, Berlin, 1972. (Cited on p. 121)

[51] T. Eisele. Nonexistence and nonuniqueness of open-loop equilibria in linear-quadratic differential games. *Journal of Mathematical Analysis and Applications*, 37:443–468, 1982. (Cited on p. 216)

[52] N. El Karoui and S. Hamadène. BSDEs and risk-sensitive control, zero-sum and nonzero-sum game problems of stochastic functional differential equations. *Stochastic Processes and their Applications*, 107:145–169, 2003. (Cited on pp. 117, 216)

[53] N. El Karoui, S. Hamadène, and M. Matoussi. Backward stochastic differential equations and applications. In R. Carmona, editor, *Indifference Pricing*, Princeton University Press, Princeton, NJ, 2010. (Cited on p. 63)

[54] N. El Karoui, C. Kapoudjian, E. Pardoux, S. Peng, and M.C. Quenez. Reflected solutions of backward SDE's, and related obstacle problems for PDE's. *Annals of Probability*, 25:702–737, 1997. (Cited on p. 63)

[55] N. El Karoui, E. Pardoux, and M.C. Quenez. Reflected backward SDEs and American options. In *Numerical Method in Finance*, pages 215–231. Cambridge University Press, Cambridge, 1997. (Cited on p. 63)

[56] R. Elliott. The optimal control of diffusions. *Applied Mathematics and Optimization*, 22:229–240, 1990. (Cited on p. 160)

[57] W. Fleming and R. Rishel. *Deterministic and Stochastic Optimal Control*. Springer-Verlag, Berlin, 1975. (Cited on p. 117)

[58] W. Fleming and M. Soner. *Controlled Markov Processes and Viscosity Solutions*. Springer-Verlag, Berlin, 2010. (Cited on pp. 117, 118)

[59] W. H. Fleming and P. E. Souganidis. On the existence of value functions of two-player zero-sum stochastic differential games. *Indiana University Mathematics Journal*, 38:293–314, 1989. (Cited on p. 216)

[60] C. Frei, S. Malamud, and M. Schweizer. Convexity bounds for BSDE solutions, with applications to indifference valuation. *Probability Theory and Related Fields*, 150:219–255, 2011. (Cited on pp. 59, 63)

[61] A. Friedman. *Partial Differential Equations of Parabolic Type*. Prentice-Hall, Englewood Cliffs, NJ, 1st edition, 1964. (Cited on pp. 117, 216)

[62] P.N. Giraud, O. Guéant, J.M. Lasry, and P.L. Lions. A mean field game model of oil production in presence of alternative energy producers. Technical report, 2010. (Cited on p. 250)

[63] O. Guéant, J.M. Lasry, and P.L. Lions. Mean field games and oil production. *Finance and Sustainable Development: Seminars Lectures*, 2009. (Cited on p. 250)

[64] O. Guéant, J.M. Lasry, and P.L. Lions. Mean field games and applications. In R. Carmona et al., editors, *Paris–Princeton Lectures in Mathematical Finance 2010*, volume 2003 of *Lecture Notes in Mathematics*. Springer-Verlag, Berlin, 2010. (Cited on p. 250)

[65] S. Hamadène. Nonzero sum linear-quadratic stochastic differential games and backward-forward equations. *Stochastic Analysis and Applications*, 17:117–130, 1999. (Cited on pp. 179, 216)

[66] S. Hamadène and J.P. Lepeltier. Zero-sum stochastic differential games and backward equations. *Systems & Control Letters*, 24:259–263, 1995. (Cited on p. 216)

[67] U.G. Haussmann. *A Stochastic Maximum Principle for Optimal Control of Diffusions*. Longman, Harlow, UK, 1986. (Cited on p. 160)

[68] M. Huang. Large-population LQG games involving a major player: The Nash certainty equivalence principle. *SIAM Journal on Control and Optimization*, 48:3318–3353, 2010. (Cited on p. 251)

[69] M. Huang, P.E. Caines, and R.P. Malhamé. Individual and mass behavior in large population stochastic wireless power control problems: Centralized and Nash equilibrium solutions. In *42nd IEEE Conference on Decision and Control*, pages 98–103, 2003. (Cited on p. 251)

[70] M. Huang, P.E. Caines, and R.P. Malhamé. Large population stochastic dynamic games: Closed-loop McKean-Vlasov systems and the Nash certainty equivalence principle. *Communications in Information and Systems*, 6:221–252, 2006. (Cited on p. 251)

[71] P. Imkeller and G. Dos Reis. Path regularity and explicit convergence rate for BSDE with truncated quadratic growth. *Stochastic Processes and Their Applications*, 120:348–379, 2010. (Cited on pp. 59, 63)

[72] B. Jourdain, S. Méléard, and W. Woyczynski. Nonlinear SDEs driven by Lévy processes and related PDEs. *ALEA: Latin American Journal of Probability*, 4:1–29, 2008. (Cited on pp. 26, 147, 240)

[73] Y. Kabanov and M. Safarian. *Markets with Transaction Costs: Mathematical Theory*. Springer-Verlag, Berlin, 2009. (Cited on p. 118)

[74] M. Kobylanski. Backward stochastic differential equations and partial differential equations with quadratic growth. *Annals of Probability*, 28:558–602, 2000. (Cited on pp. 34, 63)

[75] N. Krylov. *Lectures on Elliptic and Parabolic Equations in Sobolev Spaces*, volume 96 of *Graduate Studies in Mathematics*. AMS, Providence, RI, 2008. (Cited on p. 117)

[76] H.J. Kushner. Necessary conditions for continuous parameter stochastic optimization problems. *SIAM Journal on Control and Optimization*, 10:550–565, 1972. (Cited on p. 160)

[77] A. Lachapelle. Human crowds and groups interactions: a mean field games approach. Technical report, CEREMADE, Université Paris Dauphine, 2010. (Cited on p. 250)

[78] A. Lachapelle and J.M. Lasry. A mean field games model for the choice of insulation technology of households. Technical report, CEREMADE, Université Paris Dauphine, 2010. (Cited on p. 250)

[79] J.M. Lasry and P.L.Lions. Jeux à champ moyen. I—le cas stationnaire. *C. R. Acad. Sci. Paris, Ser. A, Mathématiques*, 343(9):619–625, 2006. (Cited on p. 250)

[80] J.M. Lasry and P.L.Lions. Jeux à champ moyen. II—horizon fini et contrôle optimal. *Comptes Rendus Mathématique*, 343(10):679–84, 2006. (Cited on p. 250)

[81] J.M. Lasry and P.L.Lions. Mean field games. *Japanese Journal of Mathematics*, 2(1):229–260, 2007. (Cited on pp. 118, 250)

[82] J.P. Lepeltier and J. San Martin. Existence of BSDE with superlinear quadratic coefficients. *Stochastics and Stochastic Reports*, 63:227–260, 1998. (Cited on p. 34)

[83] J.P. Lepeltier, Z. Wu, and Z. Yu. Nash equilibrium point for one kind of stochastic nonzero-sum game problem and BSDEs. *C. R. Acad. Sci. Paris, Ser. A, Mathématiques*, 347:959–964, 2009. (Cited on pp. 186, 216)

[84] P.L. Lions. Hamilton–Jacobi–Bellman equations and the optimal control of stochastic systems. In *Proceedings of the International Congress of Mathematicians, Warszawa, August 1983*, pages 16–24, 1983. (Cited on p. 117)

[85] P.L. Lions. On the Hamilton–Jacobi–Bellman equations. *Acta Applicandae Mathematicae*, 1:17–41, 1983. (Cited on p. 118)

[86] P.L. Lions. Optimal control of diffusion processes and Hamilton–Jacobi–Bellman equations, part 2: Viscosity solutions and uniqueness. *Communications in Partial Differential Equations*, 8:1229–1276, 1983. (Cited on p. 118)

[87] P.L. Lions and A.S. Sznitman. Stochastic differential equations with reflecting boundary conditions. *Communications in Pure and Applied Mathematics*, 37:511–537, 1984. (Cited on p. 116)

[88] D. Lukes and D.L. Russel. A global theory for linear-quadratic differential games. *Journal of Mathematical Analysis and Applications*, 33:96–123, 1971. (Cited on p. 216)

[89] T.J. Lyons. Uncertain volatility and the risk-free synthesis of derivatives. *Applied Mathematical Finance*, 2:117–133, 1995. (Cited on p. 118)

[90] J. Ma, P. Protter, and J. Yong. Solving forward-backward stochastic differential equations explicitly—a four step scheme. *Probability Theory and Related Fields*, 98:339–359, 1994. (Cited on pp. 56, 58)

[91] J. Ma and J. Yong. *Forward-Backward Stochastic Differential Equations and Their Applications*, volume 1702 of *Lecture Notes in Mathematics*. Springer-Verlag, Berlin, 2007. (Cited on p. 63)

[92] H. Markowitz. Portfolio selection. *Journal of Finance*, 7:77–91, 1952. (Cited on p. 118)

[93] H. Markowitz. *Portfolio Selection: Efficient Diversification of Investments*. Wiley, New York, 1959. (Cited on p. 118)

[94] T. Meyer-Brandis, B. Øksendal, and X.Y. Zhou. A mean-field stochastic maximum principle via Malliavin calculus. *Stochastics*, 84:643–666, 2011. (Cited on p. 161)

[95] M. Nourian, P.E. Caines, and R.P. Malhamé. Mean field analysis of controlled Cucker-Smale type flocking: Linear analysis and perturbation equations. In *Proceedings of the 18th IFAC World Congress, Milan, August 2011*, pages 4471–4476, 2011. (Cited on pp. 245, 251)

[96] J. Neveu. *Discrete Parameter Martingales*. North-Holland, Amsterdam, 1975. (Cited on p. 121)

[97] S.L. Nguyen and M. Huang. Linear-quadratic-Gaussian mixed games with continuum-parametrized minor players. *SIAM Journal on Control and Optimization*, 50:2907–2937, 2012. (Cited on p. 251)

[98] S. Nguyen and M. Huang. Mean field LQG games with mass behavior responsive to a major player. In *51st IEEE Conference on Decision and Control, Maui, HI, 2012*, pages 5792–5797, 2012. (Cited on p. 251)

[99] M. Nourian and P. Caines. ϵ-Nash mean field game theory for nonlinear stochastic dynamical systems with major and minor agents. Technical report, 2013. (Cited on p. 251)

[100] B. Øksendal and A. Sulem. Singular stochastic control and optimal stopping with partial information of Itô–Lévy processes. *SIAM Journal on Control and Optimization*, 50:2254–2287, 2012. (Cited on p. 161)

[101] B. Øksendal and A.T.K. Ta. A maximum principle for stochastic differential games with G-expectations and partial information. *Stochastics*, 84:137–155, 2012. (Cited on p. 216)

[102] E. Pardoux and S. Peng. Adapted solution of a backward stochastic differential equation. *Systems & Control Letters*, 14:55–61, 1990. (Cited on p. 144)

[103] E. Pardoux and A. Răşcanu. *Stochastic Differential Equations, Backward SDEs, Partial Differential Equations*. Springer-Verlag, Berlin, 2014. (Cited on p. 63)

[104] S. Peng. A general stochastic maximum principle for optimal control problems. *SIAM Journal on Control and Optimization*, 28:966–979, 1990. (Cited on p. 161)

[105] S. Peng and Z. Wu. Fully coupled forward-backward stochastic differential equations and applications to optimal control. *SIAM Journal on Control and Optimization*, 37:825–843, 1999. (Cited on pp. 47, 63, 179, 216)

[106] H. Pham. *Continuous-time Stochastic Control and Optimization with Financial Applications*, volume 61 of *Stochastic Modelling and Applied Probability*. Springer-Verlag, Berlin, 2009. (Cited on pp. 63, 118, 161)

[107] T. Pham and J. Zhang. Two person zero-sum game in weak formulation and path dependent Bellman–Isaacs equation. *SIAM Journal on Control and Optimization*, 52:2090–2121, 2014. (Cited on pp. 178, 216)

[108] M.C. Quenez. Stochastic Control and BSDEs. In *Backward Stochastic Differential Equations. Pitman Research Notes in Mathematics*, volume 364, pp. 83–99. Longman, Harlow, UK, 1997. (Cited on p. 161)

[109] S.T. Rachev and L. Ruschendorf. *Mass Transportation Problems I: Theory*. Springer-Verlag, Berlin, 1998. (Cited on pp. 237, 251)

[110] M. Robin. Long-term average cost control problems for continuous time Markov processes: A survey. *Acta Applicandae Mathematical*, 1:281–299, 1983. (Cited on p. 118)

[111] T. Schöneborn and A. Schied. Liquidation in the face of adversity: Stealth versus sunshine trading. Technical report, Berlin University, 2009. (Cited on p. 217)

[112] S. Shreve and M. Soner. Optimal investment and consumption with transaction costs. *Annals of Applied Probability*, 4:609–692, 1994. (Cited on pp. 98, 100, 118)

[113] A.S. Sznitman. Topics in propagation of chaos. In D. L. Burkholder et al., editors, *Ecole d'Eté de Probabilités de Saint-Flour, XIX—1989*, volume 1464 of *Lecture Notes in Mathematics*, pages 165–251, Springer–Verlag, Berlin, 1989. (Cited on pp. 26, 240)

[114] N. Touzi. *Optimal Stochastic Control, Stochastic Target Problems, and Backward SDE*, volume 29 of *Fields Institute Monographs*, Springer-Verlag, Berlin, 2012. (Cited on pp. 26, 63, 81, 83, 117)

[115] C. Villani. *Optimal Transport, Old and New*. Springer-Verlag, Berlin, 2009. (Cited on p. 143)

[116] J. Yong. Linear forward-backward stochastic differential equations with random coefficients. *Probability Theory and Related Fields*, 135:53–83, 2006. (Cited on p. 63)

[117] J. Yong and X. Zhou. *Stochastic Controls: Hamiltonian Systems and HJB Equations*. Springer-Verlag, Berlin, 1999. (Cited on pp. 63, 161)

[118] M. Yor. Sur les théories du filtrage et de la prédiction. In *Séminaire de Probabilités. XI, Lecture Notes in Mathematics* 581, Springer-Verlag, Berlin, 1975. (Cited on p. 20)

[119] X.Y. Zhou and D. Li. Continuous-time mean-variance portfolio selection: a stochastic LQ framework. *Applied Mathematics and Optimization*, 42:19–33, 2000. (Cited on pp. 118, 161)

Author Index

A
Ahmed, 160
Aiyagari, 246
Almgren, 216
Andersson, 160
Avellaneda, 118

B
Barrieu, 63
Basar, 167
Bellman, 67, 82
Benachour, 118
Bensoussan, 118, 160, 216, 251
Bertsekas, 160
Bertsimas, 216
Bismut, 160
Buckdahn, 63
Burdzy, 117
Burkhölder, 54

C
Cadenillas, 160
Caines, 245, 251
Cardaliaguet, 118
Carlin, 216
Carmona, 26, 118, 161, 216, 217, 251
Carrillo, 118
Cathiaux, 118
Chau, 251
Chriss, 216
Crisan, 63
Cucker, 244
Cvitanic, 118

D
Davis, 54, 98, 118
Delarue, 26, 57, 118, 161, 251
Dellacherie, 121
Ding, 160
Djehiche, 63, 160
Dos Reis, 63

E
Eisele, 216
El Karoui, 63, 117, 160

F
Fenchel, 91
Fleming, 117, 118
Fouque, 216
Frehse, 118, 160
Friedman, 216

G
Girsanov, 119
Gronwall, 6
Guillin, 118
Gundy, 54

H
Hamadène, 63, 117, 160, 216
Hamilton, 82
Huang, 251

I
Imkeller, 63
Isaacs, 119, 138, 216

J
Jacobi, 82
Jourdain, 26

K
Kabanov, 118
Kang, 117
Kapoudjian, 63
Karatzas, 160
Kobylanski, 34, 63
Kushner, 160

L
Lachapelle, 251
Lacker, 251
Lasry, 118, 250

Legendre, 91
Lepeltier, 34
Li, 63, 118, 161
Lions, 118, 143, 250
Lo, 216
Lobo, 216
Lukes, 216
Lyons, 118

M
Méléard, 26
Ma, 63
Malhamé, 245, 251
Malliavin, 45
Malrieu, 118
Manolarakis, 63
Markowitz, 92, 118, 154
Matoussi, 63
McCann, 118
McKean, 26
Merton, 90
Meyer–Brandis, 161

N
Nagai, 160
Nash, 167, 169
Neveu, 121
Nisio, 160
Norman, 98, 118
Nourian, 245, 251

O
Øksendal, 160

P
Paras, 118
Pardoux, 29, 63
Pareto, 169
Peng, 29, 63, 160, 216
Pham, 63, 118
Pontryagin, 119, 125, 130, 187

Q
Quenez, 63, 160

R
Ramanan, 117
Riccati, 48
Roynette, 118
Russel, 216
Rășcanu, 63

S
Safarian, 118
San Martin, 34
Schied, 216, 217
Schweizer, 63
Schöneborn, 217

Shreve, 100, 118
Skorohod, 116
Smale, 244
Soner, 100, 117, 118
Sun, 216
Sung, 251
Sznitman, 26

T
Talay, 118
Touzi, 26, 63, 117, 118

V
Vallois, 118
Villani, 118
Viswanathan, 216

Vlasov, 26

W
Woyczynski, 26
Wu, 216

Y
Yam, 251
Yang, 217
Yong, 63, 160
Yung, 251

Z
Zhou, 63, 118, 160, 161
Zhu, 26, 251

Subject Index

A

abatement, 76
action, 68, 167
adjoint
   equation, 113, 119, 128
   process, 128, 187, 190
admissible
   action, 68, 167
   control, 68
   strategy, 68, 167
affine BSDE, 28
Aiyagari's growth model, 246
American
   contingent claim, 39
   option, 67
ask price, 134
autonomous diffusion, 112

B

backward
   Euler scheme, 45
   stochastic differential equation, 27
banking, 75
barrier, 37
basis function, 45
best response map, 172
bid price, 134
borrowing, 75
BSDE
   affine, 28
   mean-field, 35
   quadratic, 34
   reflected, 37
Burkhölder–Davis–Gundy
   Inequality, 54
business as usual, 76

C

canonical process, 12
carbon regulation, 75
central planner, 93

closed loop, 68
   control, 72
   equilibrium, 170
      in feedback form, 170
   perfect state, 167
coefficient, 27
comparison principle, 103
complete market model, 33
condition
   Isaacs, 184
   min-max, 184
conditional
   cost, 123
   value function, 123
constant relative risk aversion, 89
contingent claim, 33
   American, 39
continuation
   method, 49
   region, 110
control, 68
convertible bond, 40
convex envelope, 80
cost
   conditional, 123
   functional, 69, 74, 168
   switching, 111
coupling, 16
covariable, 70, 173, 184
CRRA power utility, 98
CRRA utility function, 132
cubature, 45, 63
Cucker–Smale model, 244

D

de Finetti's law of large numbers, 248
decoupled FBSDE, 42
decoupling field, 42, 55
deterministic equilibrium, 170
Dirac measure, 220
discount factor, 34

distributed strategies, 168
Doléans exponential, 33, 74
drift, 8
driver, 27
dual variable, 70, 173, 184
duality approach, 133
dynamic programming
   equation, 110
   principle, 67, 81
dynamics
   McKean–Vlasov, 35
Dynkin game, 40

E

efficient
   frontier, 92
   portfolio, 92
electricity production, 75
elliptic, 101
ellipticity, 101
empirical
   distribution, 220
   measure, 16
envelope
   lower semicontinuous, 102
   upper semicontinuous, 102
equilibrium
   closed loop, 170
      in feedback form, 170
   deterministic, 170
   Nash, 167
   open loop, 170
equivalent martingale measure, 33
ergodic theory, 112
   control, 112
   control problem, 118
essential
   infimum, 121
   supremum, 121
Euler scheme, 45
   backward, 45

forward, 45
European option, 32
externality, 78

F

FBSDE value function, 42, 55, 80, 139
feature function, 45
feedback
　form, 174
　perfect state, 167
Fenchel–Legendre transform, 91, 98
Feynman–Kac formula, 9
fixed point, 172
flow, 6
Fokker–Planck equation, 10
form
　strong, 72
　weak, 72
forward
　backward stochastic differential equations, 42
　Euler scheme, 45

G

game
　Dynkin, 40
　lower value, 177
　of timing, 40
　upper value, 177
　value function, 177
　zero-sum, 177
Gâteaux derivative, 125
generator, 8
Girsanov theorem, 119
Gronwall's inequality, 6
growth model
　Aiyagari, 246

H

Hamilton–Jacobi–Bellman equation, 67, 82
Hamiltonian, 70, 172
　reduced, 71, 173
hedging, 39
high water mark, 106
historical measure, 77
HJB
　equation, 67, 82, 119
　　stationary, 86
　value function, 79, 80
hypoellipticity, 83

I

impulse region, 110
Inada condition, 76
independent copy, 35
infinite horizon, 85
initial endowment, 33
instantaneous cost, 69
inventory, 134
Isaacs condition, 138, 172, 173, 184
Itô obstacle, 38

K

Kolmogorov equation, 10

L

Lagrange multiplier, 212, 213
Lagrangian, 212, 213
　duality, 140
law of large numbers
　de Finetti's, 248
Legendre transform, 91
linear growth condition, 51
linear-quadratic, 179
　FBSDE, 63
　model, 136
　stochastic games, 179
Lipschitz condition, 51
local time, 96
log utility function, 132
lower
　semicontinuous envelope, 102
　value function, 177

M

major
　particle, 19
　player, 247, 248
Malliavin calculus, 45
marginal distribution, 12
market model, 32
Markov
　Nash equilibrium, 174
　property, 6
Markovian, 68
　equilibrium, 170
　strategy profile, 174
martingale representation theorem, 27
master equation, 244
matrix Riccati equation, 226
maximum principle, 119, 130
McKean–Vlasov dynamics, 35
mean-field, 219
　interaction, 11

BSDE, 35
　game, 26, 35, 172, 216
　interaction, 13
　　of order 1, 13
　　of scalar type, 13
　SDE, 26
mean-variance
　Markowitz criterion, 92, 154
　optimization, 73
　portfolio selection, 92, 154
memoryless perfect state, 167
Merton problem, 88
midprice, 134, 205
min-max condition, 184
minor
　particle, 19
　player, 247, 248
model
　Cucker–Smale, 244
monotonicity assumption, 46

N

Nash
　certainty equivalence, 251
　equilibrium, 167, 169
nondegeneracy condition, 55
nonintervention region, 95
nonlocal operator, 109

O

objective, 69
　functional, 69
obstacle, 109
　Itô, 38
open loop, 68, 167
　equilibrium, 170
operator symbol, 82
optimal
　exercise, 40
　switching, 111
optimality, 169
option
　American, 67
　European, 32
optional
　σ-field, 19
　projection, 19
order book, 205
　flat, 205
Ornstein–Uhlenbeck process, 113, 198

P

parabolicity, 101
Pareto

Subject Index

optimality, 169
  weakly efficient, 169
partial differential equation, 42
particle
  approximation, 15
  major, 19
  minor, 19
payoff
  running, 39
  terminal, 39
permanent price impact, 205
player
  major, 248
  minor, 248
pollution permit, 75
Pontryagin maximum principle, 119, 130
portfolio, 33
  efficient, 92
  self-financing, 33
price impact, 134
  permanent, 205
  temporary, 205
private state, 224
probabilistic approach
  to LQ control problems, 138
progressive σ-field, 19, 68
progressively measurable, 68
propagation of chaos, 15
propagator, 61
push-forward image, 14

Q

quadratic BSDE, 34
quasi-variational inequality, 44, 87, 109

R

reduced Hamiltonian, 71, 173, 184
reflected BSDE, 37
regime, 111
regular conditional distribution, 19
replicate, 33
representative agent, 93
Riccati equation, 48, 139
  for the decoupling field, 139
  for the value function, 139
risk
  appetite, 72

averse, 73
aversion, 72, 90
neutral, 205
premium, 33
seeking, 73
rough paths, 45, 63
running
  cost, 69, 168
  payoff, 39

S

saddle point, 177
Schauder fixed-point theorem, 235
scrap value, 39, 215
self-financing, 33
  portfolio, 33
semi-linear
  game, 181, 216
  parabolic equation, 42
setup, 23
singular
  control, 87
  control problem, 87
  measure, 87
Skorohod's lemma, 116
small investor, 32
smooth fit condition, 112
Snell envelope, 109
solvency region, 97
spike perturbation, 161
standard space, 19
state price density, 34
stationary HJB equation, 86
stochastic
  differential equation, 42
  HJB equation, 123
  maximum approach
    to LQ control problems, 138
  maximum principle, 125, 187
  partial differential equation, 222
stochastic game
  linear-quadratic, 179
strategy, 68, 167
  profile, 166
strong
  form, 72
  solution, 74
sub-game perfect, 174

switching cost, 111
systemic risk, 198

T

temporary
  impact, 134
  price impact, 205
terminal
  condition, 27
  cost, 69, 168
  payoff, 39
transaction
  costs, 205
  price, 134
transform
  Fenchel–Legendre, 91

U

uniform
  ellipticity, 83
  ellipticity condition, 55
upper
  semicontinuous envelope, 102
  value function, 177
utility function, 88, 89, 132

V

value function, 55
  conditional, 123
  FBSDE, 42
verification
  argument, 42
  theorem, 56, 123
viscosity
  solution, 43, 44, 102, 117, 177
  subsolution, 102, 177
  supersolution, 102, 177
volatility, 8

W

Wasserstein distance, 12
weak
  form, 72
  formulation, 119
  solution, 74
weakly Pareto efficient, 169
wealth maximization, 88

Z

zero-one law, 232
zero-sum game, 177